G000243013

THE MEDITERRANEAN

THE MEDITERRANEAN
Environment and Society

Edited by

RUSSELL KING

Professor of Geography, University of Sussex

LINDSAY PROUDFOOT

Reader in Historical Geography, Queen's University of Belfast

and

BERNARD SMITH

Reader in Geomorphology, Queen's University of Belfast

A member of the Hodder Headline Group

LONDON • NEW YORK • SYDNEY • AUCKLAND

First published in Great Britain in 1997 by
Arnold, a member of the Hodder Headline Group
338 Euston Road, London NW1 3BH
175 Fifth Avenue, New York, NY 10010

Co-published in the USA by Halsted Press,
an imprint of John Wiley & Sons, Inc.,
605 Third Avenue
New York, NY 10158–0012

© 1997 Arnold

All rights reserved. No part of this publication may be reproduced or transmitted in any form or by any means, electronically or mechanically, including photocopying, recording or any information storage or retrieval system, without either prior permission in writing from the publisher or a licence permitting restricted copying. In the United Kingdom such licences are issued by the Copyright Licensing Agency: 90 Tottenham Court Road, London W1P 9HE.

British Library Cataloguing in Publication Data
A catalogue record for this book is available from the British Library

Library of Congress Cataloging-in-Publication Data
A catalog record for this book is available from the Library of Congress

ISBN 0 340 65281 0 (pb)
ISBN 0 340 65280 2 (hb)
ISBN 0 470 23667 1 (pb, USA only)
ISBN 0 470 23666 3 (hb, USA only)

Typeset in 10/12pt Palatino by Saxon Graphics Ltd, Derby
Printed and bound in Great Britain by The Bath Press

CONTENTS

CONTRIBUTORS

Dr George Dardis, School of Geosciences, The Queen's University of Belfast, Belfast BT7 1NN

Professor Michael Dunford, School of European Studies, University of Sussex, Falmer, Brighton BN1 9QN

Dr Hazel Faulkner, Department of Geography, University of North London, Ladbroke House, 62–66 Highbury Grove, London N5 2AD

Dr Don Funnell, School of African and Asian Studies, University of Sussex, Falmer, Brighton BN1 9QN

Professor Brian Graham, School of Environmental Studies, University of Ulster at Coleraine, Coleraine BT52 1SA

Mr Alan Hill, Department of Geography, University of Lancaster, Bailrigg, Lancaster LA1 4YR

Dr Alun Jones, Department of Geography, University College London, 26 Bedford Way, London WC1H 0AP

Professor Russell King, School of European Studies, University of Sussex, Falmer, Brighton BN1 9QN

Professor Nurit Kliot, Department of Geography, University of Haifa, Mount Carmel, Haifa, 31905 Israel

Professor Lila Leontidou, Department of Geography, University of the Aegean, 81 100 Mytilene, Greece

Dr Allen Perry, Department of Geography, University College of Swansea, Singleton Park, Swansea SA2 8PP

Dr Jeff Pratt, School of European Studies, University of Sussex, Falmer, Brighton BN1 9QN

Dr Lindsay Proudfoot, School of Geosciences, The Queen's University of Belfast, Belfast BT7 1NN

Professor Helen Rendell, The Geography Laboratory, University of Sussex, Falmer, Brighton BN1 9QN

Dr Alastair Ruffell, School of Geosciences, The Queen's University of Belfast, Belfast BT7 1NN

Dr Bernard Smith, School of Geosciences, The Queen's University of Belfast, Belfast BT7 1NN

Professor Allan Williams, Department of Geography, University of Exeter, Amory Building, Rennes Drive, Exeter EX4 4RJ

PREFACE

This book aims to provide an up-to-date undergraduate text which examines contemporary Mediterranean issues from a perspective that recognises the littoral regions of the Mediterranean states as a discernible geographical entity. In a sense we can claim a 'first': there is no current undergraduate text which deals with the complex and diverse geographical issues facing the Mediterranean Basin as a region. It is true that some aspects of the geography of individual Mediterranean countries have been discussed in other recently published books, but these either have a regional focus which is placed elsewhere – in Europe, Africa or the Middle East; or they deal with narrowly conceived themes such as urbanisation, rural life or climate. It is also true that there is a vast amount of geographical literature on the Mediterranean region, much of it very recent and devoted to the important economic, geopolitical and environmental changes under way in this key part of the world. This geographical literature is widely dispersed in journals and other publications and one of the most important achievements of this book is its critical synthesis of much of this literature. Moreover, the literature covered is not limited to geographical writing; reflecting the trend towards interdisciplinarity in modern scholarship, the volume also embraces sections of other literatures on the Mediterranean: geological, ecological, historical, political, economic, anthropological and demographic.

Given the multi-layered nature of the identity of the Mediterranean, we do not think it appropriate to attempt to define the area of coverage of the book in any hard and fast way. Environmental and ecological analyses tend to concentrate on the Mediterannean littoral, i.e. on those regions contained within the Mediterranean watershed or on those areas whose character is conditioned by proximity to the Mediterranean Sea. On cultural, economic or political criteria, however, the spatial definition may widen, partly because of the national availability of statistics, and partly because what goes on in the Mediterranean Basin as narrowly defined may be influenced by wider-scale processes. Above all, we would argue, it is the nature and pattern of spatial interaction rather than any precise geographical boundary that define Mediterranean identity.

The issues at the heart of these regional interactions are as important as they are diverse. They range from problems of water and other resource depletion in the face of rapid and uneven modernisation, to the growing pace and scale of south–north migration and the increase in religious fundamentalism and regional nationalism. The discussions which follow in the various thematic chapters of the book stress both the shared nature of the geographic experience within the Basin and the diversity of local circumstances and expression. Recurrent emphasis is placed on the interaction between society and environment in terms of environmental management, differential regional development and associated political, demographic, cultural and economic tensions.

Aside from the introductory and concluding chapters, the volume is structured into four parts. The first group of chapters (2–4) deals with physical environments – geology, climate and earth surface processes. They are designed to establish the physical framework for the Mediterranean region and to justify its individuality. However, each of these nominally physical chapters also introduces a range of environmental issues and consequences for human activity in the region. The second section of the book sets out the historical context. These chapters (5–7) adopt a selective narrative form and summarise the major features of the historical geography of the Mediterranean from the rise of the Classical urban civilisations to the demise of the Ottoman Empire in the early twentieth century. The third part of the book is by far the largest: seven chapters (8–14) deal with current human geographical issues and stress the contested nature of Mediterranean human space. An overview of Mediterranean geopolitics and society is followed by an analysis of the evolution of uneven development within the Basin. These 'broad sweep' chapters (8 and 9) are then followed by a series of five more specific ones on the EU dimension, demography and migration, the diversity of Mediterranean urbanisation, agrarian society and tourism. The fourth section of the book (Chapters 15–17) deals with the management of the Mediterranean environment, concentrating on human interactions with, and pressures on, water, forests and soils, and the coastal zone. These chapters, as well as several of the others, call into question the long-term sustainability of current development processes in the region.

We hope this volume will appeal to a wide undergraduate audience. Many Geography Departments in British, Irish and other European universities offer courses on the Mediterranean or courses with a substantial Mediterranean content. Many departments also undertake field excursions to the region. As well as geographers, we anticipate that students studying many other courses (European Studies, Development Studies, Politics, Languages, Anthropology, etc.) will find the book a valuable source of reference. The comprehensive coverage of the literature and the research-level content of many of the chapters will make the volume a useful background text for postgraduate students contemplating research in the region; whilst the relatively non-technical language will widen the appeal to teachers and make the text accessible to upper school and junior college students. Finally, we hope readers of the book will be able to share some of the pleasure we have had in compiling it, on which note we would thank the contributors for their prompt delivery of their chapters, the cartographers at Queen's University Belfast and Sussex who drew most of the maps and diagrams, and Laura McKelvie at Arnold for her rare combination of patience and enthusiasm.

Russell King
Lindsay Proudfoot
Bernard Smith

ACKNOWLEDGEMENTS

The editors and publisher would like to thank the following for permission to reproduce photographic material:

Hum Beckers for Figures 11.2 and 13.1; George Dardis for Figure 16.5; ENIT (the Italian State Tourist Office) for Figure 12.1; Russell King for Figures 3.3, 3.5, 4.2, 4.5, 8.1, 11.3, 12.2, 13.2, 13.4, 14.2, 14.4, 16.3 and 16.10; Lindsay Proudfoot for Figures 5.3, 5.7 and 5.9; and Bernard Smith for Figures 4.1, 4.3, 4.4, 4.6, 4.7, 15.11, 17.4, 17.5 and 17.10.

1

INTRODUCTION: AN ESSAY ON MEDITERRANEANISM

RUSSELL KING

AIMS AND SCOPE OF THE BOOK

This book aims to be the first university-level text on the geography of the Mediterranean. Such a claim needs some justification, and perhaps a little qualification. First, one can point with some confidence to the paucity of geographical literature that treats the Mediterranean as a relatively homogenous regional unit. The problem here is that the traditional continental or sub-continental divisions of regional geography textbooks cut through the Mediterranean, so that bits of the Basin appear in books on Europe, Africa or the Middle East. Yet it is clear that on physical, cultural and historical criteria the Mediterranean presents itself as a more unified region than either Europe or Africa. Between these the real boundary is the Sahara Desert, which has been a more effective barrier to contact than the Mediterranean which, rather, has acted as a focus for communications and interaction.

Second, it should be stressed that this book addresses contemporary geographical issues in the Mediterranean Basin from a perspective that recognises the importance of both physical environmental characteristics and historical–cultural legacies. Indeed, it is the multi-layered interactions between physical, cultural and contemporary social and economic geographies which define the essence of the Mediterranean landscape and Mediterranean life, and which make the littoral regions of the Mediterranean states cohere as a recognisable geographical entity. The nature of these interactions will be explored in a little more detail towards the end of this chapter, and they will recur at various points throughout the book.

Third, we must recognise what *has* been published. There is undoubtedly a vast geographical literature on the region or, rather, on various areas and localities within the region. This literature is scattered in journals published in many countries and in many languages – including Japanese! Moreover, given the characteristic interdependence of physical, cultural and economic criteria in defining the essence of the region and its problems, the interest of the Mediterranean geographer also extends to literature in a range of other disciplines such as geology, ecology, archaeology, history, anthropology, economics and political science. But the quantity of literature on the Mediterranean Basin as a whole, or as a distinct ecological or geographical unit, remains curiously limited. At the textbook level, and limiting ourselves for the time being to geographical studies published in English, those that exist are either very old and hence hopelessly out-of-date (Newbigin 1924; Semple 1932); or they are too elementary to be regarded as serious university texts (Branigan and Jarrett 1975; Robinson 1970; Walker 1965); or they only treat certain parts of the basin such as southern Europe (Beckinsale and Beckinsale 1975; Hadjimichalis 1987; Hudson and Lewis 1985; Williams 1984) or certain themes such as urbanisation (Leontidou 1990), rural landscapes (Houston 1964) and mountain environments (McNeill 1992).

The above brief listing by no means exhausts the geographical library on the Mediterranean. Amongst other books of an earlier generation mention should be made of W.G. East's *Mediterranean Problems* (1940), an early essay on geopolitics, and André Siegfried's *The Mediterranean* (1949), translated from the French. Pride of place amongst French studies goes to Birot and Dresch's two-volume survey on the Mediterranean and the Middle East (1953, 1956). Perhaps the most rounded geographical survey of the region is that by Orlando Ribeiro, available both in the original Portuguese and in an updated version in Italian (Ribeiro 1968, 1983).

Yet, perversely, the two most important books on the Mediterranean are written by non-geographers. One looks back to the sixteenth century (Braudel 1972, 1973), the other forward to the twenty-first (Grenon and Batisse 1989). Fernand Braudel's *tour de force* is the classic that all Mediterranean writing must pay homage to; almost any book on the region is destined to be enriched with quotes from the great French 'geo-historian'. Grenon and Batisse look to the future of the Mediterranean. Their book is the 'Blue Plan' scenario (or set of scenarios), the UNESCO economic and social planning initiative concerned with the coastal regions of the Mediterranean that examines the environmental and natural resource problems common to all or most countries of the region. The Blue Plan will itself make many appearances in the pages that follow.

The present volume attempts what none of the others achieve: a complete and integrated treatment of the various systematic geographies of the Mediterranean Basin, treating the whole region both as a unity and a diversity, and stressing the physical–human interactions which lie at the heart of a full geographical understanding of the area. The substantive contents of the book consist of sixteen synoptic chapters, each dealing with one of the major geographical issues or themes of the Mediterranean, plus the introductory and concluding chapters. Each contributor provides an expert synthesis of recent important work relating to their topic, informed by their own perspectives and research. This combination of primary research with invaluable summaries of the extensive but widely dispersed journal and other literature enables the book to fulfil two important objectives: to make an original contribution in its own right; and to enable its readers to attain a thorough understanding of the Mediterranean region and its problems.

DEFINITION OF THE MEDITERRANEAN

To define the Mediterranean is not an easy task. In some senses it is probably a pointless exercise, for there is no single criterion which enables one to draw a line on a map which separates the Mediterranean from the non-Mediterranean. Mediterranean identity is a more nebulous, but powerful, concept that derives from environmental characteristics, cultural features and, above all, from the spatial interactions between the two. The Mediterranean is a sea, a climate, a landscape, a way of life – all of these and much more.

Various authors have defined the region arbitrarily, pragmatically, and not very satisfactorily. All three elementary geographies of the Mediterranean (Branigan and Jarrett 1975; Robinson 1970; Walker 1965) employ the definition 'the countries bordering the Mediterranean Sea'. But many of these countries extend to other, non-Mediterranean regions: France to northern Europe, Algeria to the Sahara, Egypt to the middle Nile and Syria to Mesopotamia. The break-up of Yugoslavia has produced states with variable links to the Mediterranean: Croatia has the island-studded Dalmatian coast; Slovenia, whilst claiming to be 'on the sunny side of the Alps', has a tiny foothold on the sea itself; Serbia looks more to the Danubian Plains. And then there is Portugal, Mediterranean in climate and culture, with no Mediterranean shore. Others cast the net even wider. In their discussion of Mediterranean demography Montanari and Cortese (1993) embrace Jordan and Iraq and also make out a case for Bulgaria, Romania and Georgia to be included. Figure 1.1 shows how these various countries stretch the limits of the Mediterranean region.

Much narrower definitions are proposed by the Blue Plan team. Their view is of an area 'where socio-economic activities are governed largely by their relations with the seaboard' (Grenon and Batisse 1989, pp. 15–16). This reflects the Blue Plan's overriding concern with the sea and with coastal processes such as tourism, urbanisation and water shortage. The 'proximity to the sea' definition is presented in two variants, one based on physical criteria and one on administrative units. The hydrological basin (Fig. 1.2) is proposed as the most suitable boundary of study for all matters relating to fresh water, including land-based pollution from rivers: this recognises that the Mediterranean impact of rivers such as the Rhône or the Po may originate in non-Mediterranean regions. The idea of a 'coastal zone' depends greatly on local coastal topography as well as on the nature of the phenomenon being studied. Hence the hemmed-in coasts of the French and Italian Rivieras contrast with the featureless coastal wastes of parts of Egypt and Libya; and coastal-strip tourism contrasts with the orientation of

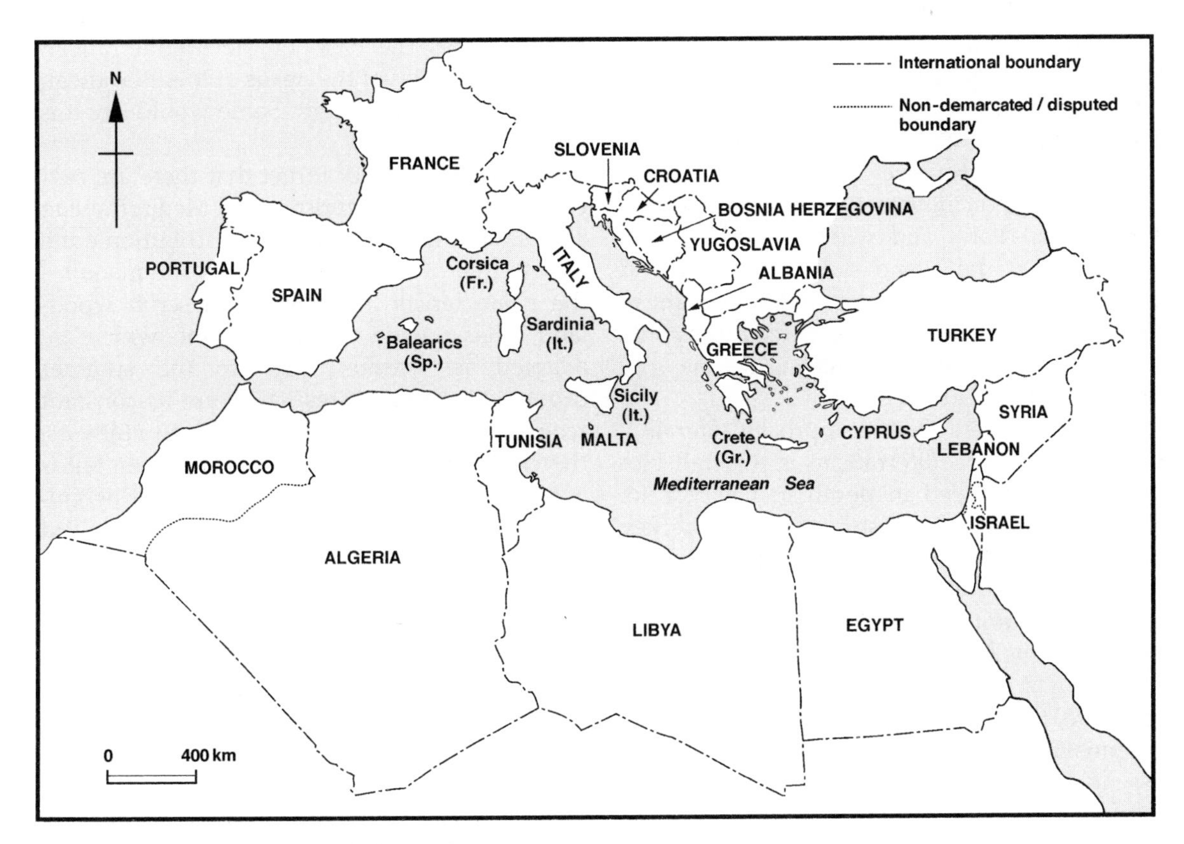

FIGURE 1.1 Countries around the Mediterranean

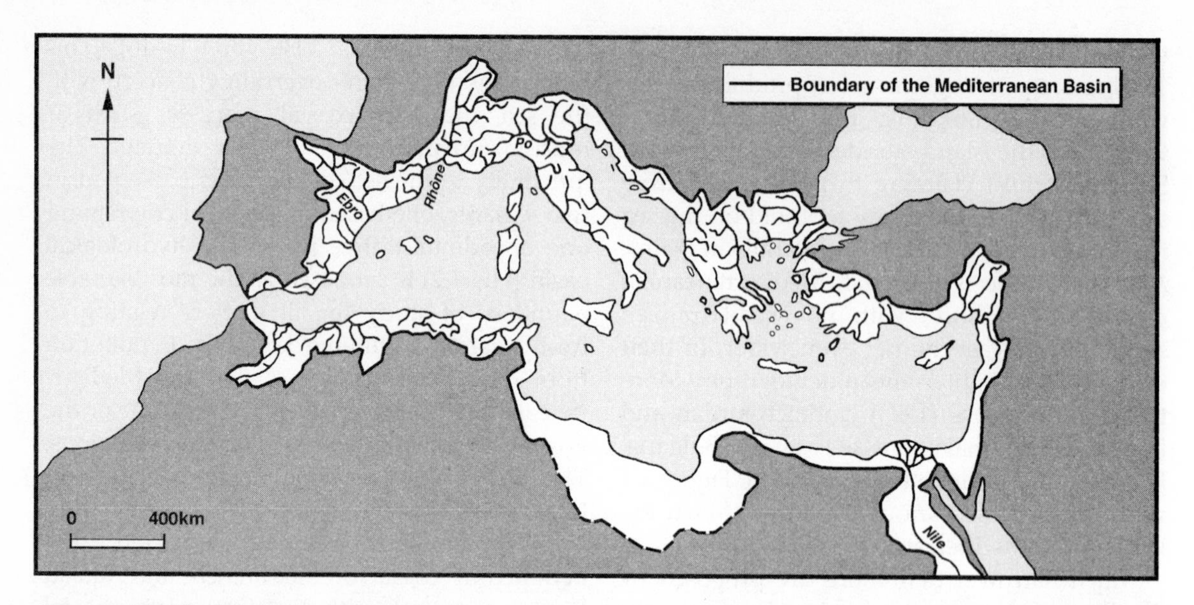

FIGURE 1.2 The Mediterranean watershed *Source:* Grenon and Batisse (1989, p. 19)

transport networks. For strictly practical reasons to do with the availability of population and economic data, the Blue Plan opted to employ the coastal administrative divisions of each nation bordering on the Mediterranean Sea (Fig. 1.3). This produces a strip of variable width depending on the size of units used, with marked variation both between countries (e.g. Tunisia and Italy) and within countries (e.g. Egypt). Only the island–states of Malta and Cyprus are delineated as wholly Mediterranean by this method; and of the larger countries only Italy and Greece succeed in having the majority of their areas included.

Perhaps a more fruitful approach (literally!) to defining the Mediterranean is through biogeography, explored in depth in Chapter 16. Figure 1.4 shows the distribution of four key plants of the region. The climax plant of the Mediterranean forest is the holm oak but this extends to Aquitaine in south-west France and misses out the eastern shores of the basin. More indicative of the Mediterranean environment is the olive, for the summer drought which is the essential feature of the Mediterranean climate is essential for the build-up of oil in the fruit, whilst the deeply penetrating roots can find moisture in even the most rugged of limestone soils. With trees that endure for centuries, the olive (Fig. 1.5) is a symbol of cultural stability and provides the basis of the Mediterranean diet. Hence it lies at the nexus of Mediterranean culture and environment. Some would say it *is* the Mediterranean.

But we should not forget that there are several other characteristically Mediterranean plants. The Aleppo pine has a distribution quite akin to that of the olive (Fig. 1.4). The maquis – the characteristic Mediterranean scrub woodland – has a wide variety of plants which are adapted in various ways to the summer drought and to the fires which are its constant scourge (Tomaselli 1977). Moving to cultivars, the vine is very widespread in the region but is not diagnostic of the Mediterranean, being cultivated as far north as the Rhine valley and even southern Britain. More coterminous with the olive are other Mediterranean tree-crops such as the fig, the carob and the pistachio, whose fruits once had great meaning to local peasant economies.

These, and the other species mentioned, both singly and in their intercropped associations, continue to embody key elements of the visible Mediterranean landscape. In saying this we are moving towards an appreciation, if not a proper

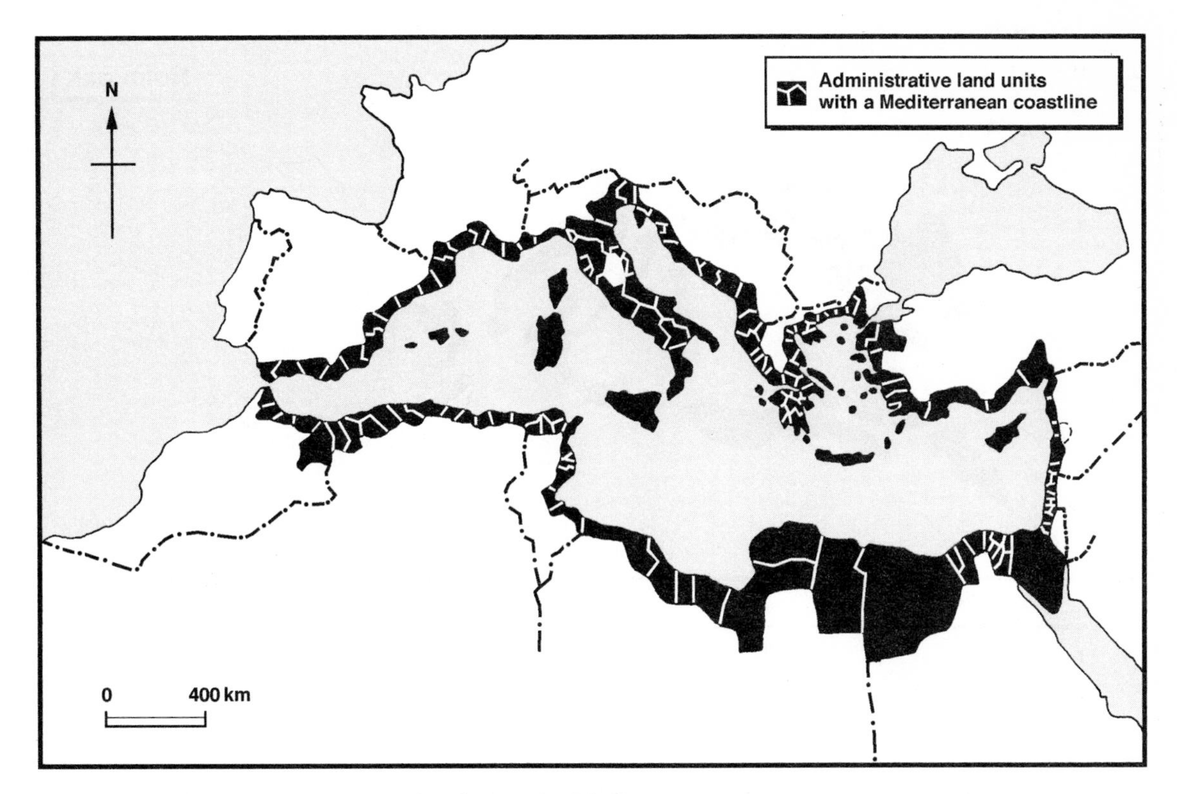

FIGURE 1.3 Administrative regions bordering the Mediterranean *Source:* Grenon and Batisse (1989, p. 18)

definition, of the Mediterranean as an *experience* rather than as some kind of objectified 'reality' to be portrayed on a map. Of all the geographers who have written on the region, J.M. Houston comes closest to this artistic appreciation of the Mediterranean when he states that the geographer must approach the landscape as a painter would (Houston 1964, p. 706). The colour, light and 'atmosphere' of the Mediterranean have long drawn artists and writers whose work constitutes the final channel by which geographers can define the essence of the region. For Lawrence Durrell, author of two marvellous books on Corfu and Cyprus (Durrell 1956, 1962), the Mediterranean 'is landscape-dominated; its people are simply the landscape-wishes of the earth sharing their particularities with the wine and the food, the sunlight and the sea'. Symbols of this landscape are 'the familiar prospects of vines, olives, cypresses ... the odour of thyme bruised by the hoofs of the sheep on the sun-drunk hills'. Never content when living away from the wine-drinking countries of the Mediterranean, Durrell yearned always to return to 'the mainstream of meridional hospitality where a drink refused was an insult given ... such laughter, such sunburned faces, such copious potations' (Durrell 1969, pp. 356–7, 369–71).

This book considers it inappropriate to define the Mediterranean in any hard and fast way. In general, its contributors focus on the Mediterranean littoral and on those areas characterised most closely by Mediterranean environment and culture. However, where common sense (or the availability of statistical data) dictates, a broader definition will be applied, as in the chapters on geopolitics (Chapter 8), economic development (Chapter 9) and demography (Chapter 11), where trends derive as much from national (or international) scale variables as from local processes.

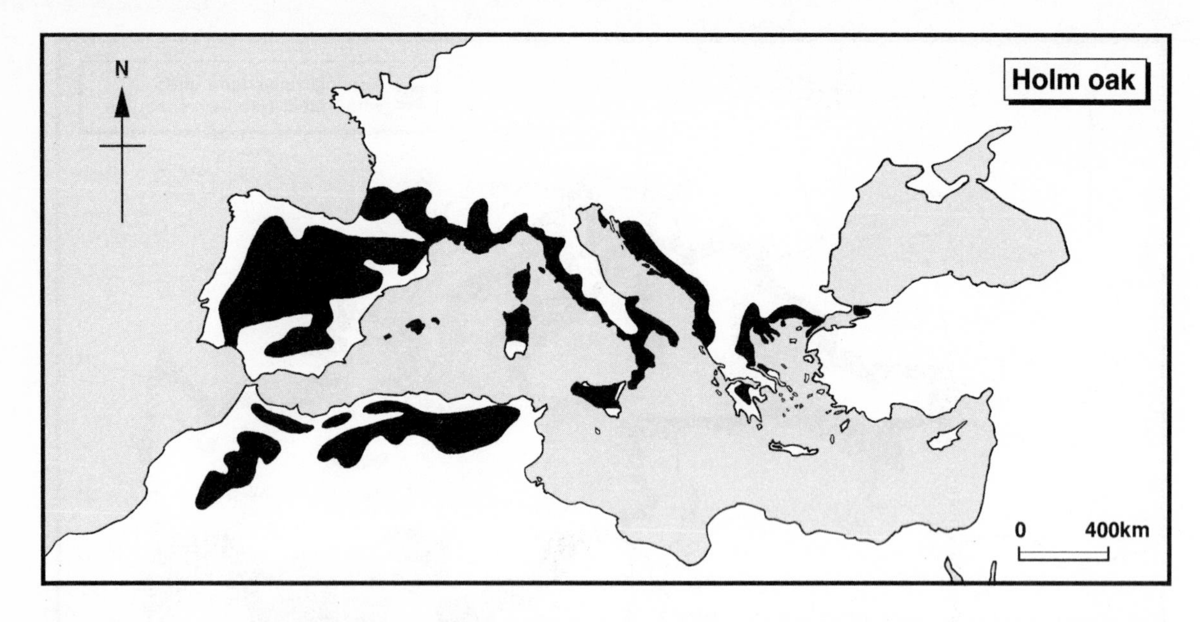

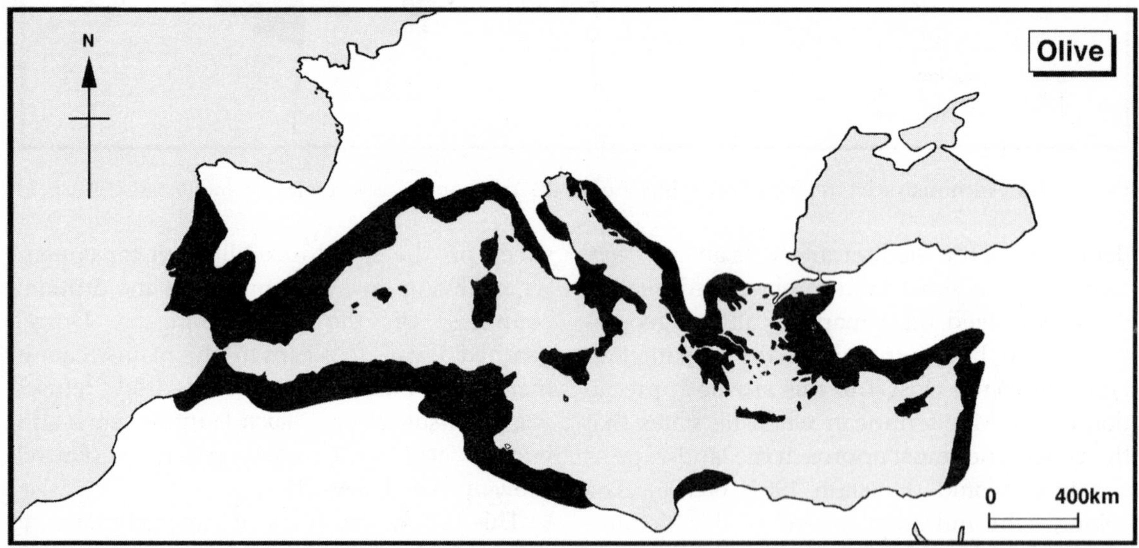

FIGURE 1.4 The limits of four Mediterranean plants *Source:* Partly after Grenon and Batisse (1989, p. 8)

MEDITERRANEANISM

The writings of Lawrence Durrell are but one example of a literary approach to understanding what is the essence of the Mediterranean – what one may call *Mediterraneanism*. One of the qualities of Mediterraneanism is undoubtedly the close interaction it represents between the physical and human realms. In fact, millennia of settled and intense occupation of the land have so humanised the landscape that physical and cultural aspects virtually blend as one. In this way terraces, irrigation systems, or estuary siltation represent landscape elements or processes which have a mixed social–physical origin and expression.

According to Houston (1964, pp. 2–7) six major, interlocking themes characterise the tra-

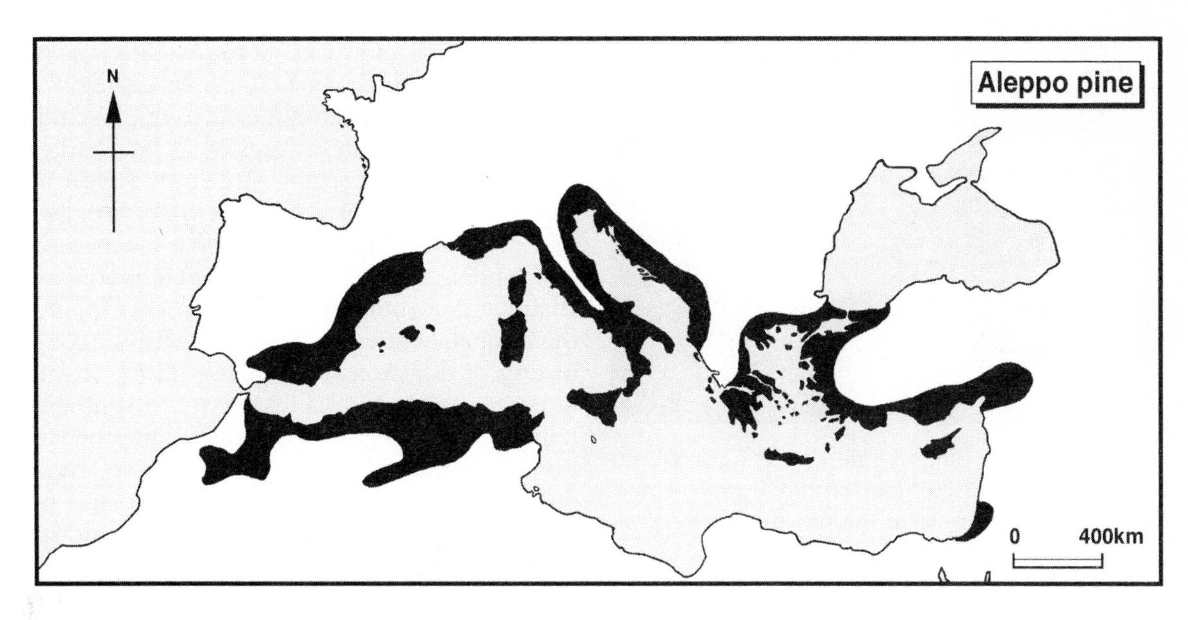

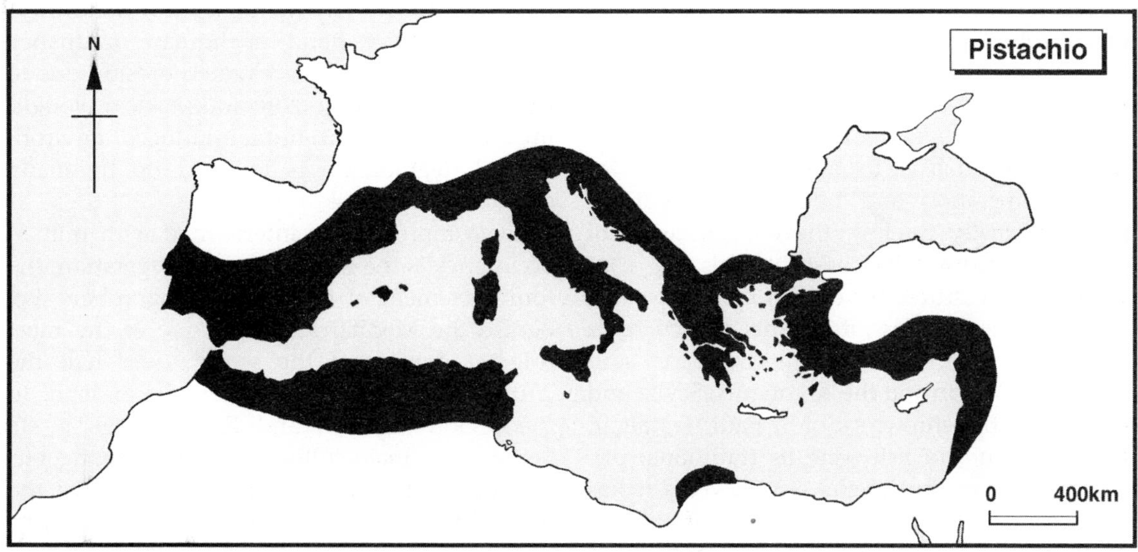

ditional Mediterranean scene – setting aside, that is, modern developments such as the transformation of coastal areas by mass tourism or the petrochemical industry. First there is the climate, more variable from place to place (and from year to year) than is generally thought (see Chapter 3), but whose unifying feature – the summer heat and drought – is perhaps the essential defining characteristic of the region. The atmosphere bathes the Mediterranean in a luminosity which makes the landscape features clear and sharp: 'there is no fusion of colours, no subordinate planes, no soft horizons enveloped mistily in cloud ... the shadows are rigidly projected and only the distant prospect shimmers frail and vague in the heat haze' (Houston 1964, p. 2). Yet this summer light should not blind us to the ferocity and damaging effects of the autumn rains – as at Larnaca in Cyprus when 192 mm of rain fell in four hours in 1981, eroding in a single cloudburst 25 times the normal annual loss of soil.

FIGURE 1.5 The olive is one of the most potent symbols of the Mediterranean environment and landscape. Olive trees in the province of Brindisi, southern Italy.

Second, there is the sea, which plays two major roles in the Mediterranean complex. It acts as a meteorological catalyst, attracting low-pressure systems in winter and acting as a kind of central heating system in summer. Second, it has functioned as a usually calm passage for human migration and trade since Neolithic times. Virtually tideless, the concentration of port–cities around its coasts reveals the ease with which cultures have been spread by this medium. Nevertheless, its storms and currents and battles feature prominently in history and mythology: Jason and the Argonauts, Scylla and Charybdis, the shipwreck of St Paul, Trafalgar, Lepanto. Salt and fish were its traditional primary resources, but in the last 30 years it has taken on new roles – attracting tens of millions of tourists, the site of large industrial complexes, and criss-crossed by oil routes leading from Suez and North Africa to the oil refineries of Europe. Heavily polluting, these late twentieth-century activities have provoked an environmental crisis in and around the sea which the Blue Plan has been trying to grapple with (see Chapter 17).

The third element is the land itself. Despite its coastal orientation, the Mediterranean is a region of mountains, of complex and fragmented relief. Even the islands are mountainous, most of them. Houston (1964, p. 51) calls the mountains and basins of the Mediterranean the Enigma Variations of tectonic geology: a symphony of the earth not easy to understand. Nevertheless, certain features recur, as Chapter 2 explains. Similar tectonic structures appear in Iberia, the Maghreb and Asia Minor: rugged fold mountains pushed up against older, rigid tablelands. The presence of unstable mountain chains and of young sedimentary rocks explains the frequency of earth tremors. There is a long history of disastrous earthquakes in the Mediterranean, and of volcanic activity in southern Italy and the Aegean. Scenically, limestone is the most characteristic Mediterranean rock, often rising steeply from flat coastal plains or intermontane basins. Such sharp relief contrasts are reflected in clear vegetation changes and in different rural economies and settlement densities. However, many Mediterranean landforms bear the heavy hand of human occupancy which has massively accelerated erosion due to deforestation, overgrazing and reckless ploughing. This theme of human-induced environmental degradation is touched on by many chapters in the book.

Even more closely intertwined with human occupancy is the Mediterranean vegetation, the fourth element of the six. Biogeographers recognise the Mediterranean as one of the most original regions of the world: over half the 25 000 plant species found there are endemic to the region (Grenon and Batisse 1989, p. 8). Some of the main cultivars – olive, vine, fig, etc. – have been mentioned already, as have the holm oak and Aleppo pine, two major tree species of the region. Undoubtedly these and other plants contribute greatly to the richly variegated personality of Mediterranean regional landscapes: the cork oaks of the Alentejo or northern Sardinia, the cedars of the Atlas and of Lebanon, the olive forests of Crete, the umbrella pines of the Roman Campagna. Exotic species have also made their impact: citrus groves in the irrigated lowlands, palms in coastal promenades and hotel gardens, eucalypts along roads and in shelter-belts, prickly-pear cactus on rocky slopes and terrace walls. In the rural economy, agriculture and pastoralism have often existed as conflicting land-use ecotypes, competing for land and investment, as some excellent studies of Sardinia have shown (Bergeron 1967, 1969; Le Lannou 1941; Weingrod and Morin 1971). Houston (1964, p. 5) stresses how these basic Mediterranean ecotypes are associated with different ecological processes. Pastoralism is regressive because of its intensive grazing and use of fire: scrubby maquis and even thinner *garrigue* are the end result. Agriculture creates its own managed ecosystem, but when the land is abandoned, progressive ecological processes will usually restore the Mediterranean forest before the soil is irreparably damaged.

The final two elements of the Mediterranean landscape derive from the human side. They are the long tradition of urban life in the Basin (see Chapter 12); and society's perception and evaluation of the resources offered by the Mediterranean environment. These resources embrace not only the natural attractions of soil, slope and microclimate but also symbolic resources: 'the sacred groves, the templed promontories, the sanctified territory of the town site' (Houston 1964, p. 6). The contrasting perceptions of the Christian and Islamic civilisations have each left a distinct impression of their own evaluation and utilisation of resources in different parts of the basin, overlapping dramatically in the Iberian peninsula where the northward advance of the Moors and the subsequent reconquest are essential historical ingredients of an understanding of the present landscape (Chapter 6). This contrast is as evident in cities as it is in rural areas with their different cropping, irrigation and tenure systems. The Islamic emphasis was on local rule and pluralistic urban societies composed of Moslems, Jews and Christians, and on privacy and enclosure in the structure of town layouts. Christian urban forms, often inherited from the Romans, reflected local autonomy and pride, with open central squares, large town halls and prominent churches. All around the Mediterranean the progressive atmosphere of the towns, long established as settlement sites and supported by strong currents of commerce and often aggressive maritime mercantilism, has contrasted with the backwardness and meagre self-sufficiency of rural life.

These six themes can be regarded as the geographer's building-blocks for an understanding of traditional Mediterranean landscapes. They are found to recur in subtle combinations throughout the region. Above all, it is their multiple relationships which provide the highest level of comprehension of what the region is about and how it functions as a human-dominated ecological system.

CONCLUSION: UNITY AND DIVERSITY IN THE MODERN MEDITERRANEAN

So far this chapter has described the regional character and coherence of the Mediterranean on traditional geographical grounds. The main factors which have been used to sustain this position of Mediterranean unity have been those of climate, ecology and cultural traditions. We have made little reference as yet to modern economic, demographic and political factors. Examination of these dimensions yields a more finely balanced situation as regards the unity-versus-diversity debate.

Tovias (1994) argues that there is such a thing as a 'Mediterranean economy'. The Mediterranean climate, together with proximity to central and northern European markets, accounts for a specialism in perishable fruits and vegetables which are traded northwards (Chapter 13). In the post-war decades, especially, such trade flows were paralleled by northward migrations of labour, another surplus product of the Mediterranean. In these respects the Mediterranean is united in being predominantly under the European sphere of economic influence. The supply of tourists also moves along a north–south axis. Another unifying feature of the Mediterranean is its strategic location between major producers and consumers of oil, giving it an important role in the transport and refining of petroleum products. In the modern cultural and political realms almost all Mediterranean countries are orientated towards Europe. European countries are admired as role models and are seen as the source of advanced technology, democratic ideals and mass media and entertainment.

On the other hand, there are several dimensions of marked differentiation within the basin. Mainly these are in the geopolitical sphere and consist of heterogenous political regimes, the coexistence and confrontation of different religions and cultures, and the fact that European political and economic union has the effect of dividing the Mediterranean states into those which are EU members (Spain, Portugal, France, Italy, Greece) and those which are not (Chapter 8). Rather than being a harmonious, inward-looking region, the Mediterranean can thus be seen as a multi-faceted fault-line separating Christian, democratic, developed Europe from Islamic, politically unstable, economically underdeveloped and demographically exploding North Africa: a frontier, in other words, between what used to be called the First and the Third Worlds.

This characterisation of the Mediterranean as a confrontation zone and an area of political fragmentation is overstated. It ignores the political efforts that are being made to unite large parts of the region via trade and cooperation agreements, the progress made in resolving the Arab–Israeli conflict, cross-Mediterranean mobility of students and workers, pan-Mediterranean cultural initiatives, and the shared concern for the environment. Conscious of the need to stimulate development in order to dampen migration flows and forestall political instability, the EU recently agreed to a $6 billion aid package to the Mediterranean states for 1995–99 – part of a 'Euro-Med' strategy which sees the Mediterranean turning more and more into a European lake (see Chapter 10). Despite the EU's close involvement with developments in Eastern Europe, political unrest in Algeria and the unpredictability of Libya mean that the EU cannot afford to shift attention away from its southern flank.

Rather than view the region as a boundary for conflict, perhaps a better analogy would be to see the Mediterranean as a geographical stage, whose spotlight is the Mediterranean sun and whose architecture is made up of the coasts, mountains, peninsulas and islands of the basin. Sometimes the stage is a theatre of war; more often it is the setting for peace, trade, and creative human movement. The actors on this stage speak many tongues and do not always understand each other; nor do we, the audience, always realise what is really going on, for the plots and story-lines are complex and not always what they seem. The Mediterranean too speaks with many voices; this book is an attempt to understand the various acts of the Mediterranean drama. Take your seat and read on!

References

BECKINSALE, R. and BECKINSALE, M. 1975: *Southern Europe*. London: University of London Press.

BERGERON, R. 1967, 1969: Problèmes de la vie pastorale en Sardaigne. *Revue de Géographie de Lyon* **42**, 311–28; **44**, 251–80.

BIROT, P. and DRESCH, J. 1953, 1956: *La Méditerranée et le Moyen-Orient. Vol. 1, La Méditerranée Occidentale; Vol. 2, La Méditerranée Orientale et le Moyen-Orient*. Paris: Presses Universitaires de France.

BRANIGAN, J.J. and JARRETT, H.R. 1975: *The Mediterranean Lands*. London: MacDonald and Evans.

BRAUDEL, F. 1972, 1973: *The Mediterranean and the Mediterranean World in the Age of Philip II*. 2 vols. London: Collins.

DURRELL, L. 1956: *Prospero's Cell*. London: Faber and Faber.

DURRELL, L. 1962: *Bitter Lemons*. London: Faber and Faber.

DURRELL, L. 1969: *Spirit of Place: Mediterranean Writings*. London: Faber and Faber.

EAST, W.G. 1940: *Mediterranean Problems*. London: Nelson.

GRENON, M. and BATISSE, M. 1989: *Futures for the Mediterranean Basin*. Oxford: Oxford University Press.

HADJIMICHALIS, C. 1987: *Uneven Development and Regionalism: State, Territory and Class in Southern Europe*. London: Croom Helm.

HOUSTON, J.M. 1964: *The Western Mediterranean World*. London: Longmans.

HUDSON, R. and LEWIS, J. (eds) 1985: *Uneven Development in Southern Europe*. London: Methuen.

LE LANNOU, M. 1941: *Pâtres et Paysans de la Sardaigne*. Tours: Arrault.

LEONTIDOU, L. 1990: *The Mediterranean City in Transition*. Cambridge: Cambridge University Press.

MCNEILL, J.R. 1992: *The Mountains of the Mediterranean World: An Environmental History*. Cambridge: Cambridge University Press.

MONTANARI, A. and CORTESE, A. 1993: South to North migration in a Mediterranean perspective. In King, R. (ed.), *Mass Migrations in Europe: The Legacy and the Future*. London: Belhaven, 212–33.

NEWBIGIN, M. 1924: *The Mediterranean Lands*. London: Christophers.

RIBEIRO, O. 1968: *Mediterrâneo: Ambiente e Tradição*. Lisbon: Fundação Calouste Gulbenkian.

RIBEIRO, O. 1983: *Il Mediterraneo: Ambiente e Tradizione*. Milan: Mursia.

ROBINSON, H. 1970: *The Mediterranean Lands*. London: University Tutorial Press.

SEMPLE, E.C. 1932: *The Geography of the Mediterranean Region: Its Relation to Ancient History*. London: Christophers.

SIEGFRIED, A. 1949: *The Mediterranean*. London: Cape.

TOMASELLI, R. 1977: The degradation of the Mediterranean maquis. *Ambio* **6**, 356–62.

TOVIAS, A. 1994: The Mediterranean economy. In Ludlow, P. (ed.), *Europe and the Mediterranean*. London: Brassey's, 1–46.

WALKER, D.S. 1965: *The Mediterranean Lands*. London: Methuen.

WEINGROD, A. and MORIN, E. 1971: Post peasants: the character of contemporary Sardinian society. *Comparative Studies in Society and History* **13**, 301–24.

WILLIAMS, A. (ed.) 1984: *Southern Europe Transformed*. London: Harper and Row.

2

GEOLOGICAL EVOLUTION OF THE MEDITERRANEAN BASIN

ALASTAIR RUFFELL

INTRODUCTION

This chapter aims to provide sufficient background information to enable readers to assess for themselves the control that geology has on both the natural landscape and human activities within the Mediterranean Basin. It is therefore selective in its treatment and cannot include detailed studies of individual areas. Instead, it synthesises and demonstrates the importance of the unique features of Mediterranean geology. To do this, it summarises the major works on Mediterranean geology, further reference to which should allow the specialist reader to enquire into any specific subject area.

Any summary of Mediterranean geology is bound to utilise information from a range of investigative techniques, published mostly over the past 50 years and concerning different geographical areas. Plate tectonic reconstructions of the Mediterranean area, for example, must use information from well beyond the area examined by this book (Fig. 2.1). Thus the reader should not be dismayed by references to locations as far from the Mediterranean Sea as the Zagros and Carpathian Mountains: geology (in the first instance) is a global subject.

Definitions and background

The present-day Mediterranean Sea forms a distinct sedimentary basin, as defined by Allen and Allen (1990). This comprises a unique area of crust, downwarped by comparison with the surrounding hinterlands to provide a place where significant thicknesses (thousands of metres) of sediment accumulate. Obvious examples of the sedimentary input to the Mediterranean are the Rhône, Nile and Ebro delta systems. In addition to sites of present-day sedimentation, ancient basins are also well documented and many modern basins exhibit evidence of previous phases of sedimentation. The Mediterranean is no exception to this dual origin. Stanley and Wezel (1985) defined the Mediterranean Basin as two closely related entities: the submerged realm and the circum-Mediterranean terranes. Broadly speaking, present-day geological basins correspond to the submerged realm; and outcrop evidence, in the form of circum-Mediterranean terranes, reflects an earlier evolution of the Mediterranean.

Early geology and structure of the Mediterranean

The oldest rocks exposed in the Mediterranean islands and borderlands are of Precambrian age

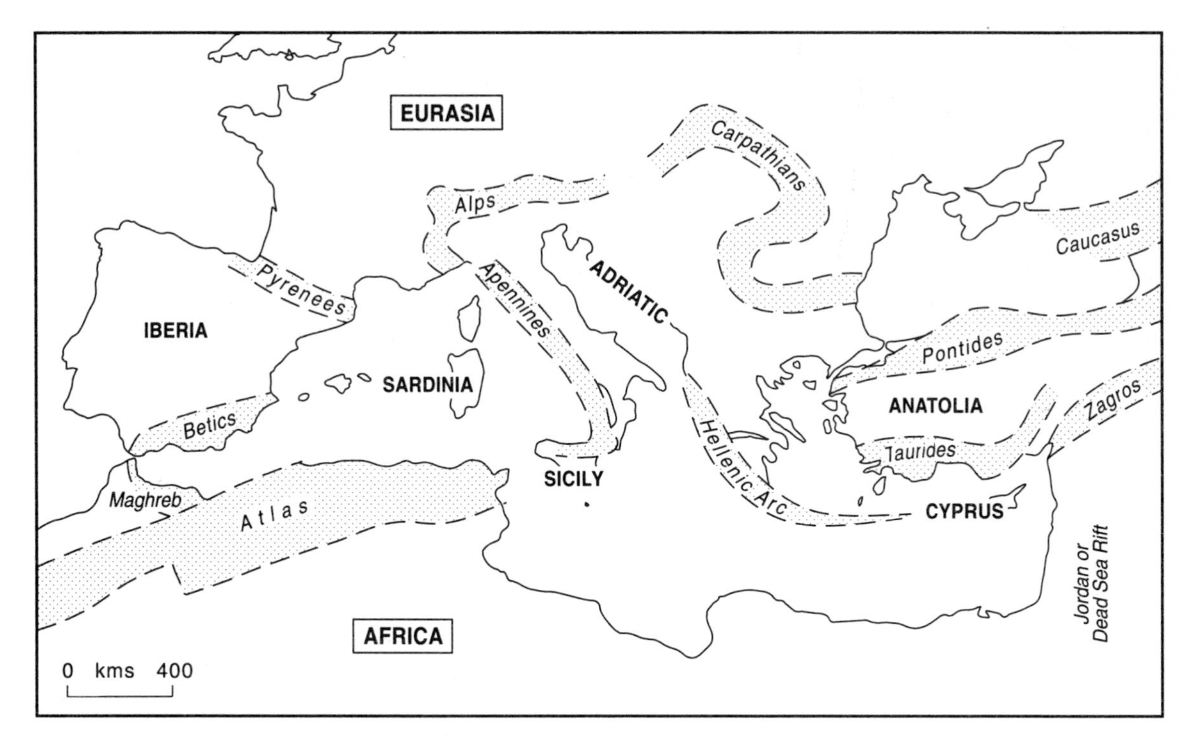

FIGURE 2.1 Distribution of volcanic chains and mountain belts around the Mediterranean (stippled). Blank areas comprise stable continental or ocean crust.

(over 600 million years old, or 600Ma) and are found in the Iberian Peninsula, Massif Central and in a belt from northern Greece to Yugoslavia. These ancient rocks have suffered numerous successive phases of deformation: none the less they can be recognised as formerly comprising shallow and deep marine sedimentary and volcanic rocks, deposited on an area of continental shelf bordering an ocean. During the Lower Palaeozoic (*ca.* 570 to 400Ma) southern Europe and Africa lay adjacent to North America, separated by an ocean from northern Europe, which lay adjacent to Greenland. This ocean closed at the end of the Lower Palaeozoic when a second (Upper Palaeozoic or *ca.* 400 to 230Ma) ocean formed to the south, roughly where the Appalachians–English Channel and Belgium–Ruhr are at present. The Mediterranean region consisted of a relatively stable platform of continental crust in contact with Africa: preserved rock strata of this age are of limited extent and comprise the typical Upper Carboniferous sandstones, coals and limestones of Europe. The Mesozoic era (*ca.* 230 to 65Ma) saw both the north–south break-up of the Mediterranean lands to form the Tethys Ocean and the initial rifting of the Atlantic. Later, the Tethys closed (in late Cretaceous times or *ca.* 100 to 65Ma) whilst rifting continued and even accelerated in the North Atlantic (e.g. the oceanic separation of Iberia from France and the opening of the Bay of Biscay).

Post-Cretaceous geology and the integrating role of plate tectonics

The geological and geomorphological evolution of the modern Mediterranean only really began in the Cenozoic (65Ma to the present), as Eurasia, separated from America and Africa, rotated anticlockwise, impacting on southern Eurasia. The Tethys Ocean still existed, but was now narrow and characterised by intense deformation and volcanicity through the Apennines, Greece and Turkey (Fig. 2.1). In this belt the emplacement of ophiolites (see below for explanation) was nearly complete and to the north and west continental impact resulted in the mountain-building (or orogenic) episodes of the Pyrenees, Carpathians and Alps. Rotation of many of the micro-continental blocks within the Mediterranean Tethys belt also occurred at this time.

Our understanding of this evolution has relied heavily upon the extensive exploration of the Mediterranean Basin since the 1960s that has paralleled the development of the theory of plate tectonics. This theory has both increased our understanding of the Mediterranean and has additionally been based on observations made in the Mediterranean area itself. Plate tectonic theory (e.g. Duff 1993; Skinner and Porter 1995) provides much of the explanation as to how the Mediterranean Basin evolved, why definitive rock types occur in certain areas and the present/past location of sites of significant sediment accumulation (depocentres).

Structure of the chapter

On the basis of the oft-quoted geological dictum that 'the present is the key to the past', it is appropriate to begin an examination of the geology of the Mediterranean by discussing the possible plate boundaries and motion within and around the basin at the present day through the examination of *neotectonics* (earthquakes and volcanoes). These have significance in explaining the geological evolution of the Mediterranean, and in helping our understanding of various human crises that have occurred throughout the Mediterranean islands and borderlands, especially in Italy, the Greek islands, Turkey, Cyprus and the Middle East. The study of present-day plate tectonics in the Mediterranean provokes a natural second question – how did this arrangement come about? In response to this, the second part of the chapter will consider certain critical outcrops of ancient oceanic crust known as *ophiolites*. Ophiolites not only offer critical evidence of past plate tectonics, but also provide the host succession to much of the ore mineralisation that is so important to the historical development of human societies in the region.

An improved knowledge of present and past distributions of continental and oceanic plates has greatly aided the development of ideas on the Tethys as a precursor ocean to the present-day Mediterranean Sea, and on the geological interpretation of the Mediterranean lands. Most especially, the common association of ophiolites and (areally) more extensive limestones are a common feature of Mediterranean geology. This limestone topography, together with the underlying *geological structure*, forms an important control on geomorphology throughout the study area. Furthermore, during the closure of the Tethys between 15 and 5Ma, the narrowing of the limited seaway between Africa and Iberia restricted the flow of marine water into the newly-formed Mediterranean. This resulted in repeated evaporation, desiccation, and the formation of thick beds of salt and gypsum (both evaporite minerals). This drying-out of the Mediterranean has been termed the *Messinian Salinity Crisis* by Hsu (1972).

Finally, the chapter will demonstrate how the *economic geology* of the Mediterranean is controlled by the interplay between all the above factors. For example, plate tectonics controls subsidence and it is in subsiding basins that significant quantities of oil and gas-bearing strata have accumulated. In addition, the ocean crust formed beneath the Tethys, and now preserved in ophiolites throughout the Mediterranean, is host to important ore minerals. Industrial minerals are also present and include ornate Tethyan limestones and marbles (used as decorative building stones), evaporite minerals (including gypsum used in plaster manufacture) and industrial bentonite clays (derived from volcanic ash).

NEOTECTONICS

Neotectonics describes present-day and recent historical plate motions evidenced by seismicity (earthquakes) and, to a lesser extent, volcanicity. The major control on seismicity throughout the Mediterranean Basin is the plate boundary between the northerly-drifting African Plate and its collision with the main Eurasian Plate to the north and a mosaic of small plates throughout the Mediterranean (Arabian, Adriatic and Iberian). These microplates are grouped with the Eurasian Plate for convenience, although some have affinities with Africa. Their movement is controlled primarily by closure of the Mediterranean but also by the opening of the Atlantic and, to a lesser extent, the opening of the Red and Arabian Seas.

Examination of the distribution of the surface earthquakes throughout the Mediterranean (Fig. 2.2 A and B) shows the clear concentration of seismic activity along the Italian peninsula and through the Greek islands. The location of deep earthquakes (Fig. 2.2 C) confirms this pattern and demonstrates the importance of the Apennine (Italian), Hellenic and eastern Carpathian seismic hazard zones. Recent seismic activity allows us to confirm the theory of Udias (1985) who suggested dividing the Mediterranean into two areas, the Western and Eastern Mediterranean Basins, separated by Italy and Sicily. The Western Mediterranean (seismic) Basin is characterised by a lack of deep earthquakes except on its eastern (Italian) margin, while the Eastern Mediterranean Basin shows abundant seismicity along its northern (Hellenic–Cyprus–Turkish) margin.

The pattern of surface earthquakes (Fig. 2.2 A) demonstrates the existence of secondary or former subsident basins beneath the Tyrrhenian, Adriatic and Levantine Seas. Conversely, the location of deep earthquakes coincides with areas of deformed and metamorphosed rock successions and volcanically active zones termed orogenic belts. These are coincident with the mountain belts of the Alps, Carpathians and Pyrenees and volcanic arcs of the Apennines and Hellenides and characterise zones of continental collision. Less abundant oceanic earthquake foci (Fig. 2.2 A) demonstrate the existence of the Azores Fracture Zone (running east–west into the Straits of Gibraltar) as well as the Mid-Atlantic Ridge.

By far the most dominant earthquake pattern of the Mediterranean is along the Hellenic Arc, stretching from the west coast of Greece to southern Turkey. It is here that the most spectacular geological activity is concentrated. Papazachos and Comninakis (1977) interpreted this dense distribution of shallow and deep earthquakes as a Benioff zone, or the place where one plate is subducted beneath another. In the case of the Hellenic Arc, the oceanic northern side of the African Plate is dipping under the Eurasian Plate, melting and causing earthquakes as it sinks. The Hellenic Arc is thus the most recent site of subduction in the Mediterranean.

Udias (1985) interprets deep earthquake focal mechanisms in terms of plate motion (Fig. 2.2 C and D). This confirms the suggestion that the Mediterranean, as a successor to a previous, larger ocean (Tethys) is still closing. Active collision is occurring between northern Africa and southern Spain, along the length of Italy, between Africa and Greece/Turkey along the Hellenic Arc, and between Africa and Arabia from Cyprus through to the Zagros orogenic belt of Iran/Iraq. In addition, Udias (1985) suggests that the east–west division of the Mediterranean Basin demonstrates the genetic relationship between the western Mediterranean and Eurasian Plate, and the eastern Mediterranean and African Plate.

The geologically recent earthquake and volcanic activity that has resulted from the African–Eurasian Plate collision has had significant effects on human civilisation. This is not only true for those populations living close to active volcanoes and active seismic zones but for all peoples of the Mediterranean when volcanic dust and earthquake-induced sea waves travel long distances. For example, one of the most violent and catastrophic volcanic explosions in human history occurred on Santorini in 1470 BC. The resultant tidal wave caused devastation to the Minoan civilisation. Similarly, the eruptions of Etna and Vesuvius are thought to

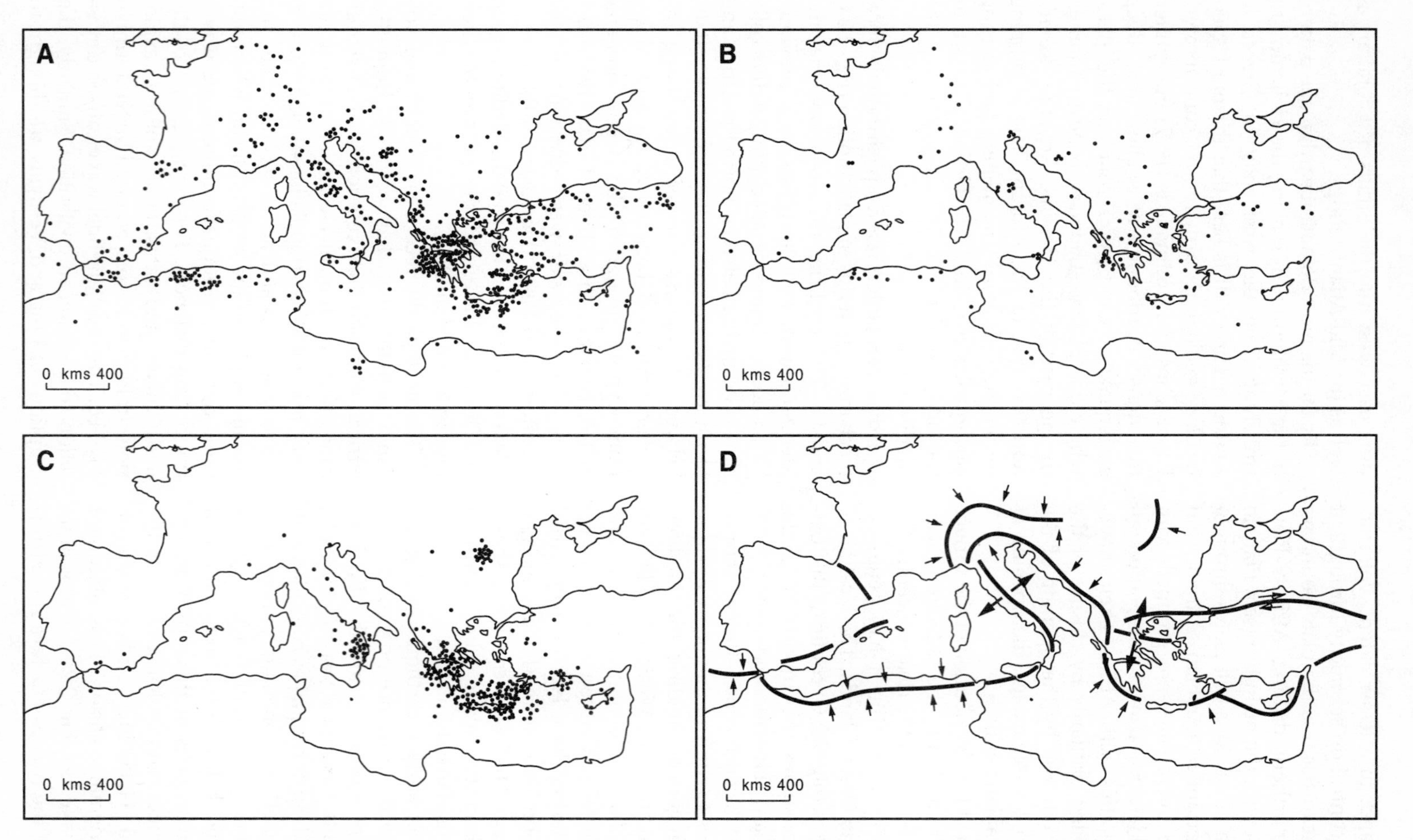

FIGURE 2.2 Location of earthquakes (1910–70) in the Mediterranean area. (A) = surface earthquakes; (B) = large surface earthquakes (greater than 6.5, Richter Scale); (C) = deep earthquakes; (D) = interpretation of present-day plate motion from earthquake data.

Source: After Udias (1985)

have caused major crises in the local population. Perhaps the most intriguing of all eruptions is the one that is thought to have occurred on the Aegean island of Thera in the Bronze Age. The story of the excavation of Akrotiri, near the Kameni volcano, is summarised in Hardy (1990), and radiocarbon dating of seeds found at the site suggests that the eruption occurred between 1680 and 1520 BC. This unfortunately long time-slot is due to the radiocarbon calibration curve being relatively flat across the seventeenth and sixteenth centuries BC. However, other palaeoclimatic evidence from around the world may help in dating the eruption as well as demonstrating its possible global significance. Baillie (1990), for example, has shown that there is a clear 'frost-ring' in American Bristlecone Pines at 1627 BC and a narrowing of European oak tree-rings at 1628 BC. Unfortunately, he also noted a growth increase in Anatolian junipers at 1628 BC, but this growth spurt could be linked to the vast quantities of magmatic water that were erupted from Thera at this time. The intriguing aspect to this story is that classical archaeology (based on Egyptian dynasties and pottery sequences) suggests a late Minoan age for the eruption (i.e. *ca.* 1500 BC), far too late for those conducting the absolute dating. Are the absolute dates wrong? Is classical archaeology using incorrect dates? Were there two eruptions? Whatever the answers, excavations at Akrotiri (Thera) and Pompeii show towns to have been obliterated and many lives lost in the eruptions and attendant earthquakes.

Ophiolites

The importance of ophiolites in modern geological studies, especially plate tectonics, is evidenced by the vast amount of literature relating to these rock successions (e.g. Duff 1993; Gass 1968; Skinner and Porter 1995). Long before their importance in plate tectonics was recognised, ophiolites were seen as unusual rock successions, commonly associated with sedimentary rocks formed in the deep oceans and exposed in orogenic belts. Many early studies of ophiolite successions were based on areas in or around the Mediterranean. The most complete ophiolite successions (from top to bottom) comprise:

- shallow marine sediments (frequently chalks)
- deep marine sediment (frequently radiolarite)
- weathered basalt (umbers); volcanic clays (bentonites)
- pillow lava (basalt)
- sheeted dykes (dolerites and basalts)
- high-level intrusives (gabbros and plagioclase granites)
- layered cumulates (dunites, peridotites, frequently altered to serpentinite).

A relatively intact succession forms the Troodos Massif on Cyprus. By contrast, in more structurally complex zones (e.g. the Alpine–Ligurian and Hellenic) the typical succession is split by faults and the ophiolite structure is harder to observe.

Submarine exploration and deep-sea drilling on mid-ocean ridges have revealed similar rock and sediment types to those found in ophiolites. The uppermost chalks and radiolarites represent the slow accumulation of the microscopic skeletons of calcareous and siliceous organisms (respectively), but Knipper *et al.* (1986) consider that overlying sediments should not be included in the formal definition of ophiolites. The sediments cover pillow lavas formed by submarine extrusive volcanism fed by dykes (intruded into one another in a tensional environment) that appear as the sheeted dyke complex in ophiolites. This tension was created by the continual splitting of the ocean floor along the mid-ocean ridge as plates moved apart (Skinner and Porter 1995). Beneath these zones of intrusive and extrusive igneous activity are one or more semi-permanent magma chambers that fed the higher level igneous activity (see Fig. 2.3). The cumulate rocks are the first-crystallised residue of these magmas, whilst the higher-level plagioclase-rich gabbros and granites represent the later crystallised melt.

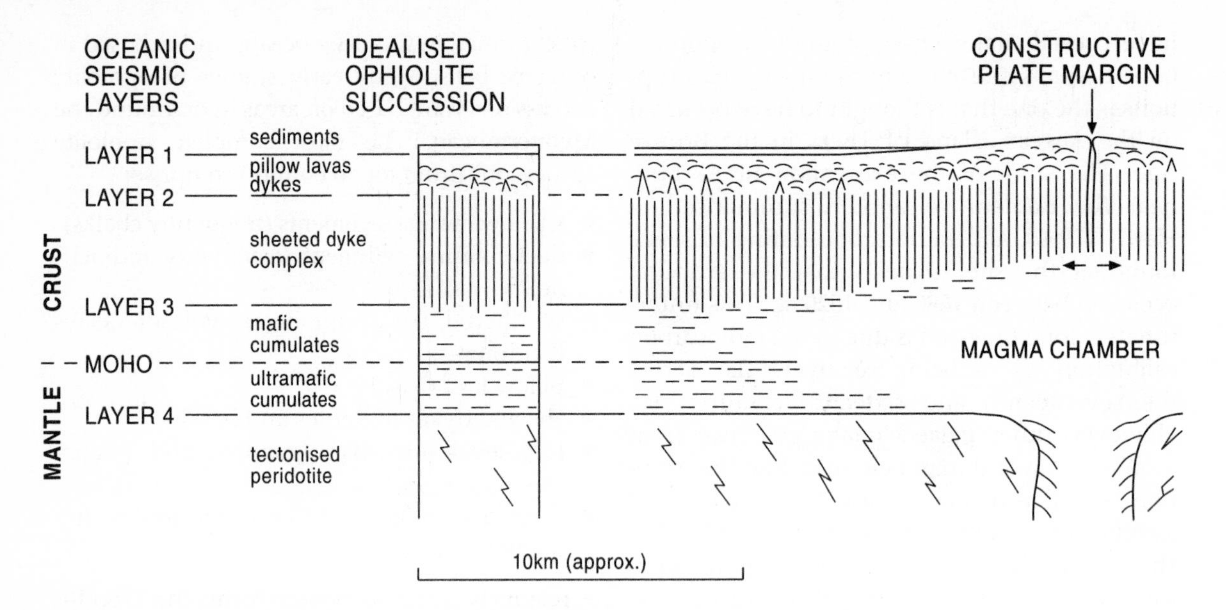

FIGURE 2.3 Greenbaum's Single Chamber Model of ocean crust generation and consequent ophiolite structure

Source: Gass (1968)

Following solidification, the entire ocean ridge would still have been hot but filled with cracks caused during cooling that allowed the circulation of sea water. This water was heated and not only aided low-temperature metamorphism of the sheeted dykes and pillow lavas but mobilised iron, manganese and copper during circulation. Later hydrothermal expulsion caused the precipitation of metalliferous ores around hot water chimneys (black smokers), both on the sea floor and in joints and veins within the solidified rock of the mid-ocean ridge. Much of the ore mineralisation found in, for example, Greece is hosted within the ancient ocean crust now exposed as ophiolites.

Following emplacement, ophiolites remain within the ocean crust for variable periods, depending on the evolution of the ocean. The ocean crust preserved in ophiolites around the Mediterranean is mostly of Cretaceous age, with some Jurassic and Cenozoic. Eventually the passive continental margins of an Atlantic-type ocean evolve to actively-subducting Pacific-type margins, when the ocean basin begins to close. During the Cretaceous this process of subduction and closure was active throughout the northern Mediterranean area. Not all ocean crust is, however, subducted. When particularly buoyant ocean crust (e.g. mid-ocean ridges) collide with similar areas of buoyant ocean crust, the oceanic slab may not sink in subduction but be accreted onto the adjacent crust. During this accretion the buoyant ocean slab may actually ride up in a process known as obduction to form a displaced mass of ocean crust.

The location of Mediterranean ophiolites is shown on Figure 2.4, where it can be seen that while the outcrops follow the major orogenic zones of the northern Mediterranean, some are currently submerged (parts of the Ligurian ophiolites), whilst still others are caught up in mountain chains (Carpathian). Following Knipper *et al.* (1986), it is clear that ophiolite emplacement occurred at different times throughout the last 100Ma, and in different tectonic environments. This complex history suggests that, prior to the present-day Mediterranean, there existed between Eurasia and

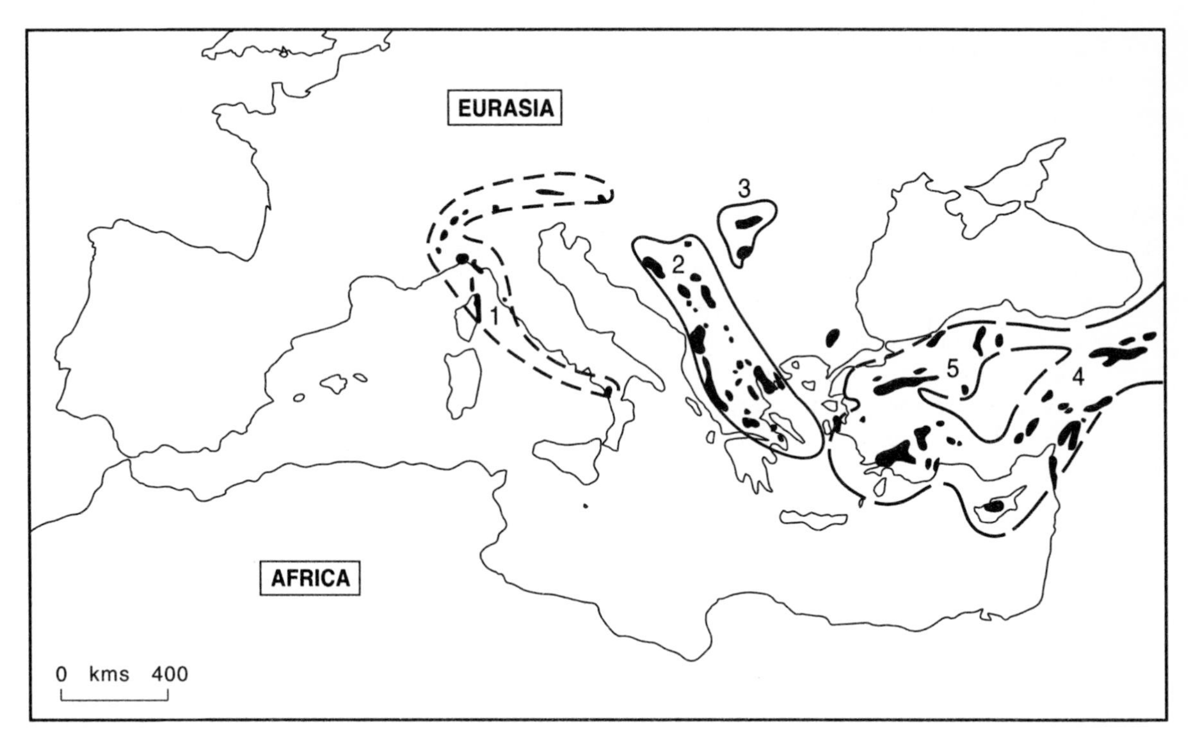

FIGURE 2.4 Distribution of ophiolites throughout the Mediterranean (in black, with inferred limits dashed). All of these are remnants of Tethyan ocean crust or back-arc basins. 1 = Ligurian; 2 = Dinaro-Hellenic; 3 = Carpathian; 4 = Peri-Arabic; 5 = Pontic – Lesser Caucasus.

Source: After Knipper *et al.* (1986)

Africa an ocean large enough to generate ocean crust, for this to be subducted, and for minor ocean basins known as back-arc basins to be formed during ocean closure. Most of the Mediterranean ophiolites were formed in back-arc basins; these contain young, hot and thus buoyant crust that is easily obducted. This (now closed) ocean was the Tethys.

Regardless of mode of emplacement, all Mediterranean ophiolites display evidence of metamorphic alteration. This may be a low-grade alteration resulting from the circulation of hot sea water, through to higher grade metamorphism associated with the intense shearing that occurred during the closure of the ocean basin and collision. Knipper *et al.* (1986) cite the occurrence of gravity slides (mudstone beds containing huge displaced blocks, termed melanges or olistostromes) above the ophiolites of the Mediterranean as evidence of obduction. Following obduction, the crust is probably depressed by the weight of the ocean slab now lying on it; this causes rapid subsidence and the formation of unstable gravity slides. These are very common along the Peri-Arabic ophiolite outcrops (Oman–Cyprus) where the similar age of former ocean crust is taken as evidence for rapid and almost synchronous obduction. The relationship between ophiolites and the existence of the Tethyan Ocean is outlined below.

THE TETHYAN OCEAN

The existence of a major ocean basin between the Eurasian and African plates from Triassic (*ca.* 200Ma) through to Cretaceous and early Cenozoic times (*ca.* 65Ma) has attracted great interest from many branches of geology (e.g. Aubouin *et al.* 1980; Dixon and Robertson 1985).

Scientists interested in plate tectonics and continental drift view the Tethys as a classic example of a closed ocean basin, and thus a model for oceanic collision belts at present. Such scientists envisage an agglomeration of the Earth's plates in Triassic time. This supercontinent is termed Pangaea. It had an early rift phase in the form of Tethys and finally dispersed with the opening of the Atlantic. Alternatively, those concerned with hydrocarbon exploration view the Tethys as the oceanic environment where rocks which now host half of the world's oil and gas reserves (Middle East) were deposited.

The evidence that the Tethys once existed and was not simply an older version of the Mediterranean comes from palaeomagnetism, palaeoenvironmental analysis and ophiolite geology.

Palaeomagnetism

Many rocks formed at or near the surface of the Earth inherit a magnetisation from the predominant global field at the time. Thus rocks formed on continents that have subsequently moved (through continental drift) show a different magnetisation when compared to the present-day polar position. By measuring the remnant magnetisation in rocks of different ages from one continent, it is possible to reconstruct the continental drift history. The movement of continental plates may be compared to both poles and other continents. Palaeomagnetic measurements from Africa and Europe indicate that these continents were in adjacent positions during the Permian and Triassic (*ca.* 300 to 250Ma), becoming progressively separated through the Jurassic and early Cretaceous, only to come closer through the late Cretaceous and early Cenozoic (Livermore and Smith 1985).

Palaeoenvironmental analysis

By examining the distribution through time and in space around the Mediterranean of different sedimentary rock types and their contained fossils, it is possible to reconstruct the distribution of past shorelines, basins, shallow-marine platforms and areas of open ocean throughout the former Tethyan Ocean.

The earliest rift phase of the Tethyan Ocean is evidenced by Triassic volcanic rocks in the Iberian, Western Mediterranean and Peri-Arabic ophiolite zones (Knipper *et al.* 1986). Although these Triassic rocks are of oceanic character, they are of different age to the bulk of rocks now preserved in the cores of the ophiolites themselves. They also differ in their chemistry, making them comparable to ocean rocks formed during the early rifting of a basin, not in a mature basin. Thus the initial opening/rifting of Tethys is considered to be Triassic in age.

Following Triassic rifting, marine waters invaded some parts of the early Tethyan Ocean (Palaeo-Tethys) to form the typical 'Alpine Triassic' dolomite–limestone successions. Further rifting and the foundering of marginal platforms/reefs are observed in the Jurassic: the rock successions contain abundant evidence for Middle Jurassic igneous activity (Finetti 1985). In the Alpine regions the distinctive 'ammonitico rosso' limestones formed on the crests of numerous fault-blocks also suggest continued rifting (Bernoulli and Jenkyns 1974). Early Cretaceous shallow-water continental margin successions are characterised by the deposition of thick limestone successions (Tithonian-type) which commonly contain warm-water fossil assemblages, typified by the Tethyan bivalves, the rudists. Deeper water successions of the Cretaceous are characterised by the deposition of dark shales. At certain discrete horizons these shales become black, laminated and organic-rich, reflecting phases of oxygen starvation on the sea floor, known as Oceanic Anoxic Events. These phases of oxygen deficiency are of interest as they not only affected Tethys, but were possibly world-wide in their distribution, suggesting a climatic or eustatic control on their development. Such shales also provide significant source-rocks for the later generation of oil and gas (see Economic geology, p. 26–8). Late

Cretaceous chalk sedimentation is dominated by open-ocean processes resulting from the Tethys becoming a mature ocean in connection with the Pacific to the east. In tandem with this, global sea levels were high in the late Cretaceous, plus the whole ocean area occupied by the Tethys continued to remain open through the opening of subduction-related back-arc basins; thus volcanic rocks and sediments are common in the late Cretaceous.

During the closure of the Tethys in latest Cretaceous and early Cenozoic times, oceanic sedimentation continued throughout most of the present-day Mediterranean. In some areas uplift (during subduction and the collision of Africa) led to erosion and re-sedimentation in newly-formed basins. Palaeomagnetic data indicate synchronous rotation of the micro-continental blocks of the Tethyan margins (Liguria, Cyprus). In orogenic zones like the Alps, continental crust was probably subducted (as a result of the force of crustal impact), to create the over-thickened crust of mountain chains (Livermore and Smith 1985). During the Cenozoic the tectonic configuration of the distinct western and eastern Mediterranean seismic zones (see section on Neotectonics, pp. 14–17) came into existence.

As already observed in the section on ophiolites, these rock successions provide direct evidence of former oceanic crust. Through the Mediterranean, ophiolites represent fragments of the rocks that once floored the now lost Tethys Ocean. Knipper *et al.* (1986) summarised the ten main groups of ophiolite outcrops of the former Tethys into three types (Fig. 2.4): Peri-Arabic, Ligurian and Mekran (Iran–Afghanistan). Of the true Mediterranean outcrops, the eastern Mediterranean Peri-Arabic ophiolites are remnants of a back-arc basin formed later than the Tethys ocean crust *sensu stricto* and emplaced simultaneously along their 3000km length. The Ligurian ophiolites are the oldest of the Mediterranean (Triassic to Jurassic) and were emplaced over a long period of time (over 70Ma). Knipper *et al.* (1986) and Ricou *et al.* (1986) both suggest that these older ophiolites may represent remnants of Jurassic ocean crust from the Tethys itself (as opposed to a later back-arc basin). This suggests that throughout the African–Eurasian collision zone most Triassic and Jurassic ocean crust was subducted during the closure of Tethys. Some of this material formed the melts that became Cretaceous–Cenozoic volcanic rock and ocean crust whilst the rest remains either melted or in partial melts in the Earth's mantle.

The Tethyan Ocean: a summary

The above evidence can be used to provide a plate tectonic model for the transition from Tethys to Mediterranean (Fig. 2.5). Triassic rifting is indicated by volcanic and shallow marine sedimentary rocks. Marine transgression occurred throughout the late Triassic and early Jurassic. By Middle Jurassic times the African Plate was separated from the Eurasian Plate by a wedge-shaped Tethys Ocean, formed by mid-ocean spreading in the centre and bordered (already) by a subduction zone at its northern margin (Liguria). Africa and Iberia were joined and the Bay of Biscay had not opened at this time (Fig. 2.5).

By latest Jurassic time the Tethyan mid-ocean ridge was being subducted northwards; to maintain rifting a new spreading centre was created in mid-Cretaceous time to the south. This process, known as 'ridge jump', causes a major reorganisation of the plate movements in the surrounding area and thus mid-Cretacous rocks may be missing in some areas and over-thickened in others. Tethys continued to form ocean crust until late Cretaceous time when the mid-ocean ridge was finally subducted, causing emplacement of parts of the Ligurian ophiolite. Cretaceous rifting in the North Atlantic occurred gradually from Africa and America up to Rockall/Ireland–Greenland, causing the opening of the Bay of Biscay at this time. As the Jurassic–early Cretaceous Tethys Ocean closed, a back-arc basin opened north of the main Africa–Eurasia subduction zone. This small, hot area of ocean crust was soon emplaced as the

Peri-Arabic ophiolite successions. Early Cenozoic continent–continent collision between Africa and Eurasia was concentrated along the Pyrenees, to be followed by collision in the Sub-Betic Cordilleras, Alps, Dinaro-Hellenic Chain and Caucausus. Latest Cenozoic collision did not include the Pyrenees (Fig. 2.1).

Cenozoic collision also caused the rotation of the numerous continental blocks formed during Tethyan rifting. This movement of robust pieces of continental crust cannot be accommodated without simultaneous compression and extension. Compression results in uplift and the erosion of mountain chains whilst extension allows the creation of basins and small areas of ocean crust. Rotation of Corsica and Sardinia (away from France) has created the Cenozoic Ligurian Basin. Similarly, the Balearic rotation formed the Gulf of Valencia and the rifting between Calabria and Sardinia opened the Tyrrhenian Sea (Fig. 2.5). This reorganisation created new, isolated areas of oceanic crust and deep rifts, allowing the accumulation of very large thicknesses of sediment (10–15kms, Tyrrhenian Sea) during rapid subsidence. Thus, from a plate tectonic perspective, the closure of Tethys, continental collision and the isolated formation of very young ocean crust and rift basins demonstrate that Tethys was in no way simply an older Mediterranean Basin.

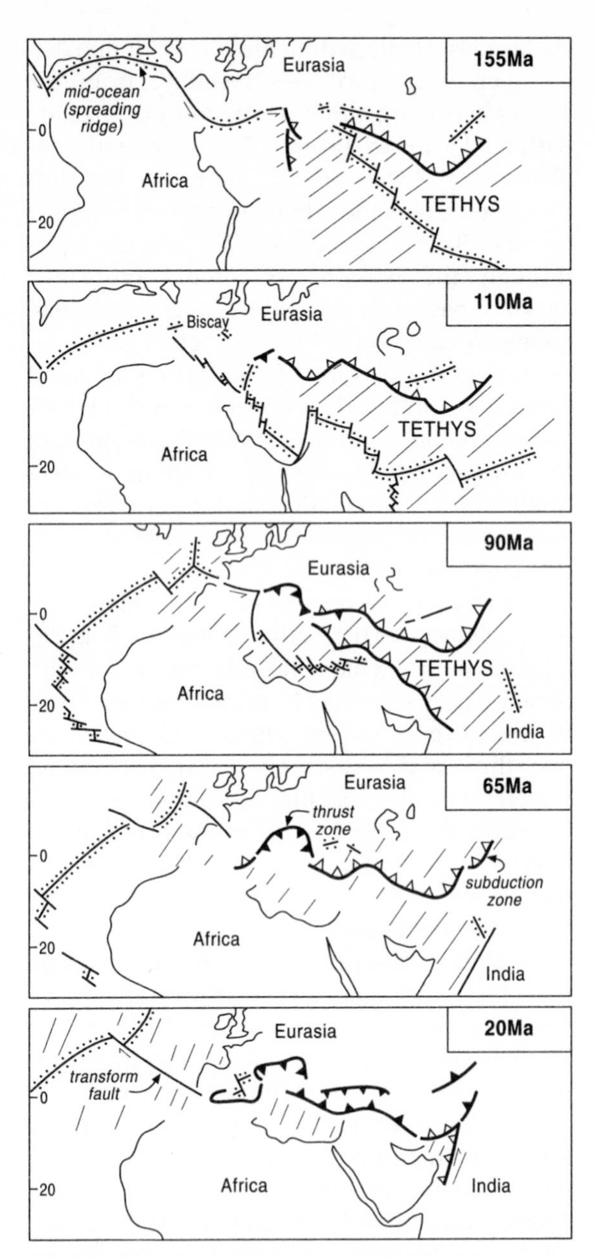

FIGURE 2.5 Plate tectonic reconstructions of the Tethys Ocean from the Middle Jurassic to the late Cenozoic. The explanation to the four main symbols is indicated on the diagram.

Source: Re-drawn from Dercourt *et al.* (1986)

GEOLOGICAL STRUCTURE AND TOPOGRAPHY

The strong link between geological structure and the topography of the Mediterranean lands and seas can be observed at a variety of scales from the macro-scale (thousands of kilometres) down to the micro-scale (hundreds or tens of metres). At the macro-scale, examination of the tectonic map of the Mediterranean (Fig. 2.1) shows the relationship between mountain belts and the distribution of island chains and mountain ranges. The present outline of the Mediterranean coast is controlled by fluvial processes such as the building of the Rhône or Nile deltas, marine processes such as the erosion of deep-water channels in the Straits of Gibraltar and Bosporus, and the erosion and marine inundation of geological structures. The greater part of the coastlines adjacent to the Hellenic Arc, Apennines, Betic Cordilleras and Atlas Mountains are controlled in their position by geological structure. Much the same observation can be made at the meso-scale of an island or coastline that is hundreds of kilometres in length. The orientation of the Greek islands of Rhodes and Crete along the Hellenic Arc and the distinctive shape of the Kyrenia Mountains in northern Cyprus are excellent examples of geological structure controlling topography. This control is apparent even at the scale of individual rock outcrops hundreds or tens of metres across. So often the sandy bays sought after by the holiday-maker are erosional features where marine and fluvial processes have exploited fault-plane weaknesses. Intervening headlands are often revealed to be undeformed rock or anticlines, especially in Tethyan limestones. Examples abound in the popular tourist destinations of the Spanish Costa del Sol, Italian Adriatic, Majorca and the Greek islands.

Whilst the shape of the coastline is a very obvious topographic effect of the underlying structure, the geologist sees the present coast as a transient feature in the Earth's evolution. The location of areas of deep marine waters and the high mountains of the hinterland have much the same structural control on their location and orientation. The deep seas of the Mediterranean are located in one of two likely structural domains. The fore-deeps of the Apennine and Hellenic Arcs originate through compressive tectonics, hence their intimate connection with earthquake and volcanic zones. This explains the location of the Tyrrhenian and Ionian Seas. The basins of the Western Mediterranean are extensional in origin, creating the broader depression of the Balearic Basin. The compressional fore-deeps, whilst still active and thus associated with neotectonics, are older features, originating with the Cretaceous closure of the Tethys some 100 million years ago. The western extensional basins are much younger, preserving the thick salt layers of the Miocene some 10 million years ago as their earliest depositional record.

The origin of the mountain belts of the Mediterranean can be viewed in much the same way as the origin of coastlines/deep marine areas. The sites of collisional tectonics, dominated by deformed/metamorphosed rock successions and current earthquake activity (Figs 2.1 and 2.2A/B) are where recent mountain-building processes dominate and high topography has ensued. These areas may be separated by resistant continental micro-plates like Iberia or by areas of extensional tectonics (Balearic, Adriatic, Dead Seas). As a result of the direct link between topography and geological structure, fluvial systems also show a clear relationship with structure and tectonics. Macklin *et al.* (1992) summarise the controls on river systems in the Mediterranean. These comprise tectonics, climate, sea level and vegetational/human change. Macklin *et al.* suggest that disrupted drainage systems are common in tectonically active areas. This may result in river reversal, capture, diversion or ponding, commonly observed by the field geomorphologist in the Greek islands, in Turkey and Italy. Conversely, mature rivers tend to drain parallel to major mountain belts (like the Po) and minor extensional basins (like the Nile and Rhône). Again, the same control on landforms at the macro-scale can be observed at the micro-scale of individual streams. The interested visitor to the interior mountains of any of the Mediterranean lands, armed with a geological and topographic map, will inevitably find a link between the geological structure, especially the faults and fractures, and the drainage pattern. Jenkyns *et al.* (1990) and Greensmith (1994) both provide guides (to the Balearics and southern Cyprus respectively) that show many examples of the geological relationship between structure and landform.

THE MESSINIAN SALINITY CRISIS

The term 'Messinian Salinity Crisis' was coined by K. Hsu (see Hsu 1972; Hsu *et al.* 1978 for summaries) to describe the widespread occurrence of gypsum and halite evaporite salts throughout surface and subsurface Miocene rocks (5–15Ma) of the Mediterranean. Sonnenfeld (1985) demonstrated how the former Mediterranean area was subjected to numerous evaporitive episodes long before the Miocene, suggesting that the erosional recycling of salts, coupled with plate tectonic configurations, may have culminated in the widespread accumulation of salts in the Miocene. Geographically isolated and rather more brief evaporitic phases have occurred since the Miocene.

The evidence for the Crisis lies in its most significant effect, namely the accumulation of thick (1000m to 1800m) salt successions. These salts are preserved beneath the waters and more recent sediments of the Balearic and Ionian seas and in the widespread outcrop of Miocene-Age evaporites in many of the Mediterranean islands and borderlands (Fig. 2.6). The deposition of such a huge volume of salt requires explanation. Although evaporitic basins are neither rare at present nor in the geological past, the Mediterranean examples, by their extent and preservation, provide a unique opportunity for the study of such deposits. In addition, the effect of the Messinian Salinity Crisis on Mediterranean palaeoenvironments was dramatic.

The existence of thick salt beds below the Mediterranean led to a critical re-evaluation of existing models describing evaporite basins. In order to appreciate the causes and consequences of the Miocene phase of evaporation, the two main models are briefly reviewed.

The Barred Basin Model (BBM)

This was first developed by Ochsenius (1876) who envisaged an enclosed seaway with limited (barred) access to open marine waters. This barrier could be closed by tectonic movement or sea level fall, whence evaporation of the basin could take place. Ochsenius realised that a single

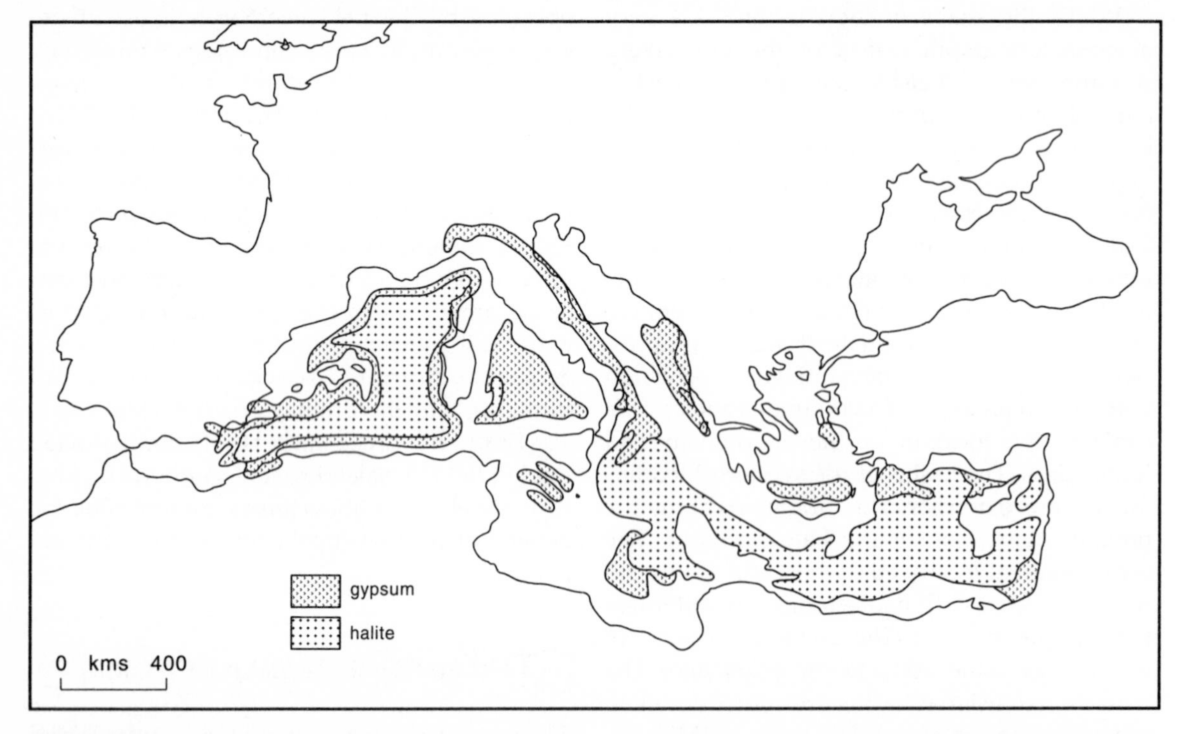

Figure 2.6 Distribution of the Messinian evaporites in the surface and subsurface of the Mediterranean. Note that halite is often presumed to cover gypsum.

Source: After Sonnenfeld (1985)

evaporative episode would not result in a significant thickness of salt, because normal ocean water rarely contains more than 3 per cent dissolved salts. To compensate for this, Ochsenius proposed repeated episodes of desiccation. Later, Krull (1917) pointed out how common gypsum/anhydrite ($Ca_2SO_4.H_2O/Ca_2SO_4$) is in the Miocene successions of the Mediterranean compared to halite/potash (NaCI/KCI). This is unusual given that sea water contains more than 30 times as much chloride as sulphate. Thus Krull modified the Ochsenius model to incorporate a non-permanent barrier, allowing the outflow of chloride-rich water and the inflow of calcium sulphate bearing marine water that later precipitated gypsum/anhydrite.

The Ochsenius–Krull model of partially barred evaporite basins has subsequently found much support and it is now recognised that all evaporite basins have some form of barrier to isolate the area of salt deposition. The extent of isolation and ultimate consequence of barring a marine basin is still the subject of some debate, the content of which often includes elements of the other models of evaporite deposition.

The Deep Basin Model (DBM)

In 1969 Schmalz observed halite precipitation occurring in the open and normal salinity surface waters of the Persian Gulf. The model developed to account for this requires a stagnant body of water (most commonly oceanic) with very little influx of fresh water. This water column, through poor circulation and evaporation, becomes stagnant (or anoxic) and hypersaline, most especially close to the sea floor. Halite may be precipitated within the water column, or at the sea floor. Although the DBM requires quite specific conditions (an open yet poorly circulated ocean in an arid climate zone), it does account for one critical observation: in many evaporite successions, evidence for complete desiccation is absent. This requires a permanent body of water and is the fundamental difference between this and the BBM.

The BBM and DBM both came under scrutiny with the discovery by the Deep Sea Drilling Project (DSDP) (Hsu 1972) of thick salt beds in the subsurface of the Mediterranean. The boreholes, sited throughout the Mediterranean, discovered evaporite beds with some clear evidence of desiccation and subaerial exposure. This conflicted with the traditional view of the Mediterranean as being similarly deep and cut by canyons in Miocene times as it is today. The conclusion drawn by the DSDP workers (Hsu *et al.* 1978) was that the Mediterranean was completely and repeatedly desiccated in late Miocene times, probably by closure of a barrier at the Straits of Gibraltar (Fig. 2.7). Marine reflux occurred through repeated and probably rapid and restricted throughflow at the straits. Hsu (1972) proposed a giant marine waterfall between Spain and Africa to account for this. Occasional freshwater input to the otherwise arid Mediterranean Basin was also recognised by the DSDP workers and may have enabled preservation of the more unstable salts by precipitation of brackish anhydrite/gypsum beds above halite (Sonnenfeld 1985). Thus occasional humid climates are also in evidence in the Miocene of the Mediterranean. Although desiccation features were recorded by the DSDP, they are by no means common and it is suggested that the bulk of Miocene evaporite sedimentation took place in brine form (Sonnenfeld 1985). This barred, brine-type sedimentation incorporates elements of both the BBM and DBM. The Mediterranean in Miocene times had occasional opening to the Atlantic via the Straits of Gibraltar, alternating with times of brine development and desiccation under a mostly arid climate. This partial opening to the Atlantic allowed the inflow of sulphate and the outflow of chloride and was maintained by a relatively stable tectonic position between Africa and Spain.

Subsequent to final deposition of the Messinian evaporites, many areas of the preexisting deep basin of the Mediterranean have continued to evolve as subsident basins. This downwarping of the crust is characterised by

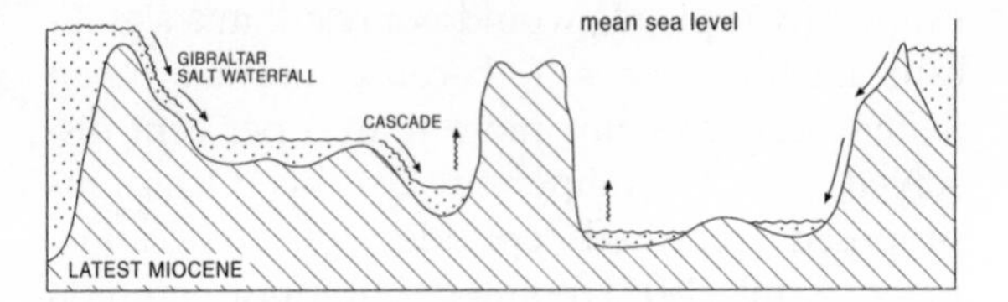

Figure 2.7 The desiccated Deep Basin Model of Messinian evaporite genesis (late Miocene age or *ca.* 6–7Ma). The left diagram shows the model at the stage of lowest sea level and complete desiccation, the right diagram shows the next stage (with rising sea levels) when Hsu's (1972) giant marine waterfall was created.

Source: After Sonnenfeld (1985)

slow and continued subsidence on the stable continental margins (Tunisia, Middle East), episodic subsidence and uplift along subduction zones (Aegean and Adriatic Seas), and rapid subsidence in newly-formed rift basins (Western Mediterranean, Tyrrhenian Sea). In these latter, unstable areas thick Messinian evaporite successions are now buried by up to 5km of sediment (Morelli 1985). The pressure associated with this burial has caused the salts to become semi-fluid and mobile. This movement of buried salt is known as halokinesis and the resultant intrusive piercement structures (diapirs, salt-walls) are common in rift basins of the Mediterranean. Whilst much of the Messinian succession remains buried beneath later sediments and waters of the Mediterranean, the successions deposited on the former basin margins and close to areas that were later uplifted (e.g. subduction zones) are now found at outcrop. In some cases the Messinian successions are now uplifted to up to 3000m altitude, attesting to the rapid and powerful processes of uplift involved in plate tectonics.

THE ECONOMIC GEOLOGY OF THE MEDITERRANEAN

Geological investigations in the Mediterranean Basin and borderlands have always gone hand-in-hand with economic interest. Exploration for precious metals and hydrocarbons has undoubtedly increased our knowledge of both ophiolite structure and the evolution of Tethys/Messinian evaporite deposition. Conversely, purely scientific-driven investigations have contributed to an appreciation of the location of economic quantities of minerals (ores, rock and hydrocarbons) and potable water.

Precious metals and ores

The genesis of ore-bearing rock in the Mediterranean may be conveniently summarised as the result of processes associated with the closure of Tethys, ophiolite emplacement, subsequent continental collision, the creation of active igneous belts and renewed subsidence. All of these processes cause igneous rock or hot aqueous fluid to migrate through rocks, dissolving some minerals and depositing them elsewhere. Where this process strips the rock of precious minerals, final deposition results in the formation of an ore-body.

As the economic wealth derived from oil and gas dictates much of the revenue-rich economy of the southern and eastern Mediterranean lands (North Africa and Middle East) at the present time (see Chapter 9), so the existence of rich metalliferous ores in the northern Mediterranean lands dictated the economies of those areas in the historic past. Copper, lead and zinc occur in abundance throughout the northern Mediterranean and include the significant deposits of Trepea (former Yugoslavia), Cyprus and Anatolia. The Anatolian deposits were likely sources for the first copper exploitation by the Hittites (*ca.* 4500 BC), who were also the first producers of iron artefacts. Worked iron is also found in late Egyptian remains; iron is currently worked in former Yugoslavia, southern France, Greece and Turkey.

The majority of metal ores in the Mediterranean were formed in the Mesozoic and had their final emplacement either at this time (roughly synchronous), or later in the Cenozoic. Significant lead and silver deposits (e.g in former Yugoslavia and at Laurium in Greece) and gold were emplaced at this time. Minor quantities of uranium are found in southern France, Spain, Italy and former Yugoslavia whilst thorium and borax (an evaporite boron mineral) occur in central Turkey. Minor deposits of tungsten and antimony occur in Sardinia; chromium and manganese in Greece and Cyprus. Other minor ores include chromite in the ophiolites of Albania, Turkey and Cyprus and asbestos in Italy and formerly exploited in Cyprus (Derry 1980). Chromite is a common constituent of the Earth's mantle, yet only ophiolite emplacement brings such material close enough to the surface to be mined. Similarly, asbestos is the hydrothermal alteration product of basic igneous rocks: the ocean environment provides the perfect location for such reaction and thus ophiolites again are the common host rock succession. Semi-precious minerals deposited during the Messinian Salinity Crisis and other evaporative episodes include significant quantities of potash (used in fertiliser) in Spain, Turkey, Italy and Israel.

Industrial (or non-precious) minerals

The rock successions outcropping in the Mediterranean borderlands and islands contain abundant rock types suitable for use as building stones, especially the white limestones of the Tethyan ocean margins and more recent Mediterranean. Weathering of Mesozoic–Cenozoic clays, especially those of volcanic origin, has resulted in the formation of bauxite and laterite (iron and aluminium-rich clay) and smectite (swelling clay). The legacy is characterised by the white buildings with red clay tile roofs of so many Mediterranean towns and villages. Cretaceous bauxites form an important iron source in southern France and Greece. Bentonites (swelling clays) are mined throughout the Greek islands for industrial fillers, oil exploration drilling fluid and for use in agrochemicals. Phosphate, an important fertiliser, is found in Tethyan successions in Morocco, Syria, Tunisia, Israel and Albania. The Messinian evaporites (as well as depositing potash) have left abundant gypsum deposits throughout the Mediterranean. These are exploited for use in plaster (most commonly plaster-board) and in industrial bondings.

Whilst the plate collision between Africa and Eurasia has caused significant earthquake and volcanic hazards to human life in the Mediterranean throughout history, earlier phases of tectonism have enabled significant ore deposition. Currently, this collisional activity is also the underlying source of geothermal energy, most especially in Italy, the world's second largest user of geothermal energy after the United States from 1970 to 1980..

Hydrocarbons

The generation and entrapment of hydrocarbons depend on the previous deposition of organic-rich source-rocks which must be buried sufficiently deeply to reach maturity and release oil or gas. This liquid hydrocarbon must then migrate into a reservoir sufficiently porous to allow later extraction from an oil- or gas-field. Lastly, the reservoir must be sealed in its upper parts in order to avoid leakage of the hydrocarbon upwards to the land surface. Typical source-rocks comprise marine shales or coastal/terrestrial coals; reservoir rocks are most commonly formed by sandstones or reef limestones, while top seals may comprise such impermeable rocks as shales or evaporites.

The existence (or former existence) of a deep (typically over 3000 metres of sediment) basin is thus a prerequisite for the maturation of source-rocks and the generation of hydrocarbons. The Mediterranean has many such deep basins, the least deformed of which occur beneath the waters of the Western Mediterranean (Ligurian, Tyrrhenian Seas; Sardo-Balearic, North Algerian basins) and onshore in north-east Spain (Ebro Basin); south-west and south-east France; Italy (Po Basin) and the Tunisian Basin. The Italian basins (on- and offshore) have traditionally been considered gas-producing basins, with Italy ranking twelfth in the list of world natural gas producers from 1975 to 1988.

One element of the prospective hydrocarbon geography of the Mediterranean that must not be forgotten is that the Tethyan Ocean was the marine area on whose borders the world's largest oil reserves were deposited. The Jurassic–Cretaceous Tethyan limestones and shales of Saudi Arabia, Iran, Iraq, Libya, Kuwait and Algeria comprise the reservoir and source-rocks. These have provided the most recent economic wealth of the southern and eastern Mediterranean borderlands (and beyond). These same rocks (especially the limestones) occur further north in the Mediterranean, where later tectonism has deformed the succession and obscured their distribution through later deposition. Major hydrocarbon discoveries of the Mediterranean to date are situated south of the Peri-Arabic, Hellenic, Ligurian and Alpine deformation zone. Thus there remains considerable hydrocarbon potential in the Mediterranean, both in the more recently formed deep basins and in the structurally complex areas of the northern tectonic belt.

References

ALLEN, P.A. and ALLEN, J.R. 1990: *Basin Analysis: Principles and Applications*. Oxford: Blackwell Scientific.

AUBOIN, J., DEBELMAS, J. and LATREILLE, M. (eds) 1980: *Géologie des Chaînes Alpines de la Tethys*. Paris: 26th International Geological Congress.

BAILLIE, M.G.L. 1990: Irish tree-rings and an event in 1628 BC. In Hardy, D.A. (ed.), *Thera and the Aegean World, Vol. 3*. London: The Thera Foundation, 160–6.

BERNOULLI, D. and JENKYNS, H.C. 1974: Alpine, Mediterranean and Central Atlantic Mesozoic facies in relation to the early evolution of Tethys. In Dott, R.H. and Shaver, R.H. (eds), *Modern and Ancient Geosynclinal Sedimentation*. Tulsa, Oklahoma: Society of Economic Paleontologists and Mineralogists Special Publication 19, 129–60.

DERCOURT, J. *et al.* 1986: Geological evolution of the Tethys belt from the Atlantic to the Pamirs since the Lias. *Tectonophysics* **123**, 241–315.

DERRY, D.R. 1980: *World Atlas of Geology and Mineral Deposits*. London: Mining Journal Books.

DIXON, J.E. and ROBERTSON, A.H.F. 1985: *The Geological Evolution of the Eastern Mediterranean*. London: Geological Society Special Publication 17.

DUFF, D. 1993: *Holmes' Principles of Physical Geology*. London: Chapman and Hall.

FINETTI, I. 1985: Structure and evolution of the Central Mediterranean (Pelagian and Ionian Seas). In Stanley, D.J. and Wezel, F.-C. (eds), *Geological Evolution of the Mediterranean Basin*. New York: Springer-Verlag, 215–30.

GASS, I.G. 1968: Is the Troodos Massif of Cyprus a fragment of Mesozoic ocean floor? *Nature* **220**, 39–42.

GREENSMITH, T. 1994: *Southern Cyprus*. London: Geologists' Association.

HARDY, D.A. (ed.) 1990: *Thera and the Aegean World, Vol. 3*. London: The Thera Foundation.

HSU, K.J. 1972: When the Mediterranean dried up. *Scientific American* **227**, 26–36.

HSU, K.J. *et al.* 1978: History of the Mediterranean Salinity Crisis. In Hsu, K.J. (ed.), *Initial Reports of the Deep Sea Drilling Project, Volume 42*. Washington, DC: National Science Foundation, 1053–78.

JENKYNS, H.C., SELLWOOD, B.W. and POMAR, L. 1990: *A Field Excursion Guide to the Island of Mallorca*. London: Geologists' Association.

KNIPPER, A., RICOU, L.-E. and DERCOURT, J. 1986: Ophiolites as indicators of the geodynamic evolution of the Tethyan Ocean. *Tectonophysics* **123**, 213–40.

KRULL, O. 1917: Beiträge zur Geologie der Kalisalzlager. *Kali* **11**, 227–31.

LIVERMORE, R.A. and SMITH, A.G. 1985: Some boundary conditions for the evolution of the Mediterranean region. In Stanley, D.J. and Wezel, F.-C. (eds), *Geological Evolution of the Mediterranean Basin*. New York: Springer-Verlag, 83–98.

MACKLIN, M.G., LEWIN. J. and WOODWARD, J.C. 1992: Quaternary fluvial systems in the Mediterranean basin. In Lewin, J., Macklin, M.G. and Woodward, J.C. (eds), *Mediterranean Quaternary River Environments*. Rotterdam: Balkema, 1–24.

MORELLI, C. 1985: Geophysical contribution to knowledge of the Mediterranean crust. In Stanley, D.J. and Wezel, F.-C. (eds), *Geological Evolution of the Mediterranean Basin*. New York: Springer-Verlag, 65–82.

OCHSENIUS, K. 1876: Über die Salzbildung der Egelnschen Mulde. *Zeitschrift Geologie Gesamt* **28**, 654–67.

PAPAZACHOS, B.C. and COMNINAKIS, P.E. 1977: Modes of lithospheric interaction in the Aegean area. In Biju-Duval, B. and Montadert, L. (eds), *Structural History of the Mediterranean Basins*. Paris: Editions Technip, 319–32.

RICOU, L.E. *et al.* 1986: Geological constraints on the Alpine evolution of the Mediterranean Tethys. *Tectonophysics* **123**, 83–122.

SCHMALZ, R. 1969: Deep-water evaporite deposition: a genetic model. *Bulletin of the American Association of Petroleum Geologists* **53**, 798–823.

SKINNER, B.J. and PORTER, S.C. 1995: *The Dynamic Earth*. New York: John Wiley and Sons.

SONNENFELD, P. 1985: Models of Upper Miocene evaporite genesis in the Mediterranean region. In Stanley, D.J. and Wezel, F.-C. (eds), *Geological Evolution of the Mediterranean Basin*. New York: Springer-Verlag, 323–43.

STANLEY, D.J. and WEZEL, F.-C. (eds) 1985: *Geological Evolution of the Mediterranean Basin*. New York: Springer-Verlag.

UDIAS, A. 1985: Seismicity of the Mediterranean Basin. In Stanley, D.J. and Wezel, F.-C. (eds), *Geological Evolution of the Mediterranean Basin*. New York: Springer-Verlag, 55–63.

3

MEDITERRANEAN CLIMATE

ALLEN PERRY

INTRODUCTION

Because the Mediterranean Basin is the world's most popular and successful tourist region its climate is perceived by many as idyllic, hospitable and delightful with unremitting warmth, endless sun and perpetual dryness. With its renowned radiance and clarity of light, it has remained seductive to north European visitors since the habit of escaping from the cold and dark of the northern winter became well established by the upper classes last century (see Chapter 14). Yeo's (1882) description of the climate of the Côte d'Azur as 'a tonic, stimulating and exciting' sums up this traditional view.

The reality, however, is very different from the perception. Extremes of summer heat as well as droughts and floods, the latter frequently destructive and sometimes catastrophic, ensure that 'harsh and capricious' are more accurate descriptions of the climate. Senequier (1993) has suggested that the meteorology of the Mediterranean is characterised by two dangerous traits, violence and rapid development. The natives are accustomed to the climate and notice the signs and portents but visitors are frequently surprised. The Mediterranean climate is characterised not only by seasonal variability but also by marked variability within and between winter seasons. While in one sense it is a transitional junction climate, representing an interactive brew of Asiatic, European, African and Atlantic influences, it is also one of the best-known and most distinctive climatic types in the world.

While the extent of 'Mediterraneanism' varies somewhat over time (Wigley and Farmer 1982), spatially it is in many respects limited and often constrained by the proximity of mountain and deserts to a narrow coastal strip. Siegfried (1948) illustrates this by reference to Alexandria (Egypt) where 'a taxi will take you in a quarter of an hour into the desert'. Hardly surprising, then, that the Mediterranean climate has been called the gift of the Mediterranean Sea. Overall, some 9 per cent of Europe has a Mediterranean climate and 46 per cent of Africa north of the Tropic of Cancer (Le Houérou 1990). The numerous bordering countries have produced a substantial, although fragmented climatic literature, written in many languages and spread through a vast array of scientific papers and journals. Only a very few comprehensive accounts of the climate exist in English: these are mostly of considerable vintage or are handbooks for weather forecasters (Meteorological Office 1962; Reiter 1975).

The classical definition of a Mediterranean climate is one where winter rainfall is more than three times summer rainfall. This strong summer–winter rainfall contrast is associated with a well-pronounced seasonal expansion and contraction of the circumpolar vortex and the consequent displacement and withdrawal of the upper westerlies from winter to summer. In the summer months the contracted circumpolar vortex ensures that the Mediterranean is a region of subsidence associated with a subtropical upper tropospheric high. In the winter, unsettled weather is prevalent at times when the westerlies are in their low index or blocked stage and the polar front jetstream exhibits a strong meandering character allowing southward transport of cold air from the north (Perry 1981). The marked biseasonality of the climate has produced a climate that is sometimes known as Etesian, from the Greek *etesios* meaning annual, referring to the summer northerly Etesian winds that affect the eastern Mediterranean each year. This seasonality imposes a unity on the Mediterranean environment. As Hughes (1988) remarks, 'from one end of the sea to the other people have found summer the time for travel and war and winter the season for agriculture and repair at home'.

While the climate acts as a uniting force in Mediterranean lands, giving them a common character where outdoor activities dominate – the open market place, temples and houses centred on courtyards – the climate has diversity too. In the case of mean annual rainfall, on the north shore of the Mediterranean totals vary from just under 200 mm in the Almeria province of Spain, the driest region in Europe (Tout 1987), to just over 1200 mm along the eastern Adriatic coast. Regionalisation of the climate on the basis of mean annual temperature records (Goossens 1986) suggests four climatic regions, while a similar study of the annual precipitation data by the same author (Goossens 1985) yields five groups of stations within the Mediterranean Basin. Mazzoleni *et al.* (1992), using data from over 400 weather stations, have recognised a hierarchical scale of climatic regions that appear to be correlated with vegetation patterns. Such studies emphasise the fact that there is considerable diversity within the broad Mediterranean climate type.

THE SEASONAL CYCLE OF CLIMATE

Mediterranean life is dominated by the climatic seasons. The disturbed regime of the cool season typically commences in mid-October. At this time the subtropical jet stream moves from its summer position over Turkey to its winter position over the northern Sudan; at the same time, there is an increase in surface pressure over the Levant. These shifts tend to occur in a series of 'jumps' and the seasonal transition assumes a pulsatory character in most years. Vigorous cyclonic disturbances, many generated within the Mediterranean Basin, typically in the Gulf of Genoa, form when southward transport of cold air occurs. These incursions of cold air from the north penetrate southwards in stages, heralding a rainy season which occurs later the further southeastwards one moves through the Basin. North of a line from north-east Spain to the north Aegean October is on average the wettest month, but in Israel it is December or January. The mean sea-level pressure distribution in winter is shown in Figure 3.1.

In some areas, for example Israel, rainfall probabilities show pronounced fluctuation during the winter months (Jacobeit 1988). Cumulus skies, indicative of heating of unstable cold air masses by the sea, which remains relatively warm and a source of energy through the winter, dominate the period from October to early May. But almost every year there are intervals of calm sunny days between the frequent storms and unsettled periods. The Greeks refer to these as 'Halcyon Days' (Dikaikos and Perry 1981); supposedly it was then that the mythical halcyon bird built its nest and hatched its eggs.

Winter in the Mediterranean is a season of extremes. Frost and snow are by no means unknown, even on the North African shore, and strong winds or gales occur as depressions move through the Mediterranean. Coasts sheltered from the north and north-west, for example in France and eastern Spain, enjoy frequent sunny weather, contributing to the 3000 hours of sun recorded by these locations each year (Pennas 1991). By March–April there is an increased risk of Atlas depressions forming over the Sahara and moving north-east through the Mediterranean (Prezerakos 1985). Unusually

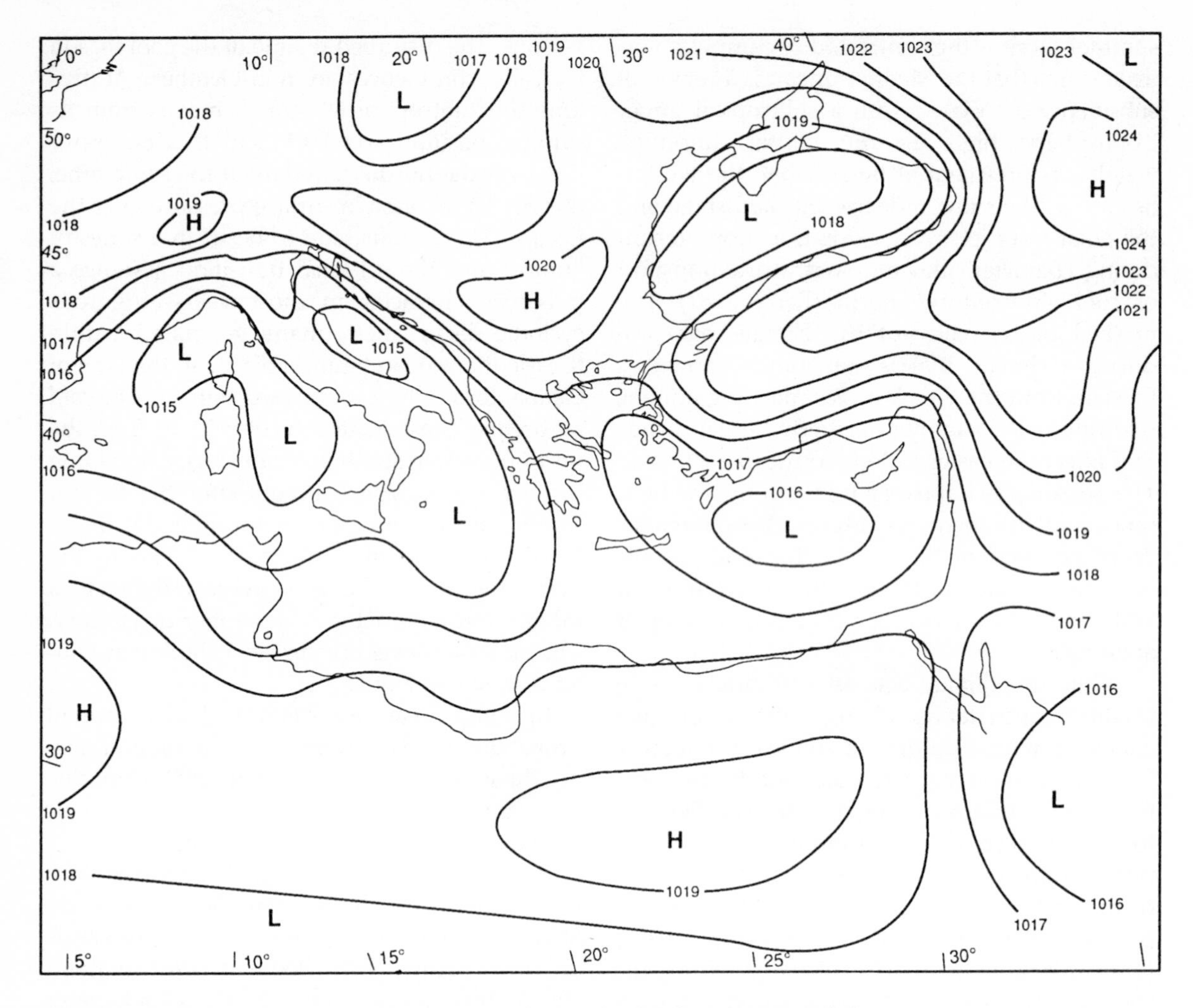

FIGURE 3.1 Mean sea-level pressure in December in the Mediterranean (mbs)

hot and dry weather in the eastern Mediterranean can occur ahead of such lows approaching from the west and these heat waves, known as *sharav* in Israel, have been documented by Winstanley (1972). Severe dust storms can occur over Egypt, Algeria and Libya behind cold fronts associated with the North Africa lows.

The change-over in spring to the settled, warm and dry conditions that dominate the July–September period is a more gradual process than the autumn breakdown. Gradually the subtropical Azores anticyclone and its associated ridge exert more influence (Fig. 3.2), and with increasing stability the rainy season ends and the heat of summer increases. Thermal low pressure areas form over Spain and North Italy and although they influence low-level airflow, they have little influence on the weather. The length of the arid season is much greater in the southern and eastern Mediterranean than elsewhere. South of 40° N only 5 per cent of total precipitation falls in summer and in most years drought conditions will be continuous for up to five months. By contrast, over the northern Adriatic summer rainfall totals are in excess of one-third of winter precipitation as a result of considerable thunderstorm activity. On many coasts a regime of sea breezes is established and if these are augmented by the normal northerly

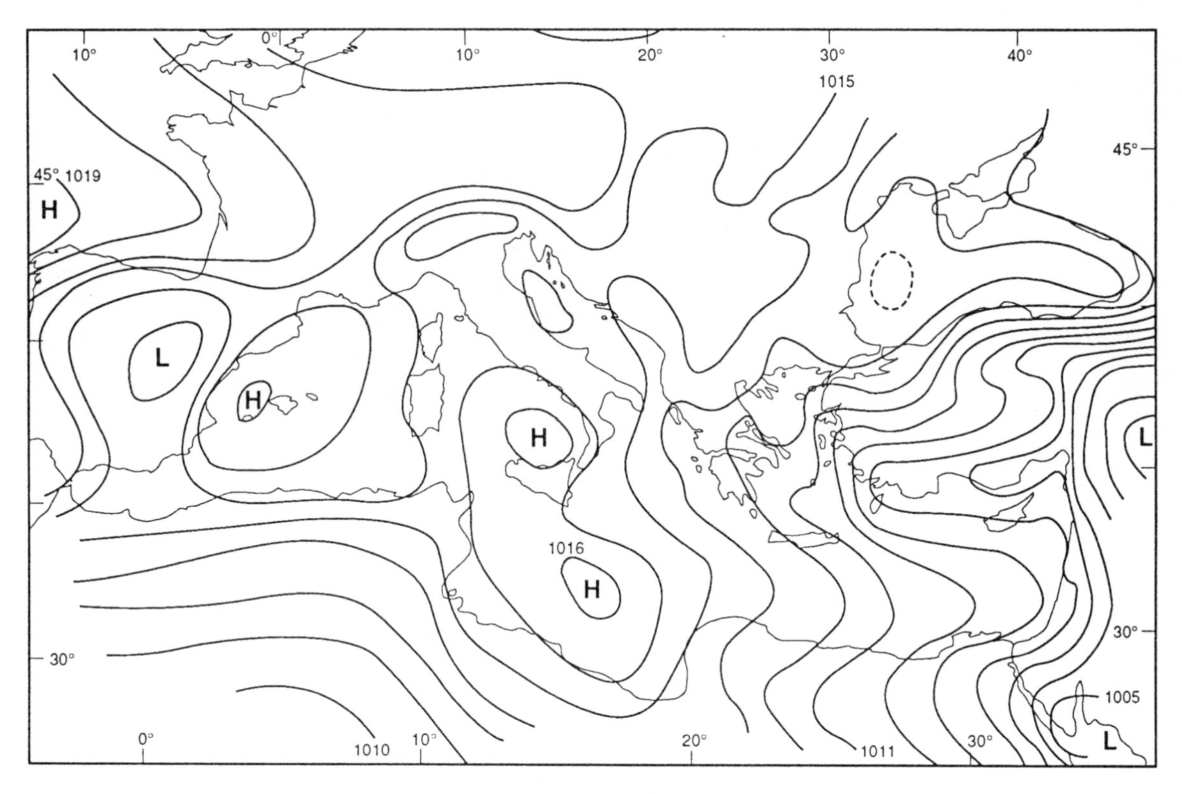

FIGURE 3.2 Mean sea-level pressure in July in the Mediterranean (mbs)

winds that dominate the eastern Mediterranean they can be particularly strong, as on the north coasts of islands like Cyprus and Crete.

In some years short but intense heatwaves exacerbate thermal stress and discomfort. For example in 1987 and 1988 Greece and neighbouring countries endured several consecutive days of temperatures over 40°C leading to a steep rise in heat-related deaths and widespread forest fires (Giles and Balafoutis 1990). Colacino and Conte (1995) suggest that in the central Mediterranean 10 per cent of the summer period is affected by heat-waves with temperatures 7°C or more above normal. Inland, in sheltered valleys the intense heat of summer is more continuous and expected. Not for nothing is Cordoba in southern Spain nicknamed *el horno* (the oven).

Urban heat islands increase the discomfort of summer in cities (Colacino 1980) along with high levels of atmospheric pollution. Athens is one of Europe's most polluted capitals and although the persistent surface inversions that lead to a build-up of pollutants and the formation of the notorious *nefos* (cloud) can occur at any time of year, it is in the heat of summer that the problem can be most distressing to urban inhabitants (Katsoulis 1988). Emergency curbs on traffic movement and industry and in 1995 the creation of a large vehicle exclusion zone in the city centre have had only a minor impact on the problems. Pollution contributed to at least 3000 deaths in 1987 and the city's daily average death rate rises by a factor of six on high pollution days.

CLIMATE AND TWO MAJOR ECONOMIC ACTIVITIES

Climate and agriculture

One of the major problems for farmers in the Mediterranean is the recurrent difficulty of providing water requirements for crops (see Chapter 13). The sporadic nature of rainfall incidence contrasts with the unwelcome dependability of the summer drought and accompanying high temperatures. Traditional crops like the vine and fig tree are well adapted to this problem, although inter-annual variability of yields, especially of the olive, can be considerable. If irrigation water is available on a regular and reliable basis, large-scale commercial agriculture is possible, as the intensive farming districts of Almeria and Murcia in south-east Spain illustrate.

South-east Spain benefits neither from Atlantic sources of moisture nor from Mediterranean depressions which form to the north-east and rarely affect this area. Tout (1987) has shown that while over a 44-year period median precipitation at Almeria was 227 mm, in several years in the 1980s it failed to reach 100 mm. Extreme winter mildness, a high percentage of possible sunshine, shelter from northerly winds and an almost complete absence of frost give the potential in this area for year-round growth and some of the earliest crops in the whole of Europe (Tout 1990). The availability of water for irrigation is all-important and this is now brought into the area from the mountains to water one of the major early salad- and vegetable-growing areas in Europe. Occasional very violent storms can do immense damage to agriculture and result in severe soil erosion. Fortunately hail, another damaging episodic weather phenomenon elsewhere in the Mediterranean, is not common. 'Red rain' containing Sahara dust washed out onto the crops and leading to staining is an occasional problem, however. Almeria province is a land of extreme contrasts, with the greatest contrast of all being that between the barren appearance of the majority of the landscape and the pockets of luxuriant growth associated with increasingly prosperous horticultural and fruit enterprises (see Fig. 3.3). Cultivation is a year-round activity on the irrigated *huertas* with multiple cropping the rule, much of it under plastic.

FIGURE 3.3 An exceptionally dry and mild climate coupled with the availability of irrigation water makes south-east Spain a land of extreme contrast between barren, eroded hills and pockets of luxuriant and intensive agriculture. Scene near Palomares in the province of Almeria.

Holiday bioclimatology

Many Mediterranean economies have a strong reliance on climate-dependent tourism, where the motive for travel is the perceived seasonal reliability of the climate (see Chapter 14). In addition, there is a growing trend for long-stay winter holidays in a climate substantially better than that in northern Europe. The major attractions are the high number of warm sunny days, particularly suitable for beach recreation, and the low inter-annual variability of the climate. Whilst bioclimate conditions are generally favourable for recreation, heat discomfort can occur on up to half of the days in July and August in a resort like Majorca (Harlfinger 1991). Figure 3.4 shows that not all resorts have identical conditions, even in high summer. At Almeria, for example, warm, close weather is much more common than in Corfu or Nice. Of course, it is extremely difficult to define 'ideal' vacation weather, owing in part to personal preferences, but according to Besançenot (1985) over 80 per cent of all days in the June–September period on the Costa del Sol fall into the 'ideal' category, decreasing to about 65 per cent on the Costa Brava.

In urban areas architectural design and town planning can also play a role in obviating the most extreme effects of summer heat. Narrow streets help to provide shade while small squares and interior courtyards, often with trees and fountains, can help to provide a pleasant microclimate. Indeed, the great Moorish

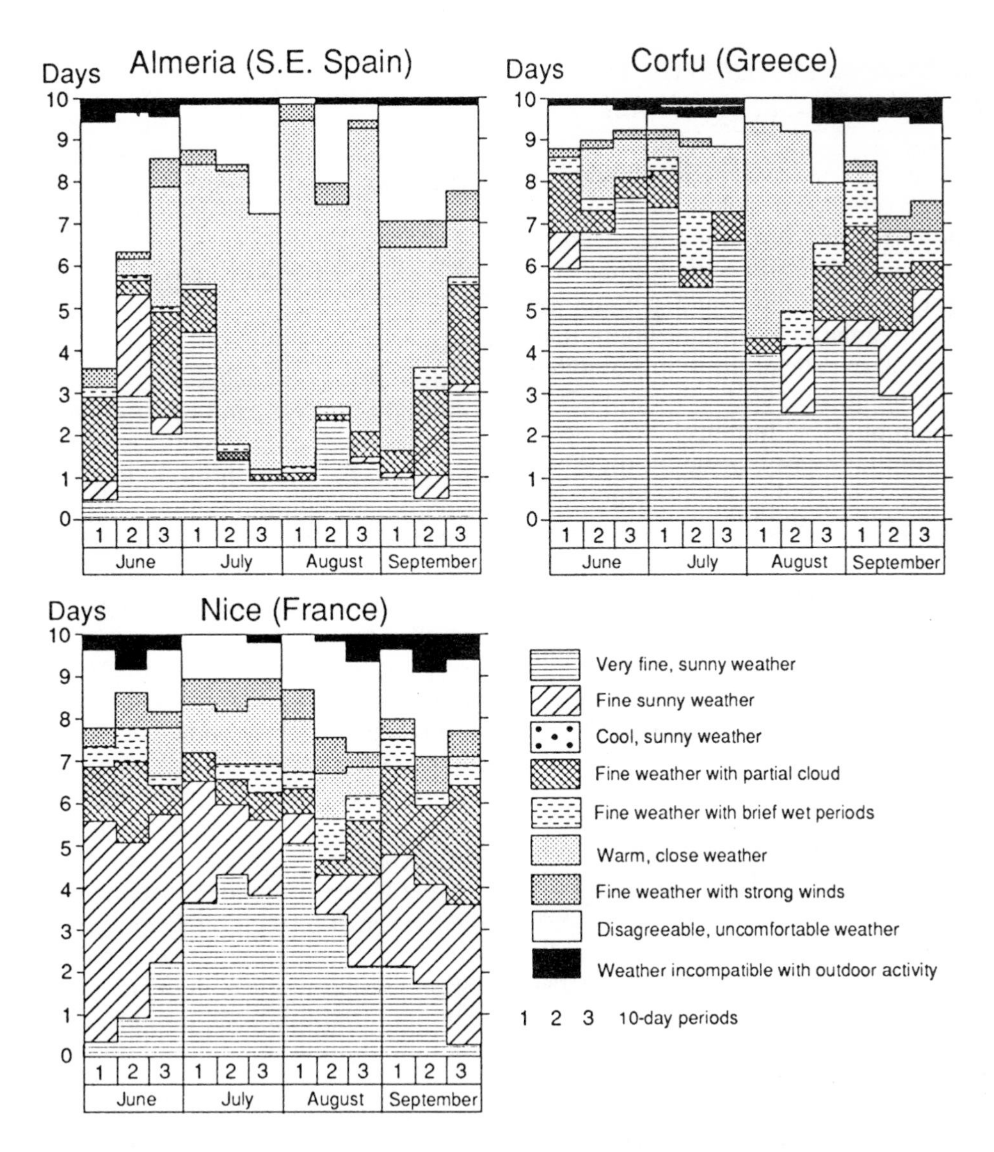

FIGURE 3.4 Examples of climate information for tourists produced by French climatologists

Source: After Besançenot *et al.* (1978)

palaces of southern Spain, like the Alhambra in Granada, provide the tourist with an excellent example of well-designed structures and environments that take account of the torrid summer climate where maximum temperatures regularly exceed 40°C.

Because of its marked geographical and seasonal concentration, tourism adds greatly to the problem of providing water supplies, stretching water resources in drought years to breaking point. These problems are often acute on islands. In 1995 the island of Majorca, with its annual influx of over 2 million visitors, used a tanker transfer scheme to bring in supplies from the mainland, while in Malta costly desalination plants have been built (see Chapter 15). The economic importance of mass tourism means that countries like Spain and Malta try to avoid adverse publicity about problems such as drought and water supply.

Floods and droughts: the price of climatic variability

Since the late 1980s, rainfall in Mediterranean Spain has been consistently below the long-term average and by the mid-1990s this long-standing drought was threatening to have a huge economic impact on agriculture and perhaps on tourism. The irregularity and variability of rainfall that are characteristic of the whole Mediterranean region frequently lead to long spells of drought interrupted by torrential downpours which can give rise to catastrophic floods. Water deficiency is likely when the normal winter rains are light, usually because deep low pressure areas persistently cross north Europe, leaving the Mediterranean under the influence of high pressure. 'This damned anticyclone' was the recent headline above the daily weather forecast in one of Spain's national newspapers.

Widespread and severe winter droughts are referred to in the contemporary weather descriptions and records of many countries, for example Greece (Grove and Conterio 1995). In a climate where temperature conditions permit year-round plant growth, low and unreliable precipitation drastically limits agricultural development. Moreover, a succession of dry years can limit groundwater availability and hence water supplies for irrigation.

One of the features of the climate of Mediterranean lands generally is the contribution that single-day rainfall totals can make to monthly and annual figures. Occasional torrential rainfall ensures that scarcely a year goes by without some area experiencing damaging floods. This is most frequent in the autumn months when sea surface temperatures are still high and eruptions of cold polar air create high evaporation rates and severe instability aloft. In November 1966 Venice and Florence were badly flooded and 112 deaths occurred throughout northern Italy, while in 1969 Tunisia and adjacent parts of North Africa were badly affected. Studies in Catalonia (Llasat and Puigcerver 1994) have identified nine occasions in the period 1960–90 with rainfall totals of 200 mm or more in 24 hours, all leading to severe material damage and loss of human life. The widespread storms of autumn 1989 in Majorca and eastern Spain have been described by Tout and Wheeler (1990) and Wheeler (1991, 1996).

Local winds

On each of the eight sides of the Tower of the Winds in Athens is a relief of a personified wind, portrayed as if flying and bearing some attribute indicating its character. Not only in Greece, but throughout the Mediterranean, local winds with particular individual characteristics have been recognised and named. The recurrence of these winds has caused them to be seen by the people living in the area affected as having a uniqueness particular to that region. These winds are not steady but occur only when appropriate synoptic conditions establish the required pressure gradient. Probably the most widely documented local winds are the *Mistral* (Barsch 1965), particularly because of the fire risk that it brings to South France (Wrathall 1985), and the *Bora* (Yoshino 1971). In Sardinia the dominant wind is the north-westerly *Maestrale*, the continuation of the Mistral. Its effects on vegetation can be seen in Figure 3.5. Three other examples also deserve mention.

The *Tramontana*, 'an ill wind in Catalonia' (Wheeler 1994), is a cool, dry, brisk wind from

Figure 3.5 The dominant wind in Sardinia is the strong and blustery *Maestrale*, which blows from the north-west, stunting and shaping the vegetation in exposed places

the north-east or north which also affects the west coast of Italy, northern Corsica and the Balearic Islands. With gusts of up to 100 km per hour, it is associated with the advance of an anticyclone from the west following a depression moving through the Mediterranean (Fig. 3.6), and it is most frequent in winter and early spring. Weather is typically fine but with instability showers. Since about 60 depressions form in the western Mediterranean Basin each year on average there are ample opportunities for this wind to blow. Wheeler (1994) has shown that there is some truth in local folklore which credits the Tramontana with producing climatically-induced stress including fatigue, depression and hypertension.

The *Sirocco* is a warm south or south-east wind blowing in advance of a depression moving eastward across the Mediterranean or North Africa. The air comes from the Sahara and is dry and dusty, but in crossing the Mediterranean it picks up moisture and reaches Malta, Sicily and southern Italy as a very enervating, hot and humid wind. It may deposit Saharan dust in these areas. In a recent study of the Sirocco in Italy (Bischoff 1992), an increase in frequency during the period 1958–88 was found, perhaps indicating a change of weather patterns in the western Mediterranean. Sivall (1957) suggested that 7–8 days of Sirocco were common in both March and April each year in the eastern Mediterranean.

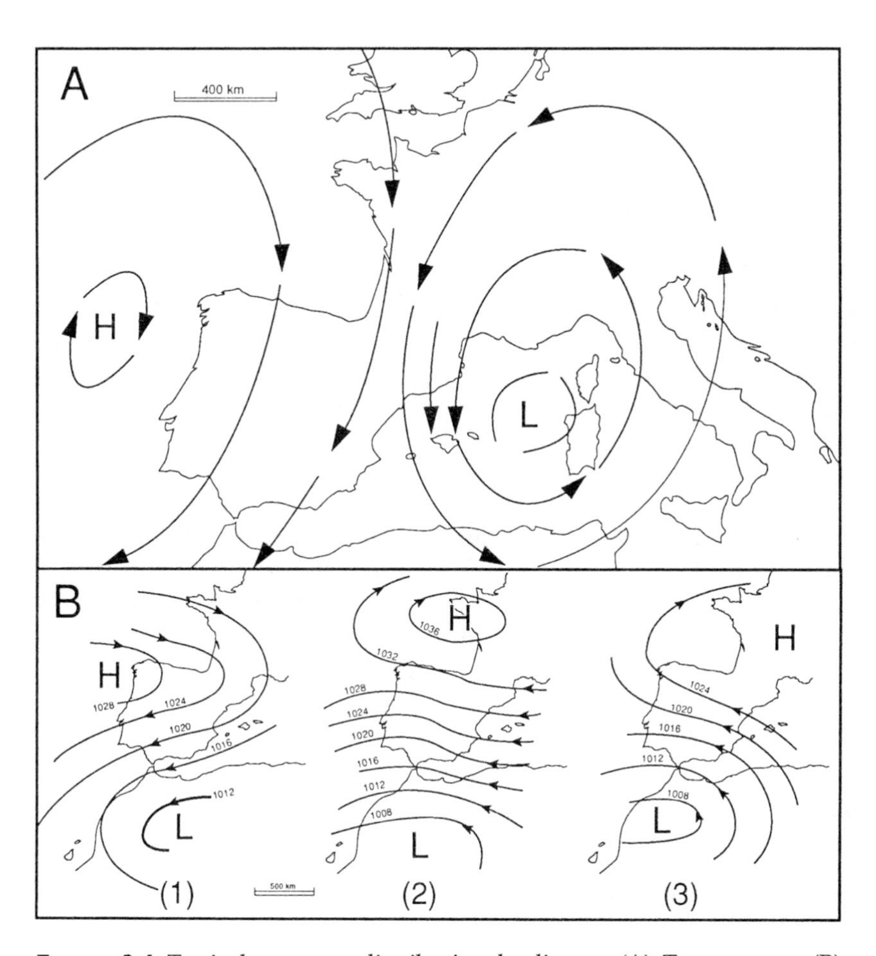

FIGURE 3.6 Typical pressure distribution leading to (A) Tramontana, (B) Levante

Finally, the *Levante* is a generally easterly wind occurring in the Straits of Gibraltar, most frequently during February–May and October–December. It is typically mild, very humid and produces overcast conditions. In particular it is responsible for producing a banner cloud that extends downwind from the rock of Gibraltar. Figure 3.6 shows the synoptic situation that produces a Levante most frequently.

TEMPERATURE AND PRECIPITATION TRENDS DURING THE PERIOD OF INSTRUMENTAL RECORD

The position of the Mediterranean between the arid climate of North Africa and the moist, temperate climate of North-West Europe suggests that it might be particularly sensitive to quite small changes of climate. Fortunately, there are a number of long-period, reliable meteorological records available to monitor climate change over the last few centuries. These include the following: Gibraltar since 1790, Marseilles since 1851, Rome since 1782, Athens since 1858, and Jerusalem since 1861. Analysis of many of these records has involved testing their homogeneity to take account of, for example, slightly different types of raingauge design or changes of site (e.g. Katsoulis and Kambetzidis 1989). Whilst many individual papers have examined historical data series at one station for one meteorological parameter – e.g. Colacino and Purini (1986) on rainfall in Rome from 1782 to 1978 – a more complete overview of historical climatic change can be obtained from the studies of larger areas and several stations and these will now be considered in more detail, looking first at temperature and then at precipitation.

Temperature

On a global scale it is well known that over the last century and a half mean air temperature has increased, probably as a result of the enhanced Greenhouse Effect produced by the substantial addition of carbon dioxide and other 'greenhouse gases' to the atmosphere by human activities. It is also known that neither the magnitude nor the rate of this increase is constant at different locations around the world. Repapis *et al.* (1992) have reported that warming in the Eastern Mediterranean region during the first half of this century was greater than the hemisphere average, and it seems that this warming trend affected all parts of the year. In the period 1950–88 there was evidence of a negative air temperature trend in the Eastern Mediterranean and a positive trend in the Western Mediterranean (Sahsamanoglou and Makrogiannis 1992). These temperature trends can be related to changes in the atmospheric circulation, with a small increase in atmospheric pressure beginning in 1960 over most of the area and a reduction in the number of depressions (Makrogiannis and Sahsamanoglou 1990). Perry (1992) has drawn attention to a European circulation type that has been prevalent in several recent winters and is associated with a see-saw of temperature between the British Isles and the eastern Mediterranean. With this circulation type, cold and often wet conditions from Cyprus eastward to Syria and Jordan occur beneath an upper cold trough while Western Europe is predominantly dry and mild. At Nicosia, for example, the warmest winter on record since 1896 was 1963, when it was very cold in Western Europe. Sea surface temperature studies suggest that minima occurred in about 1910 and the late 1970s with maxima in about 1947 and 1964 (Bartzokas *et al.* 1991). Recent years show a general increase.

Precipitation

Studies of trends and fluctuations of rainfall based on long series of instrumental records suggest that cyclical trends in precipitation are quite well marked. At Rome, Colacino and Purini (1986) detected an oscillation with a period of about a century, while at Malta the precipitation minima in 1917, 1947 and 1977 are 30 years apart (Perry 1988). Annual precipitation totals at many stations show evidence of groupings of years with similar amounts. More general studies covering the last century of central and western Mediterranean stations (Maheras 1988; Maheras *et al.* 1992) show moist and dry phases (Table 3.1). However, large-scale area-average precipitation changes are more difficult to quantify than temperature changes because of the higher spatial variability of precipitation.

A close relationship exists between the general regional atmospheric circulation and precipitation. The widespread humid period that occurred during the 1930s coincided with a clear decrease of atmospheric pressure over the Mediterranean and a decrease in the frequency of dry zonal circulation (Bárdossy and Caspary 1990). By contrast, the dominance of a meridional circulation is followed by an increase in precipitation. Because a few heavy daily falls frequently account for a large percentage of the total annual rainfall, short periods of low index meridional circulation in the wet period of the year can have a pronounced effect on annual totals and this makes forecasts of the likely trend of rainfall in future years particularly difficult.

THE PRE-INSTRUMENTAL CLIMATE RECORD

The reconstruction of climate before the era of meteorological recordings is based on a very wide spectrum of different evidence: archaeological, botanical and documentary. With their ancient civilisations, many Mediterranean countries are particularly rich in historical archives that contain much potentially useful meteorological data.

TABLE 3.1 Moist and dry phases in the last century in the Western and Central Mediterranean

	West	Central
Moist	1901–1921 1930–1941	1930–1944
Dry	1942–1954 1980–present	1980–present

Source: Maheras (1988)

Greece probably has the oldest weather observations in the world. The ancient Greek philosophers carried out observations with the purpose of drawing up forecast bulletins of both astronomical and meteorological phenomena. Aristotle's lecture notes, known as *Meteorologica*, suggest that drought contributed to the decline of Mycenean Greece in about 1200 BC (Zerefos and Zerefos 1978); and other evidence exists that severe winters were most frequent during the first half of the fifteenth century and the second half of the seventeenth century – the time of the Little Ice Age.

Recently, further reconstruction of the climate of the sixteenth and seventeenth centuries has been carried out for Crete and the central Mediterranean using Venetian and Turkish documentary sources (Grove and Conterio 1995). In some years between 1548 and 1648 weather conditions occurred which are anomalous by twentieth-century standards; for example, some winter droughts were longer-lasting and more extreme than any since instrumental measurements began. Some unusual summer rains suggest that on occasions weather situations like that in the summer of 1976 developed, when some parts of the Mediterranean had six to eight times the normal precipitation at the same time as large parts of southern England were experiencing drought (Perry 1976).

Finally, tree-ring analysis has been found to be particularly useful for reconstructing winter climate in a variety of Mediterranean regions including the northern Adriatic, south-west Europe and Morocco. This work is enabling the main outlines of climate evolution since 1500 AD to be reconstructed. It seems that not only were summers cooler during the Little Ice Age but also cool winters affected the western Mediterranean (Serre-Bachet 1994; Serre-Bachet *et al.* 1990).

THE FUTURE MEDITERRANEAN CLIMATE

The enhanced Greenhouse Effect is expected to lead to substantial changes in climate at the regional level over the next century. The Mediterranean Basin is very vulnerable to climate changes, particularly through changes in rainfall and water supply. These regional-scale changes were originally estimated using General Circulation Models (Wigley, 1992) but they are not reliable enough and high-resolution scenarios based on the statistical relationship between grid-point GCM data and observations from surface meteorological stations are now being constructed (Palutikof and Wigley 1996). In this way, a better spatial resolution can be achieved and a wealth of regional data can be extracted, including some idea of the likely frequency and severity of droughts and floods. Seasonal scenarios of the temperature change by 2030 with a 1°C global warming due to the enhanced Greenhouse Effect (Fig. 3.7) suggest an increase approaching 2°C to the north and north-east of the Mediterranean area. Precipitation changes are complex. In winter and spring the scenarios

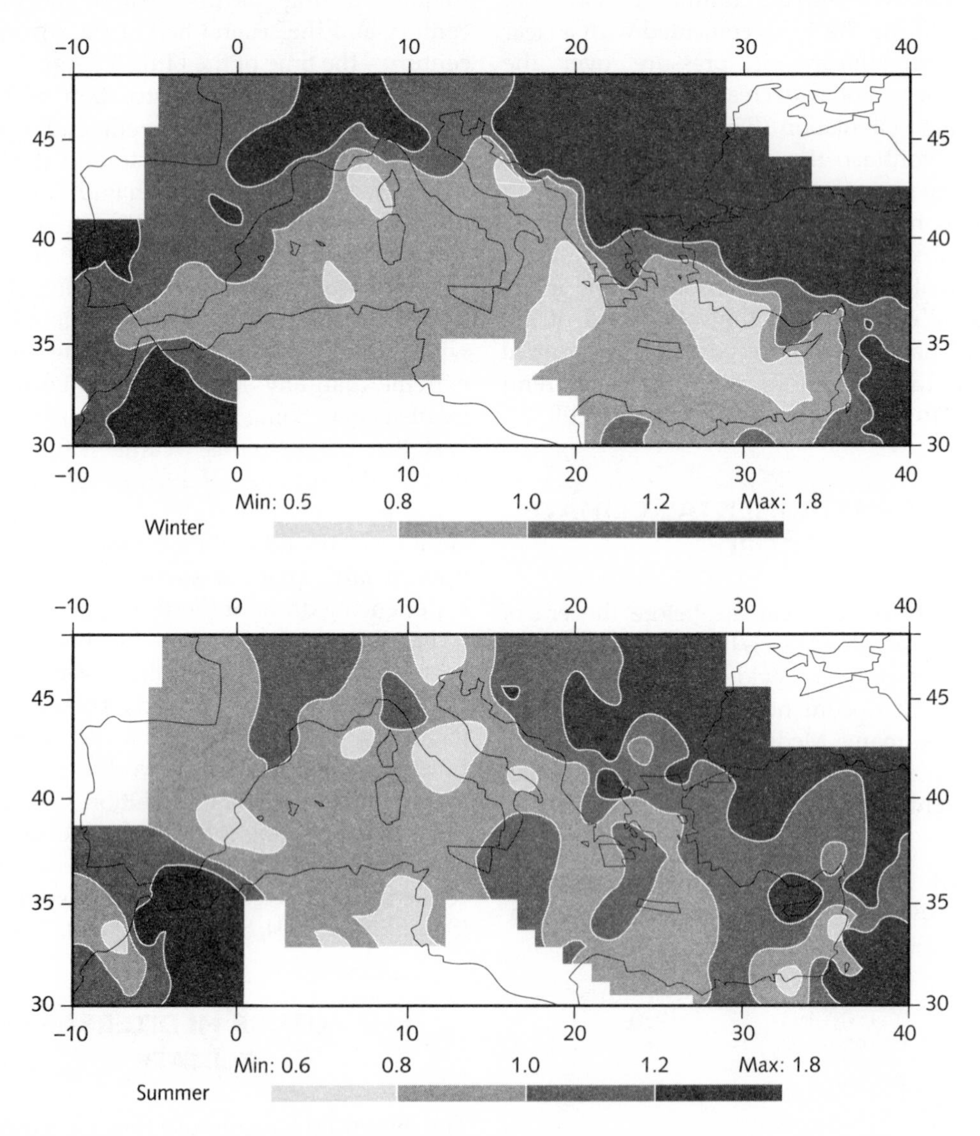

FIGURE 3.7 Predicted increase in temperature (°C) in the Mediterranean by 2030 with a general 1°C global warming

suggest higher precipitation to the north of the Mediterranean and lower precipitation over North Africa. In autumn, the contrast is between a decline in the western Mediterranean and increases over the central and eastern areas (Fig. 3.8). Potential evapotranspiration is not properly assessed by GCMs owing to their crude treatment of ground hydrology, but estimates by Palutikof and Wigley (1996) suggest that large increases of 180–200 mm may occur in summer, especially over North Africa and southern Spain.

The impacts of these expected changes are potentially great for many aspects of the environment and economies of the region. To assess them the United Nations Environment Programme (UNEP) launched, co-ordinated and financially supported a series of regional task teams. Amongst the first to report under this UNEP Regional Seas Programme was the

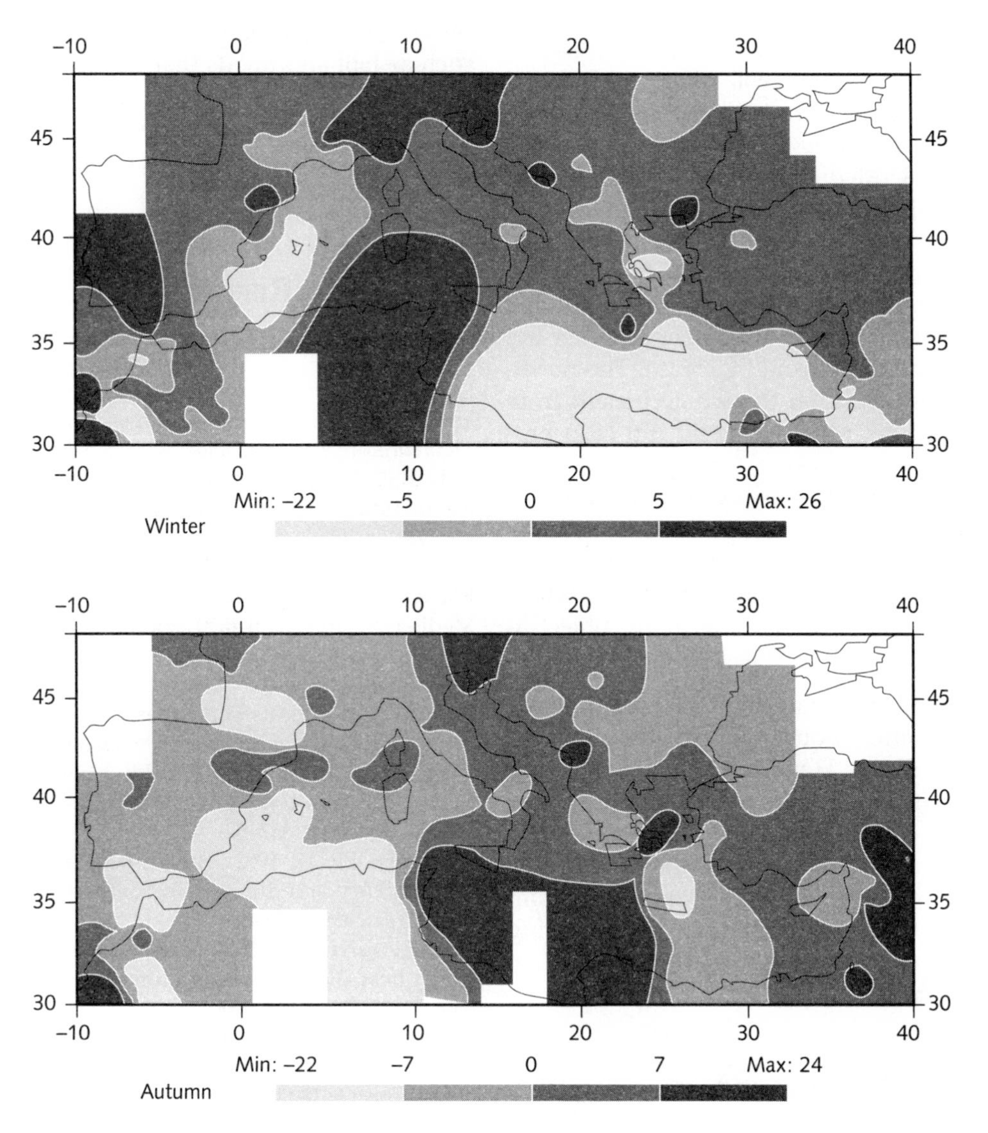

FIGURE 3.8 Predicted changes of precipitation (mm) in autumn and winter in the Mediterranean by 2030 with a general 1°C global warming

Mediterranean team (Jeftic *et al.* 1992), discussing both environmental and societal impacts in the context of a series of case studies that include the Ebro, Nile and Po deltas. A further five new case studies have been completed and published recently (Jeftic *et al.* 1996). Amongst the general findings that can be summarised from these studies the following are worth highlighting:

1. Greater soil aridity and degradation and a consequent high risk of erosion are likely to increase flood occurrence.
2. Changes of climate will affect agricultural production and forests. Some crops will be abandoned and new ones introduced as the traditional Mediterranean vegetation belt moves northwards. Citrus and other cold-sensitive crops, for example, could be grown over much larger areas of Southern Europe whilst more tropical crops might become established in North Africa. The summer fire hazard to forests would increase. The northern basin would on the whole benefit from enhanced net photosynthesis and crop productivity.
3. Associated sea-level rises could pose a major threat to low-lying coastal settlements, especially Venice. Some coastal lowlands will be lost, along with wetland resources, whilst water supplies will become more subject to saline incursions.

It is expected that many human-related activities will suffer equal or greater impacts. According to Le Houérou (1990) social and political problems will be exacerbated by major population increases in Afro-Asian Mediterranean countries and migration northwards into Europe of hungry multitudes escaping from drought, heat desertification and degrading environments. The extent to which this is an 'avoidable crisis' is evaluated in Chapter 11.

CONCLUSIONS

The Mediterranean separates dry, arid, hot North Africa from humid, temperate Europe. The whole area is vulnerable to climate change and any sustained reduction of winter rains together with increasing temperature and evaporation during summer could lead to near-desert conditions. Models of predicted future climate change suggest that the southern and eastern shores of the Mediterranean in particular could undergo deleterious changes by the middle of the next century.

The greatest variability of rainfall in the EU is observed on the Mediterranean shores but at present it is episodic destructive phenomena such as hail and floods that constitute the most acute climatic hazards. Little can be inferred on the likely future occurrence of these hazards, but drought frequency, especially in winter, seems likely to increase.

REFERENCES

BÁRDOSSY, A. and CASPARY, H.J. 1990: Detection of climate change in Europe by analyzing atmospheric circulation patterns from 1881–1989. *Theoretical Applied Climatology* **42**, 155–67.

BARSCH, D. 1965: Les arbres et le vent dans la vallée méridionale du Rhône. *Révue de Géographie de Lyon* **40**, 35–45.

BARTZOKAS, A., METAXAS, D., KATEN, M. and EXARCHOS, N. 1991: Sea surface temperature in the Mediterranean: statistical properties. *Rivista di Meteorologia Aeronautica* **55**, 52–8.

BESANÇENOT, J.P. 1985: Climate and summer tourism on the coasts of the Iberian peninsula. *Révue Géographique des Pyrénées et du Sud-Ouest* **56**, 427–49.

BESANÇENOT, J.P. MOUINER, J. and LAVENNE, F. DE 1978: Les conditions climatiques du tourisme littoral. *Norois* **99**, 357–82.

BISCHOFF, U. 1992: The scirocco: investigations on frequency and duration in Southern and Central Italy. *Erdkunde* **46**, 52–6.

COLACINO, M. 1980: Some observations of the urban heat island in Rome during the summer season. *Nuovo Cimento* **3**, 165–79.

COLACINO, M. and CONTE, M. 1995: Heatwaves in the Central Mediterranean: a synoptic climatology. *Nuovo Cimento* **18**, 295–309.

COLACINO, M. and PURINI, R. 1986: A study on the precipitation in Rome 1782–1978. *Theoretical Applied Climatology* **37**, 90–106.

DIKAIKOS, J. and PERRY, A.H. 1981: Climate during halcyon days in Athens. *Journal of Meteorology* **6**, 410–13.

GILES, B.D. and BALAFOUTIS, C.J. 1990: The Greek heatwaves of 1987 and 1988. *International Journal of Climatology* **10**, 505–17.

GOOSSENS, C. 1985: Principal component analysis of Mediterranean rainfall. *Journal of Climatology* **5**, 379–88.

GOOSSENS, C. 1986: Regionalization of the Mediterranean Climate. *Theoretical and Applied Climatology* **37**, 74–83.

GROVE, J.N. and CONTERIO, A. 1995: The climate of Crete in the sixteenth and seventeenth century. *Climate Change* **30**, 223–47.

HARLFINGER, O. 1991: Holiday bioclimatology: a study of Palma de Majorca, Spain. *GeoJournal* **25**, 377–81.

HUGHES, D. 1988: Land and sea. In Grant, M. and Kitzinge, R. (eds), *Civilizations of the Ancient Mediterranean, Vol. 1.* New York: Scribner, 89–133.

JACOBEIT, J. 1988: Intra-seasonal fluctuations of mid-tropospheric circulation above the eastern Mediterranean. In Gregory, S. (ed.), *Recent Climatic Change*. London: Belhaven Press, 90–101.

JEFTIC, L., KECKES, S. and PERNETTA, J. (eds) 1996: *Climatic Change and the Mediterranean, Vol. 2.* London: Arnold.

JEFTIC, L., MILLIMAN, J.D. and SESTINI, G. (eds) 1992: *Climatic Change and the Mediterranean, Vol. 1.* London: Edward Arnold.

KATSOULIS, B.D. 1988: Aspects of the occurrence of persistent surface inversions over Athens Basin. *Theoretical Applied Climatology* **39**, 98–107.

KATSOULIS, B.D. and KAMBETZIDIS, H.D. 1989: Analysis of the long-term precipitation series of Athens, Greece. *Climatic Change* **14**, 263–90.

LE HOUÉROU, H.N. (1990): Global change: population, land-use and vegetation in the Mediterranean basin by the mid-21st century. In Paepe, E., Fairbridge, R.W. and Jelgersma, S. (eds), *Greenhouse Effect, Sea Level and Drought*. Dordrecht: Kluwer Academic Publishers, 301–67.

LLASAT, M.C. and PUIGCERVER, M. 1994: Meteorological factors associated with floods in the NE parts of the Iberian peninsula. *Natural Hazards* **9**, 81–93.

MAHERAS, P. 1988: Changes in precipitation conditions in the western Mediterranean over the last century. *Journal of Climatology* **8**, 179–89.

MAHERAS, P., BALAFOUTIS, C. and VAFIADIS, M. 1992: Precipitation in the central Mediterranean during the last century. *Theoretical Applied Climatology* **45**, 209–16.

MAKROGIANNIS, T. and SAHSAMANOGLOU, H. 1990: Time variation of the mean sea level pressure over the major Mediterranean area. *Theoretical Applied Climatology* **41**, 149–56.

MAZZOLENI, S., LO PORTO, A. and BLASI, C. 1992: Multivariate analysis of climatic patterns of the Mediterranean basin. *Vegetatio* **98**, 1–12.

METEOROLOGICAL OFFICE 1962: *Weather in the Mediterranean*. 2 vols. London: HMSO.

PALUTIKOF, J.P. and WIGLEY, T.M. 1996: Developing climate change scenarios for the Mediterranean region. In Jeftic, L., Keckes, S. and Pernetta, J. (eds), *Climatic Change and the Mediterranean, Vol. 2.* London: Arnold, 27–41.

PENNAS, P.J. 1991: Sunshine duration study within the Mediterranean area. *Theoretical Applied Climatology* **44**, 173–9.

PERRY, A.H. 1976: Mediterranean downpours in 1976. *Journal of Meteorology* **2**, 10–11.

PERRY, A.H. 1981: Mediterranean climate: a synoptic reappraisal. *Progress in Physical Geography* **5**, 107–13.

PERRY, A.H. 1988: Trends in the rainfall of Malta. In Gregory, S. (ed.), *Recent Climate Change*. London: Belhaven Press, 125–9.

PERRY, A.H. 1992: Making sense of weather anomalies: the example of winter 1991–2. *Journal of Meteorology* **17**, 117–20.

PREZERAKOS, N. 1985: The Northwest African depressions affecting the south Balkans. *Journal of Climatology*, **5**, 654–5.

REITER, E.R. 1975: *Handbook for Forecasters in the Mediterranean*. Monterey, CA: Naval Postgraduate School.

REPAPIS, C.C., MANTIS, H.J., METAXAS, D. and KANDILIS, P. 1992: A study of the climate variability in the Eastern Mediterranean during the first half of the twentieth century. In Zeid, A. and Biswas, A.K. (eds), *Climatic Fluctuations and Water Management*. London: Heinemann, 139–47.

SAHSAMANOGLOU, H.S. and MAKROGIANNIS, T.J. 1992: Temperature trends over the Mediterranean region 1950–1988. *Theoretical Applied Climatology* **45**, 183–92.

SENEQUIER, D. 1993: A Mediterranean squall. *Meteorological Magazine* **122**, 229–38.

SERRE-BACHET, F. 1994: Middle Ages temperature reconstructions in Europe: a focus on NE Italy. *Climatic Change* **26**, 213–24.

SERRE-BACHET, F., GUIOT, J. and TESSIER, L. 1990: Dendroclimatic evidence from S.W. Europe and N.W. Africa. In Bradley, R.S and Jones, P.D. (eds), *Climate since A.D. 1500*. London: Routledge, 349–65.

SIEGFRIED, A. 1948: *The Mediterranean*. London: Cape.

SIVALL, T. 1957: Sirocco in the Levant. *Geografisker Annaler* **29A**, 114–42.

TOUT, D.G. 1987: Almeria province: the driest region in Europe. *Weather*, **42**, 242–7.

TOUT, D.G. 1990: The horticultural industry of Almeria Province, Spain. *Geographical Journal* **156**, 304–12.

TOUT, D.G. and WHEELER, D.A. 1990: The early autumn storms of 1989 in eastern Spain. *Journal of Meteorology* **15**, 238–48.

WHEELER, D. 1991: Majorca's severe storms of September 1989: a reminder of Mediterranean uncertainty. *Weather* **46**, 21–6.

WHEELER, D. 1994: La Tramontana – an ill wind in Catalonia. *Journal of Meteorology* **19**, 185–91.

WHEELER, D. 1996: Spanish climate. *Geography Review* **10**, 34–40.

WIGLEY, T.M. 1992: Future climate of the Mediterranean Basin with particular emphasis on precipitation. In Jeftic, L., Milliman, J. and Sestini, G. (eds), *Climatic Change and the Mediterranean, Vol. 1*. London: Edward Arnold, 15–44.

WIGLEY, T.M. and FARMER, G. 1982: Climate of the eastern Mediterranean and the near East. In Bintliff, J. and Van Zeist, W. (eds), *Palaeoclimates, Palaeoenvironments and Human Communities in the Eastern Mediterranean in Later Prehistory*. Oxford: Oxford University Press, 3–37.

WINSTANLEY, D. 1972: Sharav. *Weather* **27**, 146–60.

WRATHALL, J.E. 1985: The Mistral and forest fires in Provence, Southern France. *Weather* **40**, 119-24.

YEO, B. 1882: *Climate and Health Resorts*. London: Chapman and Hall.

YOSHINO, M.M. 1971: The bora in Yugoslavia: a synoptic climatological study. *Annaler der Meteorologie* **5**, 117–21.

ZEREFOS, S. and ZEREFOS, E.C. 1978: Climatic change in Mycenean Greece: a citation to Aristotle. *Archiv für Meteorologie, Geophysica und Bioklimatologie Series B* **26**, 297–303.

4

EARTH SURFACE PROCESSES IN THE MEDITERRANEAN

HELEN RENDELL

INTRODUCTION

Geomorphological processes in Mediterranean environments are conditioned by the strong seasonality and frequently steep climatic gradients that were discussed in the previous chapter. The geological setting of the Mediterranean, at complex plate junctions, is one of intense tectonic activity, marked by earthquakes and vulcanism, as we saw in Chapter 2. This interaction of climate and tectonics has produced a landscape characterised by high relative relief, which in turn has been strongly affected by a long history of human settlement and changing land-use (Fig. 4.1).

FIGURE 4.1 The long history of occupation in the Mediterranean has meant that virtually all cultivable land is exploited, as is the case on the southern slopes of the Troodos mountains in Cyprus, where the landscape has been farmed for in excess of 5000 years

Although sediment movement by wind is a significant factor in some areas, earth surface processes in the Mediterranean Basin as a whole are dominated by fluvial erosion and mass movements driven by seasonal and local moisture availability and gravitational stresses. The dominance of different processes is dictated by interactions between climatic factors, relative relief, geology and soils at particular locations. Vegetation is the most critical factor in providing protection from erosion to soils and sediments (Thornes 1990, 1995). For wind action and the overland flow of water, vegetation increases surface roughness, decreases boundary shear and reduces the likelihood of erosion. During rainfall, vegetation mediates interception and infiltration. Since agriculture dominates the Mediterranean landscape, it is impossible to discuss earth surface processes and landscapes without reference to it. Other anthropogenic impacts, such as road construction, can change rainfall-runoff relationships on slopes and may influence the development of landslides and other mass movements, while dam construction and aggregate extraction can significantly alter the behaviour of fluvial systems.

This chapter reviews the range of geomorphological and hydrological processes encountered in the Mediterranean Basin. Although the approach is generic, the material cited inevitably reflects the balance of work undertaken in various parts of the Mediterranean Basin during the last 30 years, and therefore focuses mainly on studies of contemporary processes in Spain, Italy and Israel. Earth surface processes both influence, and can be strongly influenced by, human activities. Without wishing to be too deterministic, it is clear that the vulnerability to hazard is increasing in the Mediterranean, as it is elsewhere in the world. Risks continue to be taken in the development of floodplain areas for industrial and residential use, without any thought to the probability of flooding. Landslides present a continuing hazard for urban areas, historic monuments and infrastructural links. Settlements encroach on areas of high volcanic risk, such as the Pozzuoli area of the Bay of Naples in Italy. Houses are built or rebuilt in areas of seismic risk without appropriate structural reinforcement. In all of these cases vulnerability to geological, geomorphological or hydrological processes is increased by human action. The possibility of future climate changes, with associated impacts on at least some of these processes, simply introduces a further layer of complexity.

Over and above these problems, the recognition of the comparative fragility of much of the Mediterranean area in terms of biomass production, and the potential for desertification (see Chapter 16), has triggered considerable research in recent years. Much of this has involved the definition of areas at risk and has increased understanding of contemporary processes and their interactions.

Desertification and land degradation

The issues of desertification and soil erosion have provided a major focus for work on climate change impacts funded by the European Union. As well as concentrating on developing an understanding of contemporary problems, and producing scenarios for possible climate change impacts, the EU has employed a more pragmatic approach to problems such as soil erosion. This is exemplified by the Corine Project (based in the European Commission's Directorate-General XI), which produced maps of land resources and soil erosion risk in the southern regions of the EU (Italy, Spain, Portugal, Greece and southern France). Potential erosion risk was identified using criteria that included soil erodibility and erosivity and the actual risk was then determined by taking the extent of ground cover, and therefore agricultural activity, into account. By this method areas of high and medium risk of actual soil erosion were seen to comprise 19 per cent and 36 per cent of the land area of the southern regions, respectively. Potential risk of erosion is highest in Greece, Spain and Portugal (see Table 4.1). Although they incorporate an inevitable level of generalisation, these data underline the critical role of ground cover in mitigating erosion risk and highlight the particular vulnerability of Spain and Portugal to soil erosion.

The Corine study therefore confirms that specific risks are associated with particular combinations of rainfall, slope steepness, slope length, soil type and ground cover. Since these risks result from processes involving gravitational stresses, resistances and the presence of shear forces at boundaries, it is appropriate to consider the operation of these processes in detail.

Mass movements

Mass movements occur when gravitational stresses (forces) exceed the strength (shearing resistance) of materials on slopes. They range from rock falls, topples and slides to rotational slides, debris flows and earthflows. Increases in gravitational stresses are caused by slope steepening by, for example, natural or artificial cuttings, and increases in the weight of materials on slopes. However, most mass movements

TABLE 4.1 Potential and actual soil erosion risk in the southern regions of the European Community

Country	High-risk		Medium-risk		Low-risk	
	km²	(%)	km²	(%)	km²	(%)
France						
potential	16355	9	37900	20	93443	49
actual	1693	1	22362	12	123643	65
Italy						
potential	82348	27	85211	28	122416	41
actual	30169	10	93983	31	165823	59
Greece						
potential	57414	43	27436	21	27027	21
actual	27713	19	47877	36	39287	30
Spain						
potential	202101	41	205157	41	69662	14
actual	145494	29	219908	44	111518	23
Portugal						
potential	61120	68	21890	25	4948	6
actual	26878	30	48166	54	12884	15

Source: Corine Project, Commission of the European Communities DGXI, 1993

occur either when weathering decreases material strength, which reduces cohesion and frictional resistance, or when positive pore water pressures within slope materials reduce effective normal stresses. Glacial and periglacial effects are found in the Alps and Pyrenees and glaciation in the Alps has left a legacy of steep, mostly rock-cut slopes, whereas periglacial slope deposits are extensive in the higher altitude extra-glacial areas of the Mediterranean Basin such as the Central Apennines.

Although areas of high relative relief are inevitably prone to mass movements, records of such movements tend to reflect only those landslides or rock slides which cause damage to lives, property or infrastructure. Despite the potentially biased nature of such records, those compiled by Eisbacher and Clague (1984), for mass movements in the Alps, do indicate the importance of rainfall events as triggers of several types of mass movement, especially debris flows (Table 4.2). Debris flows are particularly interesting since they often mobilise material that has accumulated within small, steep drainage catchments, and can show dynamic change during movement, as more sediment and water is incorporated into the flow. Given their nature, the products of debris flows tend to fill stream and river courses and pose additional hazards where, for example, tourist developments in the Alps and Pyrenees occupy

TABLE 4.2 Mass movement types and triggers in the Alps: data from case studies

	Earthquake-triggered	Rainstorm-triggered	Snowmelt-triggered
Debris flow: superficial		46	8
Debris flow: bedrock	1	45	17
Glacier-related mass movement			11
Rock avalanche or rock fall	8	20	21

Source: Eisbacher and Clague (1984)

valley bottoms. The role of extreme rainfall events in November 1982 in triggering mass movements on steep slopes in the eastern Pyrenees was discussed by Gallart and Clotet-Perarnau (1988). A characteristic of very large-scale mass movements, in excess of 1×10^6 m^3 of material, is their ability to temporarily or permanently dam valleys. Such mass movements require immediate remedial work to prevent the rupture of the blockage and catastrophic flooding of the valley below (see examples in Eisbacher and Clague 1984).

Seasonal moisture availability, coupled with high relative relief in much of the Mediterranean Basin, strongly influence the nature and frequency of mass movements, and failures are more often triggered by rainfall rather than by a rise in groundwater table. In the more arid parts of the Mediterranean, long-term desiccation promotes stability in areas that would otherwise (i.e. under humid temperate conditions such as those in the UK) be inherently unstable. In southern Italy, for example, slopes of in excess of 45° are cut in clays with angles of internal frictional resistance in the range 15–20° (Bromhead *et al.* 1994) which, in a humid temperate area, would have residual slope angles of 7–10°. In these more arid areas controls on mass movements are more complex. In the Plio-Pleistocene clay areas of Basilicata, southern Italy, for example, prolonged desiccation of the clays to depths in excess of 30 m below ground level are indicated (Bromhead *et al.* 1994). Thus slopes of 45° or even 60° are not uncommon because the clay effectively behaves like a rock, rather than a soil. Deep-seated mass movements occur only where perched water tables facilitate water penetration into joints and fissures deep within the clay. Such perched water tables occur at contacts between impermeable clays and overlying permeable sands and gravels. Otherwise the clay hillslopes are wetted by rainfall, rather than groundwater, and mass movements tend to be shallow. Any localised moisture concentration can produce superficial mass movements and, for example, poor quality or non-existent roadside drainage can result in the concentration of surface runoff. If this runoff is allowed to leave the road on a bend it can trigger mass movements which in turn undermine the road.

Virtually all hilltop settlements in southern Italy are subject to the effects of mass movements. Landslide damage can be so severe that towns have to be abandoned and the populace rehoused, as was the case with Craco (Del Prete and Petley 1982; see Fig. 4.2). In other cases, only the periphery of the settlement is affected by landslides, so that groups of dwellings are lost or have to be abandoned.

The Mediterranean Basin is also an area of moderate to high seismicity, and landslides have been triggered by earthquakes. Often this involves the re-activation of previous slides, as was the case for many of the landslides associated with the November 1980 earthquake in southern Italy, but sometimes large-scale mass movements have been triggered on a regional scale. Historical records show, for example, that a series of earthquakes in Calabria during the early months of 1783 triggered numerous large landslides, blocking valleys to the north of the Aspromonte. Major engineering works were required to drain landslide-dammed lakes, which not only represented a substantial physical hazard to downstream populations, but also a

FIGURE 4.2 Severe landslide damage has resulted in the complete abandonment of the hilltop settlement of Craco in Basilicata. The population have been rehoused in a new village on a more stable site.

health risk since malarial mosquitoes can breed in stagnant water.

PROCESSES ON HILLSLOPES

Background

These processes encompass rainsplash, infiltration and surface ponding, unconcentrated and concentrated surface flow (rill flow or gully flow), throughflow and piping. Gullies can affect both hillslopes and alluvial fills in valley bottoms and are steep-sided trenches which frequently end in a head-cut and which transmit flow ephemerally. However, 'gullying' does not exist as a process *per se*; instead, gullies result from the interaction of processes involving concentrated surface and subsurface flow and shallow mass movements. These processes operate effectively in poorly consolidated superficial deposits or where bedrock has a substantial silt/clay component (e.g. shales, marls or clays). Where soils have developed on a so-called hard rock such as granite or limestone, soil development may occur very slowly compared to the much shorter time scales over which a soil cover can be eroded. In areas such as southern France, southern Italy, Greece, Cyprus and former Yugoslavia, soil cover has been almost completely removed from limestone bedrock to leave hillslopes virtually bare of both soil and vegetation. In such areas further erosion is limited by the lack of material to erode (Fig. 4.3). To emphasise this point, Woodward (1995) showed that some 2.1 million ha of karst limestone in the former Yugoslavia are totally barren, an area equivalent to 38 per cent of the total limestone outcrop. Erosion does occur in limestone and sandstone areas, but Heusch and Milliès-Lacroix (1971) suggest that suspended sediment yields from catchments dominated by marls (calcareous clays/shales) are at least an order of magnitude higher than for limestone or sandstone catchments for a given value of annual runoff.

Although limestones constitute a substantial proportion of the bedrock in the Mediterranean Basin, Woodward (1995), in his review of erosion and sediment yield in the Mediterranean area, notes that they yield very little suspended sediment. He quotes the example of the predominantly limestone catchment of the Voidomatis River in north-west Greece, in which suspended sediment yield comes overwhelmingly from a small area of flysch deposits (Woodward *et al.* 1992). The dominant process in limestone areas is thought to be solution which, operating over time scales of millions of years, can produce classic karst development. The limestone areas of the Mediterranean Basin include some spectacular areas of karst scenery which frequently provide the opportunity to identify the underlying structural controls on landscape development (Fig. 4.4). Individual landforms include classic examples of caves, dolines, swallow holes and pavements (Fig. 4.5). Limestone basins are very important for water supply, since they contain substantial reserves of groundwater. Limestone is also intensively quarried, with entire hillslopes sometimes removed, primarily for use in the manufacture of cement. The value of these areas extends beyond the provision of natural resources since they also provide data on long-term landscape

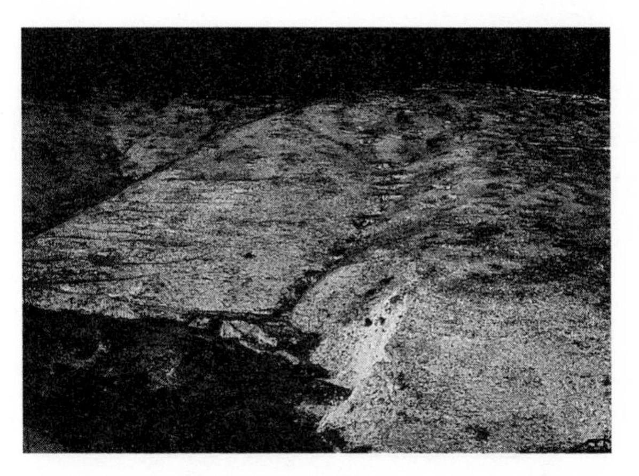

FIGURE 4.3 Bare limestone hills near Lefkara in Cyprus. Once the soil cover is lost hard limestone is exposed, sediment supply to adjoining rivers virtually ceases and eroded topsoil, which previously accumulated in valleys, is incised by hillslope runoff to form alluvial terraces that are characteristic of much of the Mediterranean.

FIGURE 4.4 High peaks of the Serra de Tramuntana in Majorca. These mountains form part of the Betic Arc at the western end of the Mediterranean, and the limestones clearly show the underlying structure of overthrusted folds that is typical of many of the mountain ranges that fringe the basin.

evolution and on human prehistory. Although the time required for the development of caves may be as long as 2.5 million years (Ginés and Ginés 1995), the coastal cave systems on the island of Majorca have been shown to have at least the potential to provide data on Mediterranean sea-level change. The well-known cave sites in the Dordogne area of southern France have also yielded valuable information concerning recent human prehistory.

Badlands

Much of the work on contemporary hillslope processes within the Mediterranean has concentrated on rainfall–runoff relationships in badland areas (see Table 4.3 for Spanish examples). Although badlands are extremely photogenic and locally extensive (e.g. in south-east Spain and southern Italy; Fig. 4.6), they are not necessarily representative of Mediterranean hillslopes as a whole. Similarly, areas of 'marginal land' characterised by scrub vegetation are not particularly representative, although in the more arid parts of the Mediterranean ground cover is often sparse. Instead, intensive cultivation, both with and without irrigation, accompanied by the seasonal removal of vegetation for grazing, is dominant.

The visual impact of badlands is often very striking and is linked to the automatic assumption that they are areas of inherently high erosion. In a detailed study of the Guadix and other Tertiary areas in south-east Spain, Gilman and Thornes (1985) demonstrated that appearances can, however, be deceptive. Instead of the anticipated high rates of erosion, not only were rates low, of the order of 25 t km^2 yr^{-1}, but the

> *in situ* archaeological deposits imply greater slope stability and less rapid erosion than the visual evidence suggests ... the sparse vegetation cover resulting from a harsh climate and

FIGURE 4.5 Doline (enclosed depression) in the limestone mountains of north-west Majorca

FIGURE 4.6 The extensively gullied landscape of Basilicata in southern Italy

TABLE 4.3 Examples of hillslope erosion studies undertaken in Spain

Author(s)	Topic	Area
Imeson and Verstraten 1988	Rills on badland slopes	Dahesas, Granada
Imeson and Verstraten 1989	Microaggregation and erodibility of calcareous soils on regolith materials	Dahesas, Granada
López-Bermúdez and Romero-Díaz 1989	Piping erosion and badland development	Guadix-Baza-Huéscar basins, south-east Spain
Gutiérrez *et al.* 1988	Piping in badland areas	Middle Ebro Basin, north-east Spain
Clotet-Peramau *et al.* 1988	Badland erosion rates	Llobregat Basin, Pyrenees
Romero-Díaz *et al.* 1988	Variability of overland flow erosion rates	Murcia, south-east Spain
Sala 1988	Slope runoff and sediment production	Montseny Mountains, Catalan range, north-east Spain
Calvo-Cases and La Roca Cervigón 1988	Slope form and soil erosion on calcareous slopes	Valencia
Cervera *et al.* 1991	Rainfall simulation from sparsely vegetated and non-vegetated badlands	Llobregat Basin, pre-Pyrenees
Francis and Thornes 1990	Runoff hydrographs from three Mediterranean vegetation cover types	Murcia, south-east Spain

from overgrazing reveals every geomorphic detail and thereby creates a misleading impression of catastrophic erosion.

(Gilman and Thornes 1985, p. 74)

Although much higher rates of erosion have been measured in badlands elsewhere in Spain and Italy, erosion estimates are notoriously scale-dependent (Roels 1985). As a general rule, the longer the time scale used and the larger the area considered, the lower the average rate of erosion. This is exemplified by data from a Plio-Pleistocene clay area in southern Italy where erosion pin measurements on bare slopes yielded values of 22 mm yr^{-1} compared with rates of 1.0–0.6 mm yr^{-1} at the subcatchment scale over time periods of 5–8 kyr, and a base-line incision rate for the area of 0.43 mm yr^{-1} over a period of 700 kyr. In a study in Murcia, south-east Spain, Romero-Díaz *et al.* (1988) demonstrated a good correlation between rainfall and sediment yield at the catchment scale, but a much poorer relationship with sediment yields measured at a smaller scale using Gerlach troughs to trap eroded sediment.

Soils, vegetation and erosion

A number of studies have focused on interactions between the physico-chemical properties of the soil and processes of surface wetting, rainsplash, crust formation and surface flow. Many Mediterranean soils exhibit both slaking and dispersion during wetting. Slaking is a physical process whereby soil aggregates are burst apart by the escape of air trapped by advancing wetting fronts. Dispersion, on the other hand, is a physico-chemical process that is driven by the presence of sodium cations in the exchange complex of montmorillonite clays.

Dispersion can be suppressed by the presence of small amounts of humic materials (1–2 per cent organic matter) or by the use of various soil conditioners. Many Mediterranean soils are calcareous, and organic matter appears to be particularly important for stabilising micro-aggregates in these soils (Imeson and Verstraten 1989).

Saline or sodic soils also present potential difficulties, and in some areas saline intrusion into coastal aquifers already affects soils and the quality of irrigation water used on them (see Chapter 15). In other areas soils are naturally sodic, with high values of exchangeable sodium percentage, the key index property for dispersive soils. Since sodic soils are prone to swelling and dispersion, they are also highly erodible. These soils are also subject to piping through the sectional enlargement of desiccation crack networks (Gerits *et al.* 1987).

In the context of potential short-term changes in climate, Imeson and Emmer (1992) have emphasised the importance of changes in the salt balance, carbonate precipitation and the supply and breakdown of organic matter within soils. They postulated that changes in precipitation and evapo-transpiration have the potential to increase the areal extent of saline or sodic soils. They commented that:

> in those regions with high evaporation rates capillary rise is accelerated and salts accumulate residually where drainage is nearly absent. Particularly serious would be a decrease in winter precipitation to the extent that seasonally accumulated salts are not flushed from the soil. ... Particularly in Spain and Italy, changes in salt accumulation could lead to an increase in the areas of soils affected by vertic conditions, irrigation and crop management would become more difficult or expensive, and the need for crops with higher salt tolerance would increase.
>
> (Imeson and Emmer 1992, p. 103)

Accumulations of calcrete (secondary carbonate), initially as Bt horizons within Mediterranean soils, are fairly common and lead to important changes in water-holding capacity. Calcrete tends to have an extremely low permeability but, once water penetrates the calcrete, it also exhibits high water retention characteristics. The combination of these two properties in a semi-arid environment has important implications for plant growth and, in particular, water supply to plants during periods of summer drought. Studies of the impact of surface and buried stones on rainfall–runoff relationships also reflect a recognition of the need to deal with complicating factors associated with soil erosion. Data appear to indicate that a complex interaction may take place depending upon both stone size and areal coverage (Abrahams and Parsons 1994).

The impact of vegetation cover on soil erosion has only relatively recently attracted serious attention, although in simple terms a vegetation cover is the most effective protection against erosion. Work has tended to concentrate on vegetation conditions on semi-arid hillslopes, dominated by scrub vegetation, rather than vegetation and erosion interactions on agricultural fields. With vast areas of the Mediterranean Basin given over to agriculture, human impact on soil erosion is hard to ignore. Removal of vegetation for the planting of vines, almonds and other tree crops leaves large areas of bare soil open to rill erosion if slopes are moderate or steep (Fig. 4.7). Cultivation of grain

FIGURE 4.7 Rill erosion by spring rains of a vineyard in the hills of south-central Spain, from which the protective ground vegetation has been removed

involves the removal of surface cover between harvesting and sowing. Moreover, moves towards the cultivation of autumn-sown cereals have effectively optimised erosion, with the thunderstorms of early autumn coinciding with a lack of vegetation on arable fields. In arid and semi-arid parts of the Mediterranean, grazing can also result in the temporary removal of surface vegetation cover. Fire, both controlled (as in stubble burning) or uncontrolled, can also have a major, if local, impact on surface vegetation cover (see Chapter 16).

Francis and Thornes (1990) investigated the commonly held belief that the optimal erosion-resistant vegetation cover is a tree cover by studying rainfall–runoff relationships under degraded matorral, shrub matorral and high matorral with pines. They used small 5 m × 2 m plots and different intensities of simulated rainfall. The results are complex and inevitably biased by plot size, but they demonstrate that although high matorral offers the best protection from erosion, scrub or bush matorral provides extremely effective ground cover compared to degraded matorral. These authors suggest that encouraging the development of a good cover of natural scrub vegetation presents a rapid solution to erosion problems.

Some agricultural practices, such as terracing, may put a brake on hillslope erosion by controlling sediment and water movement and by effectively decoupling hillslopes from channels. The problem with terrace systems is that they require constant maintenance, and it has been argued that hillslope erosion and valley bottom aggradation in the Mediterranean are often associated with terrace systems falling into disuse and disrepair.

CHANNEL PROCESSES

Annual rainfall within the Mediterranean ranges from 200–400 mm yr^{-1} to in excess of 1200 mm yr^{-1}. Rainfall seasonality leads to a dominance of ephemeral flow over perennial flow in drier parts of the Mediterranean where inputs from base-flow are minimal or non-existent. These ephemeral flow regimes are 'torrential'; that is, they are dominated by a few relatively high magnitude events and for much of the year channels may be dry. The spatial and temporal resolution of rainfall predictions/scenarios (Palutikof and Wigley 1996) is too poor to allow reliance on them; but any change from perennial to ephemeral flow conditions within catchments – as a result of changes in rainfall supply, groundwater conditions and base flow – is likely to be associated with some fundamental changes in flood hydrographs and sediment transport. Recently published studies by Laronne and Reid (1993) and Reid and Laronne (1995) for the ephemeral Nahal Yatir, in Israel, show that, in comparison with a similar-sized perennial system, the bedload mobilised during a flow event was several orders of magnitude higher. Mobilisation of bedload, even under conditions of very low flow, is associated with the lack of armouring on the bed of the ephemeral torrent. From the minimal information currently available, it would appear that flow in ephemeral systems is characterised by very steep rising and falling stages, and by an extremely rapid rise to peak discharge, of the order of 15–30 minutes for smaller catchments of a few km^2 in area.

Conversion from perennial to ephemeral systems will also have an impact on, and be affected by, systems of water management. In areas of predominantly bedload transport, the higher mobility associated with ephemeral flow conditions will profoundly affect any dams within the catchment. Data for a small ephemeral catchment in the Pyrenees (Clotet-Perarnau *et al.* 1988) show that a rockfall-dammed natural sediment trap accumulated some 2100 m^3 of sediment from a catchment 0.031 km^2 in area over a period of 40 years. The fill comprised 13 fining-upwards sedimentary units, each representing a particular flow event within the catchment. At a different scale, records show that, in 1962, 30 m of sediment accumulated behind a landslide dam on the torrential Tagliamento river in northern Italy in the two months before the landslide dam was breached (Eisbacher and Clague 1984).

Data from Central Italy point to the existence of substantial changes in the behaviour of rivers and the character of their floodplains during the late Holocene (Cilla *et al.* 1994). These changes included a switch of channel planform geometry from braided to meandering and are thought to be due in part to changing patterns of landuse.

In much of southern Italy, sediment fluxes within catchments have been radically altered by dam construction in recent years. Sediment yields of 1159 t km^{-2} yr^{-1} and 2458 t km^{-2} yr^{-1} were recorded for the Bradano and Sinni catchments, respectively, prior to dam construction (Morandini 1962). Not only is the economic life of the San Giuliano and Senise Dams on these rivers now threatened, but the disruption of coastal sediment systems has led to erosion of beaches along the Ionian coast which poses problems for tourist developments. Sediment inputs from rivers and movement alongshore and onshore/offshore are critical to the maintenance of wide beaches, and the reduction in this sediment supply has directly affected beach volumes. Problems of reservoir siltation and beach starvation are also common in other areas of the Mediterranean (see Woodward 1995).

Another contemporary impact on channel processes is the lowering of groundwater levels through extraction for irrigation and water supply. In the Guardiana and Zancara catchments, in the Manchega area of central Spain, pumping of aquifers has led to an inexorable fall in groundwater levels at a rate of several metres per year and, consequently, the drying up of these groundwater-fed rivers (Pérez González 1994).

Aeolian processes

Aeolian processes only dominate in certain parts of the Mediterranean Basin. Sediment movement by wind requires the conjunction of strong turbulent winds with a supply of sediment suitable for entrainment. In many parts of the Mediterranean these conditions are only satisfied along coasts, where sandy beaches provide sediment for coastal dunes. Large active dune systems are developed in parts of the Sahara and in the Sinai and Negev Deserts, but very much smaller, seasonally active dune systems are found, for example, in the Manchega Plain of central Spain (Pérez González 1994).

Other aeolian processes involve dust entrainment, transport and fall. Dusts come primarily from arid areas in the Mediterranean Basin and deflation occurs from the surfaces of playas, alluvial fans, ephemeral stream beds, exposed bare surfaces and active dune fields (Pye and Tsoar 1987). The major dust sources are the Sahara and the Sinai/Negev Deserts and significant deposition of dust occurs in the peripheral areas of these deserts. Examples of this are dust infiltration and the development of biogenic crusts on linear dunes in the northern Negev in Israel and dust inputs into soils and soil development in both Israel (Yaalon and Ganor 1979) and Greece (Macleod 1980). In the northern Negev, dust has become trapped on gravelly surfaces, on vegetated surfaces at the desert fringe and on stabilised sand dunes (Gerson and Amit 1987). The interaction of dust deposition and dune dynamics is particularly interesting and has been investigated by Pye and Tsoar (1987). They show that dust deposition can lead to the formation of a surface crust which, in turn, enables algae and lichens to colonise the dune surface. This process appears to be most effective where the dune forms are linear, since linear dunes tend to have stabilised flanks but mobile crests.

Conclusions

A wide range of processes, operating at different scales, has shaped the Mediterranean landscape. The combination of high relative relief and a strongly seasonal climate means that landsliding and flooding are facts of life in many parts of the Mediterranean Basin. Over the whole of the Basin, however, the operation of geomorphological processes has been influenced to a greater or lesser extent by human activities, including agriculture and the management of water resources. These impacts are developed in Chapters 15–17.

References

ABRAHAMS, A.D. and PARSONS, A.J. (eds) 1994: *Geomorphology of Desert Environments*. London: Chapman and Hall.

BROMHEAD, E.N., COPPOLA, L. and RENDELL, H.M. 1994: Geotechnical background to problems of conservation of the medieval centre of Tricarico, southern Italy. *Quarterly Journal of Engineering Geology* **27**, 293–307.

CALVO-CASES, A. and La ROCA CERVIGÓN, N. 1988: Slope form and soil erosion on calcareous slopes (Serra Grossa, Valencia). *Catena Supplement* **12**, 103–12.

CERVERA, M., CLOTET, N., GUARDIA, R. and SOLE-SUGRAÑES, L.I. 1991: Response to rainfall simulation from scarcely vegetated and non-vegetated badlands. *Catena Supplement* **19**, 39–56.

CILLA, G., COLTORTI, M. and DRAMIS, F. 1994: Holocene fluvial dynamics in mountain areas: the case of the River Esino (Appennino Umbro-Marchigiano). *Geografia Fisica Dinamica Quaternaria* **17**, 163–74.

CLOTET-PERARNAU, N., GALLART, F. and BALASCH, C. 1988: Medium-term erosion rates in a small scarcely vegetated catchment in the Pyrenees. *Catena Supplement* **13**, 37–47.

DEL PRETE, M. and PETLEY, D.J. 1982: Case history of the main landslide at Craco, Basilicata, South Italy. *Geologica Applicata e Idrogeologia* **17**.

EISBACHER, G.H. and CLAGUE, J.J. 1984: *Destructive Mass Movements in High Mountains: Hazards and Management*. Ottawa: Geological Survey of Canada Paper, 84–16.

FRANCIS, C.F. and THORNES, J.B. 1990: Runoff hydrographs from three Mediterranean vegetation cover types. In Thornes, J.B. (ed.), *Vegetation and Erosion*. Chichester: Wiley, 363–84.

GALLART, F. and CLOTET-PERARNAU, N. 1988: Some aspects of the geomorphic processes triggered by an extreme rainfall event: the November 1982 flood in the Eastern Pyrenees. *Catena Supplement* **13**, 79–95.

GERITS, J., IMESON, J.M., VERSTRATEN, J.M. and BRYAN, R. 1987: Rill development and badland properties. *Catena Supplement* **8**, 141–60.

GERSON, R. and AMIT, R. 1987: Rates and modes of dust accretion and deposition in an arid region – the Negev, Israel. In Frostick, L.E. and Reid, I. (eds), *Desert Sediments: Ancient and Modern*. Oxford: Blackwell, Geological Society Special Publication No. 35, 157–69.

GILMAN, A. and THORNES, J.B. 1985: *Land-use and Prehistory in South-East Spain*. London: George Allen and Unwin, The London Research Series in Geography 8.

GINÉS, J. and GINÉS, A. 1995: Speleochronological aspects of karst in Mallorca. In *Karst and Caves in Mallorca*. Palma: Monografies de la Societet D'Història Natural de les Balears 3, 99–112.

GUTIÉRREZ, M., BENITO, G., and RODRÍGUEZ, J. 1988: Piping in badland areas of the Middle Ebro Basin, Spain. *Catena Supplement* **13**, 49–60.

HEUSCH, B. and MILLIÈS-LACROIX, A. 1971: Une méthode pour estimer l'écoulement et l'érosion dans un bassin. Application au Maghreb. *Mines et Géologie (Rabat)* **33**, 21–39.

IMESON, A.C. and EMMER, I.M. 1992: Implications of climatic change on land degradation in the Mediterranean. In Jeftic, L., Milliman, J.D. and Sestini, G. (eds), *Climatic Change and the Mediterranean, Vol. 1*. London: Edward Arnold, 95–128.

IMESON, A.C. and VERSTRATEN, J.M. 1988: Rills on badland slopes: a physico-chemically controlled phenomenon. *Catena Supplement* **12**, 139–50.

IMESON, A.C. and VERSTRATEN, J.M. 1989: The microaggregation and erodibility of some semi-arid and Mediterranean soils. *Catena Supplement* **14**, 11–24.

LARONNE, J.B. and REID, I. 1993: Very high rates of bedload sediment transport by ephemeral desert rivers. *Nature* **366**, 148–50.

LÓPEZ-BERMÚDEZ, F. and ROMERO-DÍAZ, M.A. 1989: Piping erosion and badland development in south-east Spain. *Catena Supplement* **14**, 59–73.

MACLEOD, D.A. 1980: The origin of red Mediterranean soils in Epirus, Greece. *Journal of Soil Science* **31**, 125–36.

MORANDINI, G. (ed.) 1962: *L'Erosione del Suolo in Italia, Vol. II: Aspetti Geografici*. Padua: CNR.

PALUTIKOF, J.P. and WIGLEY, T.M.L. 1996: Developing climate change scenarios for the Mediterranean region. In Jeftic, L., Keckes, S. and Pernetta, J.C. (eds), *Climatic Change and the Mediterranean, Vol. 2*. London: Arnold, 27–56.

PÉREZ GONZÁLEZ, A. 1994: Depresión del Tajo. In Gutiérrez Elorza, M. (ed.), *Geomorfología de España*. Madrid: Editorial Rueda, 389–434.

PYE, K. and TSOAR, H. 1987: The mechanics and geological implications of dust transport and deposition in deserts with particular reference to loess formation and dune sand diagenesis in the northern Negev, Israel. In Frostick, L.E. and Reid, I. (eds), *Desert Sediments: Ancient and Modern*. Oxford: Blackwell, Geological Society Special Publication No. 35, 139–56.

REID, I. and LARONNE, J.B. 1995: Bedload sediment transport in an ephemeral stream and a comparison with seasonal and perennial counterparts. *Water Resources Research* **31**, 773–81.

ROELS, J.M. 1985: Estimation of soil loss at a regional scale based on plot measurements – some critical considerations. *Earth Surface Processes and Landforms* **10**, 587–95.

ROMERO-DÍAZ, M.A., LÓPEZ-BERMÚDEZ, F., THORNES, J.B., FRANCIS, C.F. and FISHER, G.C. 1988: Variability of overland flow erosion rates in a semi-arid Mediterranean environment under matorral cover, Murcia, Spain. *Catena Supplement* **13**, 1–11.

SALA, M. 1988: Slope runoff and sediment production in two Mediterranean mountain environments. *Catena Supplement* **12**, 13–29.

THORNES, J.B. 1990: The interaction of erosional and vegetational dynamics in land degradation: spatial outcomes. In Thornes, J.B. (ed.), *Vegetation and Erosion*. Chichester: Wiley, 41–53.

THORNES, J.B. 1995: Mediterranean desertification and the vegetation cover. In Fantechi, R., Peter, D., Balabanis, P. and Rubio, J.L. (eds), *Desertification in a European Context: Physical and Socio-Economic Aspects*. Brussels: European Commission EUR 15415EN, 169–94.

WOODWARD, J.C. 1995: Patterns of erosion and suspended sediment yield in Mediterranean River Basins. In Foster, I.D.L., Gurnell, A.M. and Webb, B.W. (eds), *Sediment and Water Quality in River Catchments*. Chichester: Wiley, 365–89.

WOODWARD, J.C., LEWIN, J. and MACKLIN, M.G. 1992: Alluvial sediment sources in a glaciated catchment: the Voidomatis Basin, northwest Greece. *Earth Surface Processes and Landforms* **17**, 205–16.

YAALON, D.H. and GANOR, E. 1979: East Mediterranean trajectories of dust-carrying storms from the Sahara and Sinai. In Morales, C. (ed.), *Saharan Dust*. Chichester: Wiley, 187–93.

5

THE GRAECO-ROMAN MEDITERRANEAN

LINDSAY PROUDFOOT

The cultural landscapes of the Mediterranean Basin are complex, ancient and diverse. Long regarded as one of the major cultural hearths of world civilisation, the 'greater' or 'historical' Mediterranean defined by Braudel (1972) has been home since at least the eighth millennium BC to a succession of cultures of varying origins, ethnicity, social complexity and technological attainment. Braudel's historical Mediterranean extended far beyond the physical littoral which forms the focus of this book, south to the southern margins of the Sahara; east to the Syrian desert and the Asian steppes, and finally north and west to the Alps and the Atlantic. But the defining unifying core was the Mediterranean Sea itself, and around its shores peoples and ideologies have collided throughout history in processes of assimilation, integration and eradication which have left numerous, frequently enigmatic, traces in the contemporary human landscape.

The clarity and decipherability of these imprints depend on the nature and extent of the cultural discontinuities which separate them from the present day, and these are partly a function of time. Some of the most profound discontinuities occurred as a result of the Arab conquests and the spread of Islam in the southern and eastern Mediterranean in the period following the Prophet Mohammed's death in 632 AD. Thus the conquest of the Maghreb by the Arab Umayyad dynasty in the late seventh century AD imposed an Islamic hegemony over a region of complex classical settlement. Although the fate of the earlier Roman cities such as Dougga (Tunisia) or Leptis Magna (Libya) varied (Lepelley 1992), there is little doubt that by the eleventh century AD, most had experienced a terminal, if in some cases protracted, decline, as Islamic Arab society evolved its own patterns of settlement articulated on centres such as Qayrawan (Tunisia), Fez and Marrakesh (Morocco). Thereafter, during the years of the Mamluk and Ottoman Empires, the by-now generally abandoned or disfunctional Roman settlements survived as the disregarded legacy of a Classical polity which the Arabs had sought to replace, not preserve (Lewis 1995, p. 245).

Arguably, despite such cultural displacement, the Hellenisation and Romanisation of part or all of the Mediterranean Basin constituted two of the most important cultural developments in its history. Only the spread of Christianity and the growth of Islam had a similarly extensive impact. The legacy of these Classical civilisations has been immense, both in terms of European history, through the Renaissance, and more immediately in the Mediterranean as a source of political inspiration for the reunification of Italy and the assertion of Greek independence during the nineteenth century (Clogg 1992; Lewis 1995, pp. 315–19; Riall 1994). They continue to resonate in the present day, both as part of what is perceived to be a common 'European' cultural heritage, with all the attendant pressures of heritage tourism which this engenders, and as historical justification for the assertion of contested nation–state identities in Macedonia, Cyprus and the eastern Aegean (see Chapter 8).

Accordingly, any attempt to understand the human origins of the contemporary Mediterranean demands some consideration of the nature of the Hellenisation and Romanisation which unified part or all of the region between the eighth century BC and the fourth century AD. This chapter explores the complex processes involved in the geographical growth and subsequent transformation of these civilisations. The discussion does not purport to provide an exhaustive history of the period. Rather, it emphasises some of its major geographical manifestations, particularly the emergence of the concept of the city-state or *polis*, the growth of the Roman territorial state, and the nature of Roman urbanism.

The chapter concludes with a brief consideration of the factors which may have contributed to the fragmentation of the Roman Empire from the fourth century AD onwards.

GREECE, HELLENISM AND ROME: TOWARDS A GEOGRAPHY OF THE MEDITERRANEAN, *CA* 800 BC–395 AD

The emergence of the Greek *polis*

By the early eighth century BC, and for reasons which are still unclear, the islands and peninsulas of the Aegean witnessed the emergence of new forms of social and spatial organisation which differed in significant respects from anything that had gone before. In the Cyclades and Dodecanese, along the western coast of modern Turkey, and above all in eastern and southern districts of Macedonia, Thessaly and the Peloponnese in modern mainland Greece, numerous, generally small, *poleis* developed (Fig. 5.1). Any precise definition of a *polis* is made difficult by the ambiguity of the term and the great variety of the territorial units and socio-political structures it described. The word is widely translated to mean city–state, but it was originally used to describe the defended citadel or *acropolis* around which many Greek cities were built, and only later to refer to the cities themselves and the territories or states which were dependent on them (Austin and Vidal-Naquet 1977, p. 51).

Nevertheless it is still easier to describe at least some of the more frequently recurring characteristics of the early *poleis* as they developed during the Archaic period (*ca.* 750–500 BC) than to explain their origins. Their total number is hard to determine. By 600 BC there were about 200; two hundred years later there were over 340 in the Cyclades, the Dodecanese, Macedonia and the Aeolian and Ionian coasts of Asia Minor alone. Their size and importance varied, but with the notable exception of Athens and Sparta, and later, Corinth, the vast majority were small in both territory and population. Thus whereas the Athenian *polis* extended to over 1000 square miles, and in the fifth century BC supported a population of perhaps 300 000, most city–states appear to have been less than one-tenth of this (Pounds 1990, p. 28; Smith 1967, p. 58).

Although recent archaeological research has raised doubts over the conventional assumption that the growth of these city–states was synonymous from the outset with urbanisation, even the most radical revisionist critique accepts that by 600 BC, most *poleis* had an urban central-place as their core (Morris 1991; Snodgrass 1991). Sparta remained an exception. There, the *polis* centred not on a city but on a group of five villages, a situation which reflected the state's unusual social organisation based around peer fratries or *Homoioi*. For the rest, even the smallest cities were equipped with the accoutrements of civilised urban life.

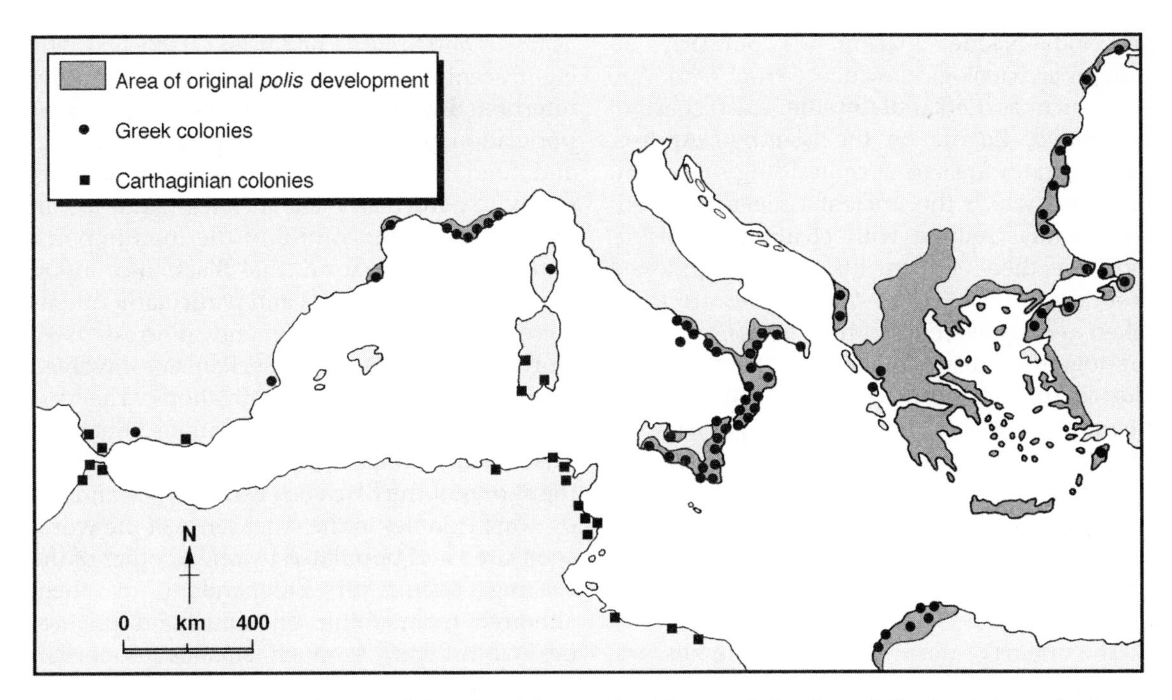

FIGURE 5.1 Distribution of original Greek *poleis ca.* 750–550 BC and Greek and Carthaginian colonies *ca.* 500 BC

Source: After Pounds (1990); Snodgrass (1991)

Most were fortified, all possessed a market or *agora*, a place of assembly (frequently in the *agora*), and a seat of justice or government. Many, as noted, sheltered under a fortified citadel.

This concern over the provision of public amenities left a legacy of sophisticated monumental architecture which until recently has distracted archaeologists from the important rural–urban relationships linking each city to its territory. Nevertheless, these built forms accurately reflected the social abstractions which lay at the heart of the city–state. Although varied in social and political organisation as much as in size and shape, each *polis* was a defined community of citizens and non-citizens (slaves and resident foreigners), occupying a recognisably discrete area and living under an established constitution, independent in theory at least of outside control, and with a clear sense of its own autonomous identity (Forrest 1986). Citizens acquired their political rights by virtue of birth, but as a group were economically disparate. The main criterion of poverty was whether or not one had to work for a living. Land ownership was the most socially prestigious form of wealth, but despite their origins as a peasant–citizenry, by no means all citizens owned equal amounts of land. Consequently, a recurrent theme in the history of many *poleis* was the periodic demand for land redistribution among the citizen class (Austin and Vidal-Naquet 1977, pp. 20–3).

The question remains, however, of how and why the *poleis* developed as and when they did, and it must be stressed that there are no clear answers. One central issue is the extent to which the *poleis* originated in the Greek 'Dark Ages', the four hundred years or so of migration and cultural dislocation which followed the final collapse of the Mycenaean palace cultures by the early twelfth century BC. The association between some *poleis* and the distribution of late Mycenaean palace sites in eastern and central Greece and in the islands of the Aegean has long been recognised, but does not in itself demonstrate continuity of occupation (Austin and Vidal-Naquet 1977, p. 51). Similarly, although archaeological evidence from Dark Age sites such as Lefkandi on the west coast of Euboea, or Zagora on the island of Andros, demonstrates the sort of centralising settlement processes which the Ancient Greeks' own origin-legends credited with creating the *poleis*, most of these settlements were abandoned around *ca.* 700 BC. Thus if these sites are to be taken as early evidence of the *synoikismos* ('coming together') made famous by Aristotle's conclusion (in the fourth century BC) that 'when several villages are united in a single community, large enough to be nearly or quite self-sufficing, the state comes into existence', they clearly also demonstrate that this was sometimes short-circuited by the emergence of other, more powerful, neighbouring city-states (Snodgrass 1991, pp. 7–9).

The origins of the *poleis* thus remain obscure. For the present, it may be best simply to envisage their early growth as an evolutionary process, involving social polarisation and a restatement of communal religious beliefs within societies which had already reached a certain threshold of wealth and sophistication (Coldstream 1983). This may have involved the occasional re-use of much earlier proto-urban sites, but in general it was a process in which, in the central and western Aegean at least, urban growth was a more protracted and variable component than was once thought (Morris 1991; Rihill and Wilson 1991). What is clear is that once the idea of *polis* had begun to emerge, it very quickly became an extremely efficient vehicle for the transmission of Hellenic culture beyond mainland Greece and the islands of the central Aegean, east to Asia Minor and the coasts of the Black Sea, and west to southern Italy, Sicily and southern France and Spain, where Greek colonists came into conflict with Carthaginian cities of earlier, Phoenician, origin (Fig. 5.1).

Formal Greek colonisation reached its peak between *ca.* 750 and 550 BC, but had been preceded in Asia Minor by Greek migration and settlement from at least the ninth century onwards, when a number of fortified sites such as Iassos and Smyrna had been established. The eighth-century colonisation has been variously interpreted as a means of disposing of surplus population and thus relieving pressure on land and food resources; as a search for new resources, particularly metal ores; and as an attempt to ensure control of the food imports, particularly corn from the Black Sea, upon which many Greek cities and particularly Athens increasingly came to depend (Forrest 1986; Hornblower 1986). The mechanisms involved were as diverse as the motivations. The vast majority of the new foundations, such as Theodosius, Sinope or Odessos, all founded on the shores of the Black Sea between 600 and 550 BC, were colonies in the strict sense of the word, sponsored and populated by one or other of the existing *poleis*, yet independent of them although maintaining economic and political ties with them. A much smaller proportion were simple trading-stations or *emporia*, such as Al Mina on the River Orontes in northern Syria, or Naukratis on the River Nile.

The establishment of a colony was a formal affair. An *oikistes* or coloniser was appointed by the sponsoring city, and he was responsible for exploring the best site (after ritual consultation of the oracle), locating a water supply, building the defences, and allocating land to the colonists (Owens 1991, pp. 30–73; Ward-Perkins 1974, pp. 14–17). Defensible hillside or peninsular sites often located near estuaries or rivers were preferred, as for example at Istrus, founded in 657 BC on the Black Sea, and Syracuse and Segesta, founded in the early eighth century BC in Sicily.

These processes of state- formation, urbanisation and colonisation were thus intimately bound up with the opportunities and constraints offered by the physical environment, and the relationships between them have long absorbed geographers. Early explanations favoured a straightforward environmental determinism. Thus Semple (1932) sought to explain the individually small-scale but, in aggregate widespread, proliferation of *poleis* in terms of the characteristically fragmented insular and peninsular geography of the Aegean, the general poverty of its land resources and the relative ease of its maritime communications. Subsequent writers have been quick to point out that the 'fit' between the distribution of *poleis* and 'natural' geographical units was far from perfect: small islands were often divided between several city states (Lesbos had six), while large parts of Attica and the Peloponnese were ruled by single cities (Smith 1967, p. 58). Even so, we do not have to ascribe to the extremes of environmental determinism to recognise that the ecological fragility and geological instability of parts of the eastern Mediterranean imposed their own rhythms and limitations on the pattern of human activity. The suggestion that the catastrophic explosion of the volcanic island of Thera *ca.* 1500 BC may have been partly responsible for the destruction of late Minoan palace-culture in Crete is well known, but earthquakes and coastal deposition (as at Ephesus, Miletus and Priene, in western Turkey), and the constant need to manage water supplies also posed continuing and widespread problems for urban communities throughout Antiquity (Bayhan 1994; Erim 1993, Thomas 1993). Conceivably, therefore, the intense fragmentation of the *poleis* system may have reflected the constraints and opportunities inherent in the physical environment, on the one hand, and, on the other, the technological and social capacity of the city as a social construction to cope with these.

Urbanism and society in the later Greek world, *ca.* 500–29 BC

By the fifth century BC, the Greek urban ideal had in many ways reached its apotheosis. In a regionally uneven process of political evolution, the monarchies and aristocracies of the Archaic period had given way to various forms of oligarchic government, and, following the reforms of Cleisthenes in 508 BC, to democracy at Athens. Externally, the threat of Persian conquest was at least temporarily removed with the defeat of Xerxes by sea at Salamis in 480 BC and on land at Plataea a year later, achieved by a confederacy of city–states led by Athens and Sparta. This united action in the face of an external threat could not mask the underlying political instability of the city–state system. Arguably, this arose not simply from the political rivalry engendered by its extreme fragmentation, but also from the uneven growth in demand for food and other resources and the attempts by the largest cities, notably Athens, to secure control over these (Hornblower 1986, pp. 133–5). Accordingly, during the next 150 years, until Philip of Macedon defeated an Athenian-led alliance at Chaeroneia in 338 BC and established what was in effect an imperial monarchy ('The League of Corinth') in Greece, the Hellenic world witnessed the emergence of a series of relatively short-lived city alliances or confederacies, which sought to establish either their own independence from more powerful neighbours, or hegemonic control over their rivals. The leading contenders were Athens, Thebes and Sparta, and although Athens succeeded in imposing a (frequently contested) form of federal authority over most of the Aegean between 478 and 404 BC, neither it nor its competitors were sufficiently powerful to achieve more permanent control.

With the absorption of the city–states into the Macedonian kingdom in 338 BC, the context for urbanism changed. The city–state, territorially limited and dependent on trade or alliance for much of its food and other needs, gave way to the territorial state, able to marshal a much wider range of resources in pursuit of its interests. Under Philip's son, Alexander (336–323 BC), the Aegean's resources were mobilised in a campaign against the Persians which extended Macedonian authority and Hellenic culture as far as the Punjab, Afghanistan and the Asian steppes in the east and Egypt in the south. Following Alexander's death, the Empire he had created quickly fragmented politically but remained as a zone of Hellenistic culture. By 300 BC, three major successor kingdoms had emerged, each led by the descendants of one or other of Alexander's generals. The Ptolemies ruled Egypt and parts of the Levant; the Seleucids Turkey, Syria and Afghanistan; and the Antigonids, Macedonia. Although each state was to vary in power and extent over the next 300 years, with the Seleucids in particular quickly losing territory to the new Attalid kingdom in western Turkey and to the Parthians and Bactrians further east, all three dynasties survived as major powers in the political geography of the eastern Mediterranean until their incorporation within the Roman Republic and Empire between the late second century BC and first century AD.

In all of this, the role and character of urbanism also changed, but perhaps to a lesser extent than might have been expected. The structures of urban life established during the Archaic period, and elaborated during the fifth and fourth centuries BC, were replicated by Alexander and his successors throughout the Middle East (Owens 1991, pp. 74–92). Alexander is credited with establishing at least 35 cities (it may have been double this), and the Seleucids with founding or refounding a further 60 in Turkey and Iran (Price 1986). In contrast to the ancient *poleis* in Greece, whose accretive origins were frequently reflected in their irregular ground-plans and disordered use of urban space, most of these Hellenistic cities were planned foundations. Many, such as the third-century BC Seleucid foundation at Dura-Europus in Syria (Fig. 5.2) were laid out on the rigidly orthogonal plan traditionally associated with the work of the fifth-century BC planner, Hippodamos of Miletus. Others, such as the hilltop Attalid capital at Pergamum, to the north of modern Izmir in western Turkey, experimented with the varied topography of their sites to create massively monumental urban vistas, which departed radically from the formality of the Hippodamian concept.

This willingness to adapt traditional urban forms was also typical of Hellenistic architecture. The two great Classical orders, the Doric and Ionic, continued to be used but in a much

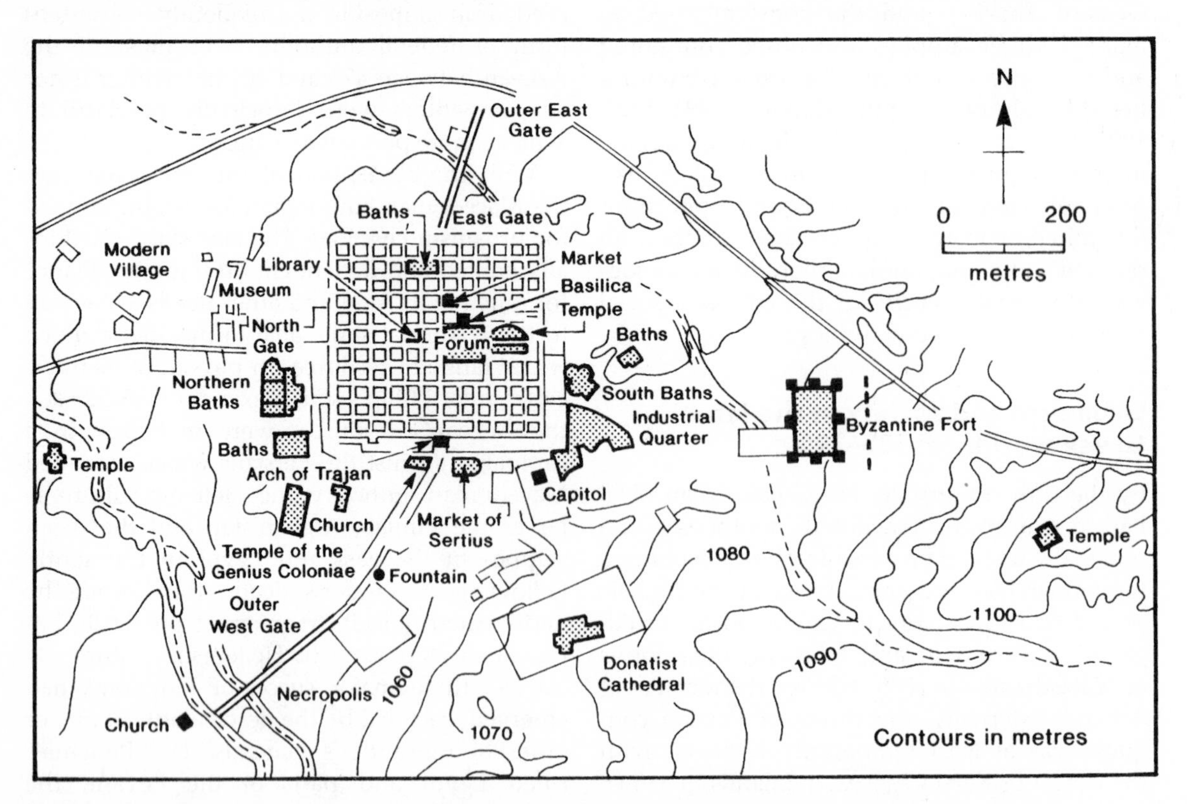

FIGURE 5.2 Planned third-century BC Seleucid foundation at Dura-Europus, Syria

Source: After Finlay (1977)

less formal fashion than before, and were supplemented by the more flamboyant Corinthian style, which was favoured by the Seleucids (Fig. 5.3). The towns themselves continued to be equipped with similar arrays of public buildings to those which had characterised Archaic and Classical cities. Temples proliferated, their dedications to gods or goddesses such as Apollo or Athena reflecting each community's religious loyalties and self-image. Town councils continued to meet in the *bouleuterion*, or council chamber, where their deliberations might reflect the delicate balance between the community's residual autonomy and the overarching authority of the king, most frequently expressed in the demand for taxation. The one truly innovative development was the appearance of the *gymnasium*, used for both physical recreation and artistic purposes, and a sign, in the Hellenistic world, of the adoption of Greek culture.

The changing morphologies of Greek cities thus reflected the changing political ideologies which underpinned the Classical and Hellenistic worlds. In economic terms, however, Greek cities remained what they always had been, centres of consumption rather than production in an overwhelmingly agrarian economy, where land and its ownership remained the essential test of elite status. Functional specialisation certainly existed, both within and between towns (Murray 1986), and petty commodity production certainly grew, at least until the second century AD, but neither of these things, nor the inter-regional patterns of trade they gave rise to, could distort the fundamental dependence of towns upon the countryside (Greene 1986; Hopkins 1983).

FIGURE 5.3 Corinthian architectural decoration, Didyma, Turkey

The rise of Rome

As in the case of the Greek *poleis*, the most intriguing questions concerning the growth of the Roman state relate not so much to the chronology of its foundation and subsequent territorial expansion, which is well known, but rather to why its power expanded as rapidly and extensively as it did. Tradition rather than history dates Rome's foundation to 753 BC, but for the first 200 years of its existence it was a relatively small and unimportant monarchy, probably without a defined urban core, and simply one of many more or less autonomous communities in a peninsula characterised by extreme ethnic, political, religious, social, economic and linguistic diversity. To the north, centred in Tuscany, were the Etruscans, a linguistically distinctive urban culture. Once thought to have originated in Asia during the Dark Age migrations, they are now considered to have developed in Italy under Greek influence (Pounds 1990, p. 31). Around Rome and in the mountainous interior to the east and south, various Iron Age tribal groups, including Latins and Samians, spoke one or other of a series of related 'Italic' languages. Also to the south but on the coast, lay the 30 or so independent *poleis* of Magna Graecia, such as Cumae and Tarentum on the mainland, or Naxos and Syracuse in Sicily, which had been founded as part of the eighth-century BC Greek colonial movement and which exerted tremendous cultural influence over pre-Roman Italy and over early Rome itself (Smith 1967, pp. 62–5).

Although Rome was never a simple Etruscan dependency, its early development occurred under their influence, and it was not until the late sixth century BC that it finally established its independence and secured control over Latium. Rome's subsequent political evolution first as a

Republic (from 509 BC), and latterly as an Empire (from 29 BC), has been thoroughly rehearsed elsewhere (see, for example, Crawford 1978; Wacher 1987; Wells 1984). We need only note here that the internal social and political tensions which characterised the Republic – particularly the rivalry between the patrician and plebeian orders – and which eventually created the conditions in which the figure of an emperor could emerge, were themselves at least partly a result of Rome's territorial expansion and the problems this created in the social allocation of wealth.

Figure 5.4 demonstrates that the greater part of this expansion occurred prior to 14 AD, and was particularly rapid under the later Republic and early Empire, when the Roman polity broke through the confines of the Mediterranean world to establish itself in north-west Europe and, in the east, in the Danube Basin. Prior to this, the early Republic had taken some 300 years (509–218 BC) to establish control over peninsular Italy and its contiguous regions, and a further century to consolidate its authority in Spain, Greece and Carthaginian North Africa. The pace of expansion slowed during the first and second centuries AD. By the peak of the empire in the early second century AD, the only significant gains had been the annexation of lowland Britain in the far north-west, and Armenia and Kurdistan to the south-east of the Black Sea.

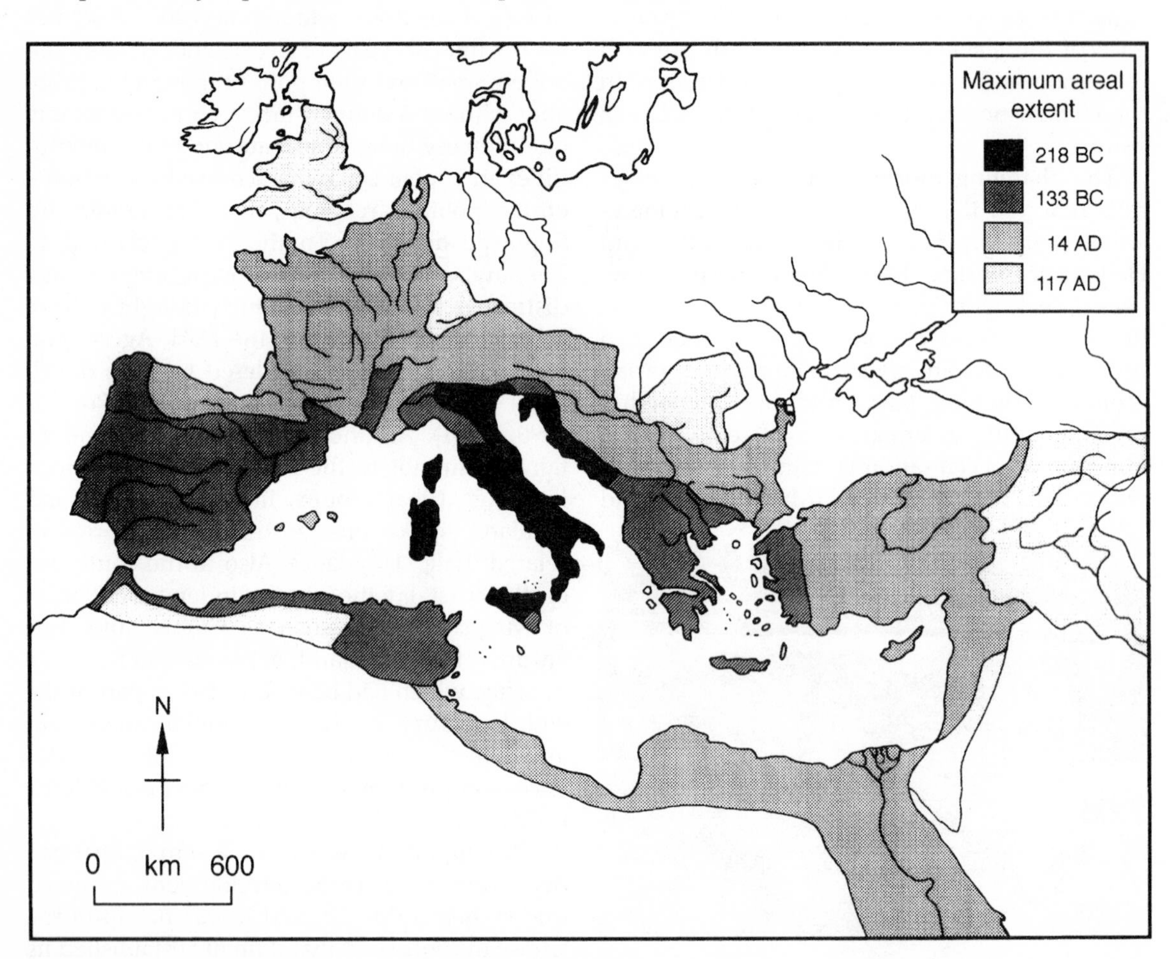

FIGURE 5.4 The territorial growth of the Roman Republic and Empire

Source: After Boardman *et al.* (1986)

The trajectory of Roman expansion was thus far from uniform, and its causation was undoubtedly complex. Previous explanations of the city–state's growth under the monarchy and during the early Republic have stressed the importance of its relatively rich resource endowment and strategic location. This was at the lowest fording point over the River Tiber on what was, in effect, a cultural frontier between the relatively advanced Etruscan and Greek cultures respectively to the north and south, and the Iron Age Latin tribes in the interior (East 1966, pp. 5–13; Smith 1967, pp. 62–5). While it is easy to conceive that these locational attributes may well have played some part in securing Rome's initial growth, they can hardly be held to account for its subsequent expansion throughout the peninsula and beyond. Political and social factors may have been of more importance here. Crawford (1978) stresses the success of the Republic in mobilising the resources, particularly in manpower, of the Latin and other states it conquered or allied with in Italy, and incorporating them within its own social organisation and polity, either through the grant of Latin and eventually (in 91 BC) Roman citizenship to the existing population, or by founding its own colonies (*colonae*) in the newly acquired territory. These 'allied' citizens were required to fight in Rome's interests, and one consequence was the city's growing military prowess and success, which was also reflected in the essentially militaristic nature of Roman society (Rawson 1986).

Underlying all of this may have been other, more fundamental, expansionist dynamics. Delano Smith (1979, p. 9) has stressed the 'unusual strength of demographic levels' in Roman Italy – by 100 AD its population may have approached 5 million – and argues that it was this that made possible the political expansion, conquest and colonisation. In turn, the negative demographic consequences of such rapid expansion – losses of manpower during war, for example – may have helped set in train the third-century AD decline by curtailing the population's capacity for self-replacement. In a not dissimilar argument, Marxist historians have suggested that it was the Roman reliance on chattel slavery – itself a borrowing from the Hellenic world – that was the prime cause both of their expansion and decline (see for example, Anderson 1974). Although this view has been criticised for over-simplifying a complex socio-economic phenomenon (Greene 1986), the argument has an appealing symmetry. With every successful war of aggression, whether against the Samnites in 327–304 BC, in Spain in 197–133 BC, or against Carthage in the Third Punic War (149–146 BC), the Republic acquired additional territory and slaves. In the Marxian view, the slaves were the more important of these two commodities, and were in fact the prime object of these wars. Rome's slave population was not self-reproducing, and the state's economic dependence on them as a low-cost labour force demanded that they be continually replenished from external sources – by conquest. Yet these wars also inevitably brought new territory to be exploited, which in turn demanded yet more slaves. Arguably, a self-perpetuating cycle of expansion was created which only ended when the frictions of distance within the expanded Empire became great enough to distort and destroy the core–periphery linkages upon which it depended, and the wars of aggrandisement – and the external replenishment of the slave supply – accordingly ceased.

Roman settlement

Within this system, Roman urbanism played a major role in articulating the structures of social, economic and political authority, but in fact formed only part of a pattern of settlement whose extensive rural base has only recently begun to be understood (Corbier 1991; Millet 1991). Recent surveys of Roman rural settlement in Etruria, Biferno and near Cosa in Italy; in the Guadalquivir valley in southern Spain, and in Libya, Crete and northern Syria, have demonstrated that this was much more extensive than had previously been realised. In some regions, particularly in the more marginal environments of North Africa and the East, it was also supported by complex environmental management systems, particularly of water supplies (Fig. 5.5). More generally, however, the long-term environmental impact of both Roman and Greek farming appears to have been negative. Recent research has suggested that it was agricultural activity rather than climatic change which was responsible for the widespread soil erosion and 'Younger Fill' of late Classical times (Greene 1986, pp. 84–7, 98–141).

The importance which the 'new' archaeology has placed on Roman rural settlement has to some extent simply confirmed what we already knew. Namely, that the Roman economy remained predominantly based on agriculture, and that the survival of the Empire as a geopolitical unit depended in part at least on the efficiency with which diverse regional economies were integrated in the service of the whole. This integration was achieved by trade and taxation articulated through the provincial urban network, and geared towards meeting the demands of provincial garrisons and the imperial administration alike (Hopkins 1978, 1980, 1983; Osborne 1991). As in earlier periods, therefore, Roman towns and cities remained centres of consumption rather than production, places where the power relationships inherent in the Republic and Empire were mediated and the arts of civilisation practised (Greene 1986; Patterson 1991). Global estimates of the total number of Roman towns and cities are rendered difficult by problems of definition and functional change. Nevertheless, in a partial survey, Pounds tentatively identifies over 600 in Greece and Italy alone by the second century AD,

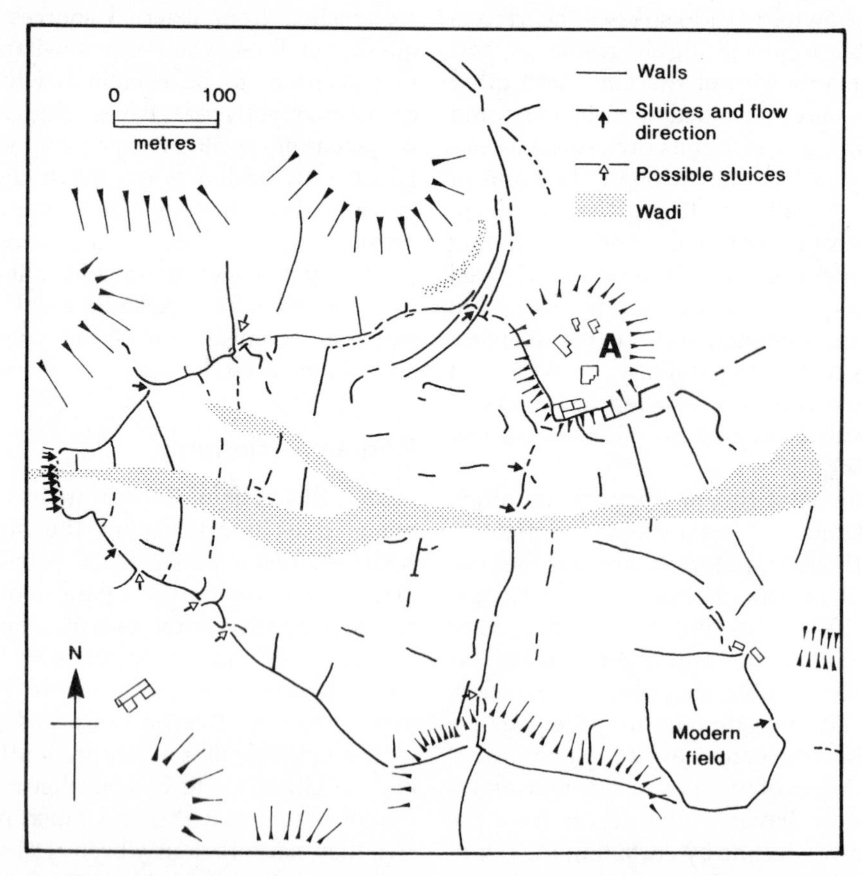

FIGURE 5.5 A typical Roman farm complex (A) in the Libyan pre-desert

Source: After Greene (1986)

together with perhaps 150 in Iberia (particularly in Andalusia), and a further 50 or so in Mediterranean France and 60 in Dalmatia (Pounds 1990, pp. 56–7).

Rome was the epitome and focus of this urbanism. The city provided political leadership, inspired morality and religious conviction, and established conceptions of taste and fashion. These were imitated by provincial cities throughout the Republic and Empire, and were necessary in the fusion of the diverse cultural traditions of the Hellenic, Hellenistic, Near Eastern, Eastern and European worlds into a homogeneous entity. Until the second century AD, Rome pursued this role as chief city in a federation of city–states, each technically equal in their social and political construction. Thereafter, as the social, economic and political stresses inherent in the slackening pace of Empire became more evident, successive Emperors shed their previous status as 'first citizen' and became increasingly autocratic and interventionist (Marchese 1983). Thus imperial authority, seated in and synonymous with Rome, began to assume much more immediate and direct control over virtually every aspect of the organisation and administration of the provincial urban network and regional economies.

There was every need to do so as far as the city of Rome itself was concerned. Already by the first century AD a megapolis of perhaps 1 million inhabitants, imperial Rome had no other rival in terms of its size and importance. Alexandria and (Roman) Carthage may have had populations of around 300 000 at their height, but 20 000 or less was more normal, even for major regional centres (Wells 1984, p. 213). Rome was thus dependent on a wide range of regional economies – and efficient communications and transport technologies – for the resources it required to feed and maintain its growing population (Greene 1986). The grain trade is a case in point. Corn was the single most important food item required by the urban populations of the Republic and Empire, and Garnsey (1983) estimates that at its maximum, Rome required 200 000 tonnes of corn a year. The sources of supply expanded as Rome and its territorial authority grew. During the early Republic, supplies came mainly from Campania, Etruria and Umbria, but were gradually supplemented by corn from Sardinia, Sicily and North Africa. In each case, the sources of supply became permanent and regular once these territories were annexed by Rome. By the first century AD, North Africa was the single most important source, providing perhaps 50 000 tonnes of corn a year, along with olive oil (Garnsey 1983, pp. 119–20). The corn was sent as payment in kind for imperial taxation, as rent from public land, and by way of private trade, although the latter became progressively less important with the growing assertion of centralised imperial control.

In Rome, the citizens who received these supplies lived in a congested and stressful urban environment, plagued by problems of pollution which would be instantly recognisable by twentieth-century urban man (Ramage 1983). Like Athens before it, Rome's accretive origins, complex topography and lack of centralised planning even after the disastrous fire of 64 AD had created a highly irregular plan and morphology (Fig. 5.6). Coherent planning on an admittedly grandiose scale appears to have been confined to the periodic extension by successive emperors, particularly Augustus and Trajan, of the imperial fora and the imperial palace (on the Palatine). By contrast, many of the residential and commercial districts were characterised by often poorly built high-rise tenements, prone to collapse and risk of fire, and lacking adequate sanitation and water supply, despite the extensive aqueducts which supplied the city with water from up to 30 miles away (Pounds 1990, pp. 58–9). There is some evidence that in response to these problems, services within the city decentralised, and those who could afford to, moved out to the surrounding countryside (Potter 1991; Ramage 1983), but in general Rome retained its centripetal attraction – and consequent problems – as the hub of the later Classical world.

In the provinces, Roman towns and cities fulfilled various roles as administration and market

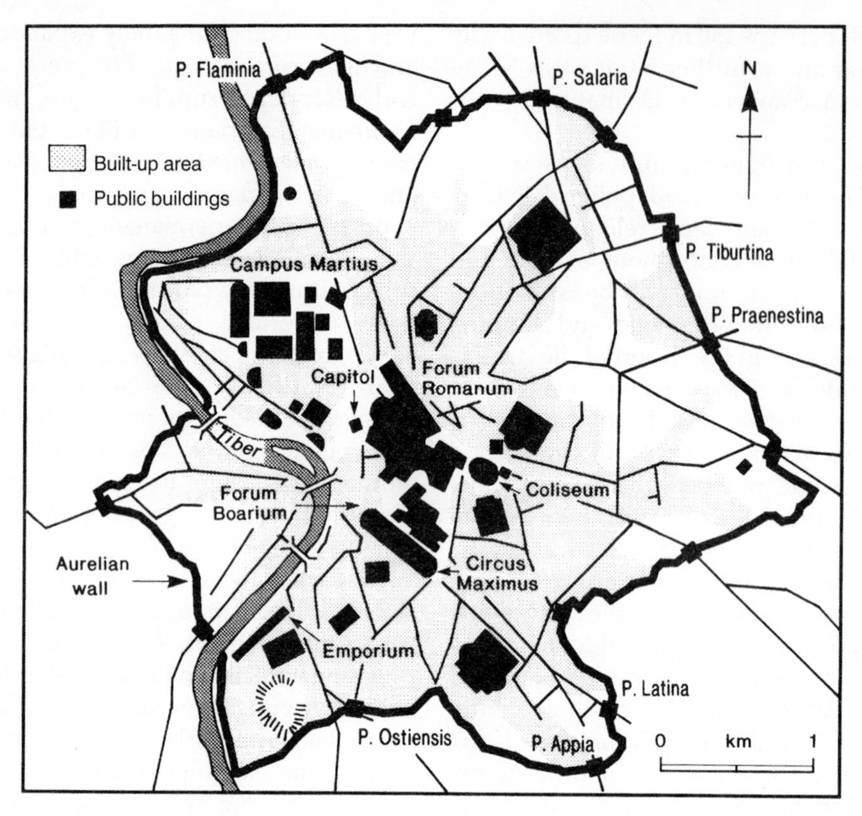

FIGURE 5.6 Imperial Rome

centres, garrisons, and centres of petty commodity production. Pre-eminently, they were places where wealthier subjects could enjoy the benefits of civilised urban life – *Romanitas* – and where their energies and wealth could be directed towards the service of Rome. All were geared towards promoting the continuing functional unity of the Republic and Empire, but clearly developed on the basis of very diverse earlier urban traditions. We can usefully distinguish, therefore, between the Roman urbanism which developed on the basis of the sophisticated Hellenistic and Carthaginian urban cultures of the eastern and southern Mediterranean, and that created in the less advanced tribal societies of Atlantic Europe. In the Hellenistic and Carthaginian worlds, urbanism was more ancient and just as complex as that of the Romans themselves. Consequently, in Egypt and other parts of North Africa, the Levant, Asia Minor and Greece, the major problem for the Roman administration was to ensure the continued loyalty of existing towns and cities and their efficiency as redistributive centres in the trade and taxation nexus, rather than necessarily to found new ones.

The differences between Roman and Greek cities were essentially those of scale and elaboration rather than function or purpose. Existing Hellenistic cities were almost invariably extensively refurbished by the Romans, but in ways which emphasised the underlying functional continuity between them. Thus, the Greek *agora* might be enlarged as a Roman *forum*, and the *bouleuterion* replaced by a *basilica*, but each served the same purpose as its predecessor. For example, at Ephesus, an ancient Hellenistic city on the Ionian coast of western Turkey, virtually all the major buildings were constructed or rebuilt during the first and second centuries AD. In line with the usual practice throughout the Roman Empire, this work was funded both by private benefaction by individual citizens (the

Library of Celsus, the Gate of Mazeus and Mithridates, Fig. 5.7), and by corporate investment (the Temples of Hadrian and Domitian, the Dea Roma and Divus Julius Caesar). The latter were imperial cult temples, and were built by the city in honour of the eponymous emperors, only after they had been granted the right to do so, itself a great privilege and a mark of the city's importance as the provincial capital of Asia. In contrast, at Aphrodisias, in western Anatolia, the Romans appear to have been instrumental in founding a town at a proto-urban religious sanctuary of considerable antiquity. Originally dedicated to Astarte, the sanctuary appears to have been rededicated to Aphrodite in a conscious attempt to enhance its attraction as a place of pilgrimage following the Roman annexation of the area in 132 BC (Erim 1993).

FIGURE 5.7 Roman refurbishment at Ephesus, Turkey: the Library of Celsus and the Gate of Mazeus and Mithridates

Settlement adjustments of this sort are likely to have been commonplace during the process of Romanisation, but, arguably, may have been more widespread in the western Mediterranean, where the urban tradition was less extensive than in the Hellenistic east. Here, the Romans supplemented the existing but relatively limited Greek and Carthaginian urban networks by founding *colonae*, or veterans' settlements, much as they did in Britain and Gaul. For example, although most Roman cities in North Africa were founded at existing urban centres, a significant number of *colonae* were also established in what is now Algeria during the early second century AD. These formed a strategic military screen designed to defend the fertile coastal plain and its settlements from attack via the Aures mountains. Places such as Timgad and Lambessa were laid out according to the rigidly orthogonal principles of Roman military planning, and their plans contrasted sharply with the irregular accretive plans of older cities such as Constantine (Algeria), Dougga (Tunisia) or Volubilis (Morocco) (Fig. 5.8). These cities and others appear to have developed – sometimes on much earlier prehistoric foundations – as the capitals of independent pre-Roman kingdoms, although they invariably came under Punic influence (Finlay 1977). In fact, many of the Roman cities in modern Tunisia and Algeria were of Carthaginian origin. Once they had finally defeated Carthage in 146 BC, the Romans incorporated her dependent cities within their own polity, and rebuilt Carthage itself in 44 BC.

In the western provinces of Atlantic Europe, on the other hand, the urban tradition was less well founded. Here, Roman towns were much more likely to be entirely new, colonial, foundations. Various functional and legal classifications of Roman urbanism have been proposed which aptly reflect the imposed and recent character of urban life in these western provinces. At the apex of these hierarchies, planned *colonae* such as Colchester in Essex, or Merida in Extremadura (Spain), performed a major role as centres of

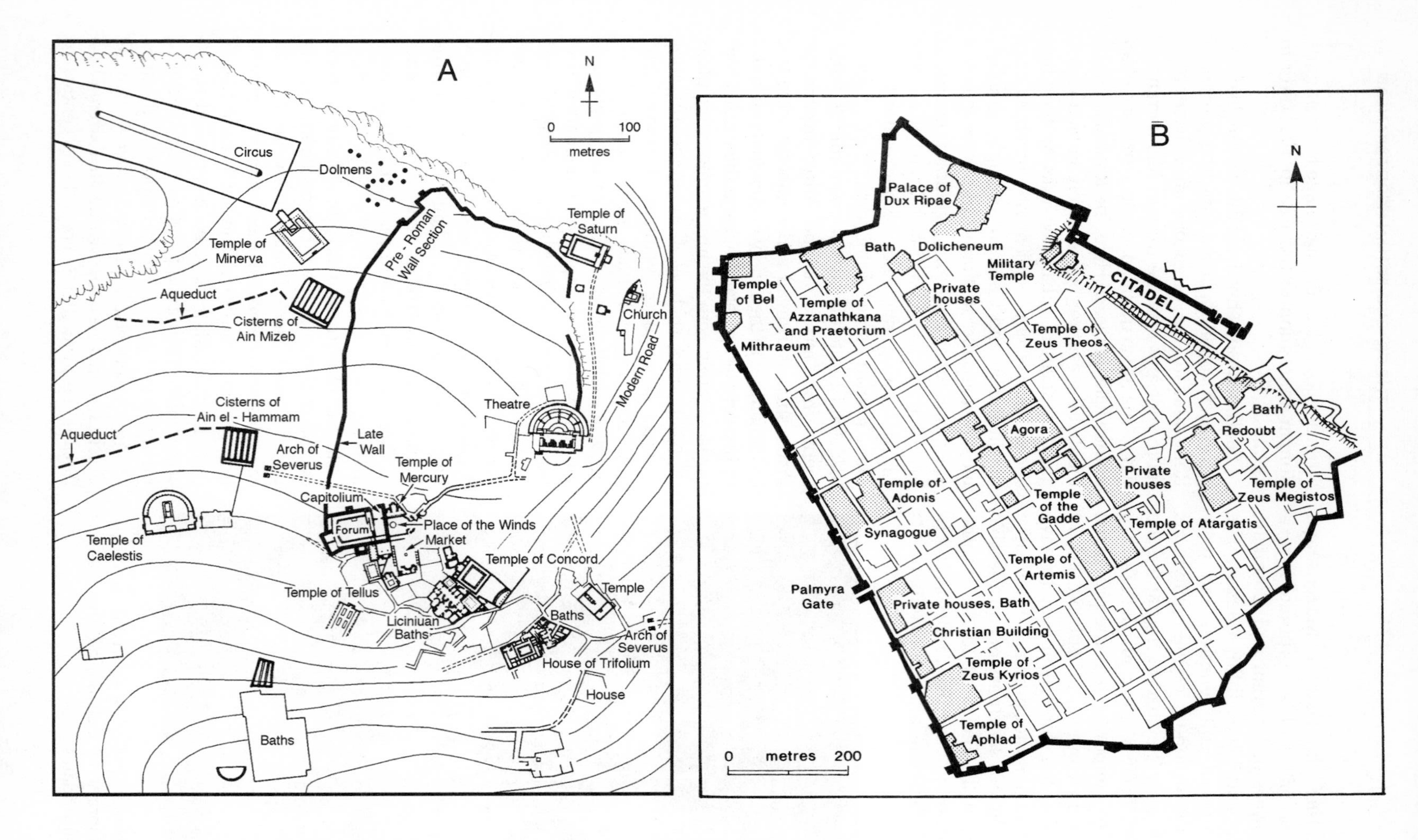

FIGURE 5.8 Accretive and planned urban foundations in North Africa: (A) Dougga, (B) Timgad

Source: After Finlay (1977)

acculturation. Populated by citizen veterans of the Roman army, they demonstrated to the indigenous population the advantages of loyalty to Rome, and were equipped, accordingly, with all the amenities of civilised urban life. At Merida, for example, founded as the capital of the Augustine province of Lusitania (northern Portugal) in 25 BC, Augustus' lieutenant, Agrippa, built a luxurious theatre two years later as part of the complex array of public and private facilities the town enjoyed (Fig. 5.9). Below the *colonae* in legal status were the various *civitates* or tribal capitals, some of the more successful of which acquired Latin rights of citizenship as *municipia*. Arguably, the success of the acculturation process was to be measured in terms of the number of these *municipia* which developed under the aegis of the various provincial administrations (Wacher 1987). Smaller again and functionally simpler than most of the *municipia* were the *vici*, or small market centres, which formed the base of the urban hierarchy, but could prosper sufficiently to acquire *civitas* status.

THE END OF EMPIRE

The second century AD represented the peak of the Roman imperial achievement. Thereafter, the fabric of empire, and its extraordinary political and social cohesion, epitomised by the network of provincial cities, gradually began to fragment into a series of more or less autonomous successor states. The causes of this fragmentation remain obscure. The Marxist emphasis on the importance of a possible decline in the supply of slaves has already been noted, but whether this was a cause or a consequence of the political fragmentation remains unclear. It is possible that the Empire simply expanded beyond the capacity of its technology to maintain effective links between the core and the periphery. This might explain the attempted assertion of centralised control in the third century AD. As the friction of distance increased, so the trade and taxation nexus would become progressively less able to supply the demands for food and revenue emanating from Rome (the core), giving rise in turn to increasing social and political instability, as the currency was devalued and taxation demands increased. As Rome's ability to control the periphery weakened, so the provinces tended to become more autonomous, but also more vulnerable to the growing external threat posed by the Germanic and other tribes who swept across the northern and western provinces in increasing numbers from the mid-third century AD onwards.

FIGURE 5.9 Conspicuous consumption in the west: the Theatre of Agrippa, Merida, Spain

All of this is speculative, but this type of spatial model would go some way towards explaining the evidence for monetary inflation and increasingly onerous taxation in the third century AD, as well as the increasing signs of social and political instability and the attempts by successive emperors to rationalise the imperial administration. The first of these occurred *ca.* 284 AD under Diocletian; the last and most effective in 395 AD, when the Emperor Theodosius I died, and the Empire was divided between his two sons. Thereafter, the Eastern Empire was ruled from Constantinople, itself refounded as an imperial city in 330 AD on the site of the old Greek city of Byzantium, while the Western Empire was ruled from Rome.

The subsequent history of the two empires was dramatically different. The Western Empire rapidly collapsed as a coherent geopolitical entity in the face of repeated invasions by Lombards, Ostrogoths, Visigoths and Vandals.

Rome was sacked by Visigoths in 410 AD, and the last Roman emperor ruling from the city was deposed in 476 AD and replaced by a barbarian leader. Although the various petty states which subsequently emerged in Italy and elsewhere in the west owed much to their non-material Roman cultural heritage, not least in terms of language and law, they owed at least as much to the more recent Germanic influence (Wickham 1981). Within this increasingly autonomous and localised world, much of the existing Roman settlement and communications infrastructure lost its imperial rationale and withered away. Where towns and cities survived, such as Toledo in Castile or, famously, Rome itself, they did so on the basis of the continued exercise of newly acquired forms of social, cultural or political authority, seated within the emerging post-imperial social formations. Pre-eminent among these was the institution of the Christian Church. In many regions, particularly in Italy, southern France and Spain, the urban-centred territorial dioceses established as the basic unit of religious administration following Constantine's declaration of Christianity as the official state religion in 312 AD, provided both the framework and rationale for urban survival.

In the Byzantine world of the eastern Mediterranean, the changes were neither as rapid nor as uniformly degenerative (Cameron 1993). The eastern imperial administration had proved to be more securely rooted than in the west, and consequently the Empire survived as a recognisable geopolitical entity comprising Egypt, the Levant, Asia Minor and Greece until the Arab conquest of the late seventh century. Indeed, in the early sixth century AD, the Emperor Justinian succeeded in temporarily re-establishing Byzantine control over Italy and parts of the Maghreb, the latter region having been settled by Vandals moving south from Spain in the early fourth century AD. One consequence of this continuing imperial control was the further erosion of urban and provincial autonomy. Thus, while there is evidence of accelerated urban and rural settlement growth in various parts of Syria, Cyrenaica, Turkey and the Negev as late as the sixth century AD (Liebeschuetz 1992, pp. 25–36), this was no longer in the context of a 'classical' federation of city–states. Instead, it reflected the ability of a centralising imperial monarchy to create – at least temporarily – the conditions of socio-economic stability necessary for such growth. The corollary was that once this imperial authority eventually began to decline, as it did in the face of the seventh-century Arab conquest, these rejuvenated settlement systems proved to be just as vulnerable to fragmentation and decay as their counterparts in the west. By 650 AD, Egypt and the Levant as far as Antioch had fallen into Arab hands, and this signalled the start of the protracted but progressive Islamic conquest, which culminated in the capture of Constantinople and the extinction of the Byzantine Empire by the Ottoman Turks in 1453 AD. The assertion of Islamic and Turkic authority over what had been the heartland of the Hellenistic world extinguished the last geopolitical structure surviving from the Graeco-Roman world, and created a series of cultural discontinuities whose consequences are still apparent today.

References

ANDERSON, P. 1974: *Passages from Antiquity to Feudalism*. London: Verso.

AUSTIN, M.M. and VIDAL-NAQUET, P. 1977: *Economic and Social History of Ancient Greece: An Introduction*. London: Batsford.

BAYHAN, S. 1994: *Priene, Miletus, Didyma*. Istanbul: Keskin.

BOARDMAN, J., GRIFFIN, J. and MURRAY, O. (eds) 1986: *The Oxford History of the Classical World*. London: Oxford University Press.

BRAUDEL, F. 1972: *The Mediterranean and the Mediterranean World in the Age of Philip II*. 2 vols. London: Collins.

CAMERON, A. 1993: *The Mediterranean World in Late Antiquity, A.D. 395–600*. London: Routledge.

CLOGG, R. 1992: *A Concise History of Greece*. Cambridge: Cambridge University Press.

COLDSTREAM, J.N. 1983: The meaning of the regional styles in the eighth century B.C. In Hagg, R.

(ed.), *The Greek Renaissance of the Eighth Century B.C: Tradition and Innovation*. Stockholm: Skrifter utgivna ar svenska Institutet i Athen, 4th ser., 30.

CORBIER, M. 1991: City, territory and taxation. In Rich, J. and Wallace-Hadrill, A. (eds), *City and Country in the Ancient World*. London: Routledge, 211–40.

CRAWFORD, M. 1978: *The Roman Republic*. London: Collins.

DELANO SMITH, C. 1979: *Western Mediterranean Europe*. London: Academic Press.

EAST, W.G. 1966: *An Historical Geography of Europe*. London: Methuen.

ERIM, K.T. 1993: *Aphrodisias*. Istanbul: NET.

FINLAY, M.I. 1977: *Atlas of Classical Archaeology*. London: Chatto and Windus.

FORREST, G. 1986: Greece: history of the Archaic Period. In Boardman, J., Griffin, J. and Murray, O. (eds), *The Oxford History of the Classical World*. London: Oxford University Press, 19–49.

GARNSEY, P. 1983: Grain for Rome. In Garnsey, P., Hopkins, K. and Whittaker, C.R. (eds), *Trade in the Ancient Economy*. London: Chatto and Windus, 118–30.

GREENE, K. 1986: *The Archaeology of the Roman Economy*. London: Batsford.

HOPKINS, K. 1978: Economic growth and towns in Classical Antiquity. In Abrams, P. and Wrigley, E.A. (eds), *Towns in Societies*. Cambridge: Cambridge University Press, 35–79.

HOPKINS, K. 1980: Taxes and trade in the Roman Empire, 200 BC–400 AD. *Journal of Roman Studies* **70**, 101–25.

HOPKINS, K. 1983: Introduction. In Garnsey, P., Hopkins, K. and Whittaker, C.R. (eds), *Trade in the Ancient Economy*. London: Chatto and Windus, ix-xxv.

HORNBLOWER, S. 1986: Greece: the history of the Classical Period. In Boardman, J., Griffin, J. and Murray, O. (eds), *The Oxford History of the Classical World*. London: Oxford University Press, 124–55.

LEPELLEY, C. 1992: The survival and fall of the classical city in Late Roman Africa. In Rich, J. (ed.), *The City in Late Antiquity*. London: Routledge, 50–76.

LEWIS, B. 1995: *The Middle East: 2000 Years of History from the Rise of Christianity to the Present Day*. London: Weidenfeld and Nicolson.

LIEBESCHUETZ, W. 1992: The end of the ancient city. In Rich, J. (ed.), *The City in Late Antiquity*. London: Routledge, 1–49.

MARCHESE, R.T. 1983: Introduction. In Marchese, R.T. (ed.), *Aspects of Graeco-Roman Urbanism: Essays on the Classical City*. Oxford: BAR International Series 188, 1–9.

MILLET, M. 1991: Roman towns and their territories: an archaeological perspective. In Rich, J. and Wallace-Hadrill, A. (eds), *City and Country in the Ancient World*. London: Routledge, 169–90.

MORRIS, I. 1991: The early polis as city and state. In Rich, J. and Wallace-Hadrill, A. (eds), *City and Country in the Ancient World*. London: Routledge, 25–58.

MURRAY, O. 1986: Life and society in Classical Greece. In Boardman, J., Griffin, J. and Murray, O. (eds), *The Oxford History of the Classical World*. London: Oxford University Press, 204–33.

OSBORNE, R. 1991: Pride and prejudice, sense and subsistence: exchange and society in the Greek city. In Rich, J. and Wallace-Hadrill, A. (eds), *City and Country in the Ancient World*. London: Routledge, 119–46.

OWENS, J. 1991: *The City in the Greek and Roman World*. London: Routledge.

PATTERSON, J.R. 1991: Settlement, city and elite in Samnium and Lycia. In Rich, J. and Wallace-Hadrill, A. (eds), *City and Country in the Ancient World*. London: Routledge, 147–68.

POTTER, T.W. 1991: Towns and territories in Southern Etruria. In Rich, J. and Wallace-Hadrill, A. (eds), *City and Country in the Ancient World*. London: Routledge, 191–210.

POUNDS, N.J.G. 1990: *An Historical Geography of Europe*. Cambridge: Cambridge University Press.

PRICE, S. 1986: The history of the Hellenistic period. In Boardman, J., Griffin, J. and Murray, O. (eds), *The Oxford History of the Classical World*. London: Oxford University Press, 315–37.

RAMAGE, E.S. 1983: Urban problems in Ancient Rome. In Marchese, R.T. (ed.), *Aspects of Graeco-Roman Urbanism: Essays on the Classical City*. Oxford: BAR International Series 188, 61–92.

RAWSON, E. 1986: The expansion of Rome. In Boardman, J., Griffin, J. and Murray, O. (eds), *The Oxford History of the Classical World*. London: Oxford University Press, 417–37.

RIALL, L. 1994: *The Italian Risorgimento: State, Society and National Unification*. London: Routledge.

RIHILL, T.E. and WILSON, A.G. 1991: Modelling settlement structures in Ancient Greece: new approaches to the *polis*. In Rich. J. and Wallace-Hadrill, A. (eds), *City and Country in the Ancient World*. London: Routledge, 59–96.

SEMPLE, E.C. 1932: *The Geography of the Mediterranean Region: Its Relation to Ancient History*. London: Christophers.

SMITH, C.T. 1967: *An Historical Geography of Western Europe before 1800*. London: Longman.

SNODGRASS, A.M. 1991: Archaeology and the study of the Greek city. In Rich, J. and Wallace-Hadrill, A. (eds), *City and Country in the Ancient World*. London: Routledge, 1–24.

THOMAS, C.M. 1993: *Ephesus*. Istanbul: DO.GU.

WACHER, J. 1987: *The Roman Empire*. London: Dent.

WARD-PERKINS, J.B. 1974: *Cities of Ancient Greece and Italy: Planning in Classical Antiquity*. London: BCA.

WELLS, C. 1984: *The Roman Empire*. London: Collins.

WICKHAM, C. 1981: *Early Medieval Italy: Central Power and Local Society 400–1000*. London: Macmillan.

6

THE MEDITERRANEAN IN THE MEDIEVAL AND RENAISSANCE WORLD

BRIAN GRAHAM

CONTEXT AND THEMES

In his magisterial work, *The Mediterranean and the Mediterranean World in the Age of Philip II* (1972), Fernand Braudel envisages a unified and coherent region surviving until as late as the sixteenth century, one in which the almost timeless realities of terrain, climate, agriculture, cities, trade, transport and population transcended cultural and political fragmentation (Prince 1975). Braudel's vision is a powerful mental construction, one of the most telling examples of *géohistoire*. In this epistemology, history is conceptualised in slow motion, revealing permanent values that can be set against a relatively unchanging physical environment (Butlin 1993).

None the less, one of the principal difficulties in writing about the medieval and Renaissance Mediterranean stems directly from the political and social fracturing, fragmentation and heterogeneity of the region. There is a symmetry to the historical geography of the period which begins and ends in crisis. Between the fifth and eighth centuries, the classical legacy of the sea as a unifying axis was slowly lost. The seventh- and eighth-century spread of Islam from its Middle East heartland across North Africa into Iberia, juxtaposed with European Christianity to the north, transformed the Mediterranean into a theatre of ideological conflict in which the sea became one of the great divides of history. Whereas the classical Mediterranean had unified Eastern and Western Europe with the Levant and North Africa, the links binding the Byzantine Empire – the imperial eastern Mediterranean centred on Constantinople – to the Germanic successor kingdoms to Rome in the west were broken by the Islamic advance. Braudel regarded Christian–Moslem conflict as constituting the defining characteristic of the Mediterranean Basin until as late as the sixteenth century.

At first, this ideological hostility might be envisaged as having separated the northern and southern elements of the Mediterranean world but, by the late sixteenth century, following the emergence of the Turkish Ottoman Empire in the east and the eventual completion of the Christian reconquest of Iberia, the axis of confrontation was transformed to one between east and west. By then, however, both the Christian Spanish Empire in the west and the Muslim Ottoman Empire in the east were in comparative decline, eclipsed in European power politics by the rise of north-west Europe, a process epitomised by the emergence of centralised states ruled by absolute monarchies, the Reformation and the early stages of an eventual capitalist world-economy (Fenech 1993).

However, even during the high medieval period that separated these two extended crises, we need to be aware of the deterministic dangers that underlie the conceptual model of a Mediterranean *world* defined by Christian–Islamic conflict. In the first place, neither medieval Christianity nor Islam could be regarded as monolithic ideologies. While the former was characterised by the slow evolution of the western Latin liturgy and the eradication of heresy, a process that continued throughout the medieval period, it was also distinguished by internal schism, most notably that between Latin and Orthodox faiths. The Muslim world was equally divided upon itself, not least through the failure of its founder, Mohammed, born in Makkah (Mecca) *ca.* 570, to designate a successor and thereby prevent the dynastic warfare and schism which followed his death in 632.

Second, the component regions of the Mediterranean were subject to many different, opposing and often contradictory forces, not simply political in origin but also reflecting complex economic and demographic forces. In the west, five centuries of slow, stuttering reconquest – the *Reconquista* – caught Iberia between the opposing forces of north African Islam and the political struggles of Christian north-west Europe. During the eleventh and twelfth centuries in particular, the reform of the Latin Church carried out by Pope Gregory VII (1073–85) and orchestrated through the network of abbeys controlled by the great Burgundian house of Cluny, brought much of northern Iberia within the orbit of French ecclesiastical and political ambitions. Again, the thirteenth-century Albigensian Crusade, launched against the Cathar Church in Languedoc in south-west France, was as much concerned with the emergence of an embryonic but northern French state, centred on the Île-de-France, combined with the power politics of the Roman Church, as it was about suppression of heresy (de Planhol 1994). East of the Rhône, Provence and – to some extent – northern Italy looked across the Alps towards the shifting and fragmented world of the Holy Roman Empire. By the eleventh century, however, the Italian city–states, most notably Venice, Genoa and Pisa, were effectively independent entities, deriving enormous wealth from straddling – and often controlling – the trade routes between north-west Europe and the Levant.

East again, the confused geopolitics of the Balkans were eventually resolved through the development of the Islamic Ottoman Empire. Originating only in the mid-thirteenth century in north-west Anatolia, this occupied the vacuum created by the collapse of the Byzantine Empire, Constantinople falling to the Turks in 1453. The late medieval Balkans were successfully subdued but the ineffectual siege of Vienna in 1529 marked the limits of Islamic expansion into south-eastern Europe. It is important to remember, however, that the Islamic world of the eastern Mediterranean also looked towards the wider Muslim territories of Arabia, Persia and the steppelands of central Asia. The desert cities and their oases were linked by caravan routes that connected the Mediterranean economy to Indian Ocean trade and also led into the Asian heartland. This Muslim hegemony was briefly interrupted by the Crusades of the eleventh to thirteenth centuries and the doomed attempt by Latin Christianity to establish Frankish kingdoms in the Holy Land. Finally, Islamic North Africa did indeed face across the sea to southern Europe, particularly in the west where the Moroccan dynasties often controlled, or were influenced by, events in al-Andalus – Moorish (Islamic) Spain. But just as significant were the mountain and desert lands to the south. Not only did important trade routes stretch across the Sahara, but successive nomadic tribes sporadically irrupted northwards to seize control along the Mediterranean littoral. In Morocco, for example, the High Middle Ages were dominated by two Berber dynasties, the desert Almoravids, who briefly controlled an empire that stretched from Ghana to Algeria, and the Almohods who originated in the High Atlas Mountains.

In conceptualising the Mediterranean world, Braudel was thus careful to define it as a wide zone that extended in all directions well beyond the shores of the sea. As he observes, Mediterranean history has felt the pulls of both its desert and European poles. He stresses, too, the fragmenting effects of the sea that has attracted southern Europe to its shore 'and contributed in no small measure to prevent the unity of that Europe' which 'it divided to its own advantage' (1972, p. 188). Inevitably, the deterministic hue of these statements invites qualification. The construct of a Mediterranean *world* is a powerful one but it must be treated with some caution as an explanatory framework for the region's medieval and Renaissance geography. The Mediterranean may well have possessed enduring physical and even human characteristics but it was also distinguished by heterogeneous social, cultural, political and economic processes that applied differently to its disparate component regions.

In attempting to make selective sense of this diversity – both across space and through time – the remainder of this chapter is organised around three distinct themes. The first – ideology, conquest and social structure – examines the conflict and interaction between Christianity and Islam. Second, the complex and varied nature of Mediterranean medieval economy, trade and urbanisation is examined while, finally, the chapter concludes with a brief discussion of the enigmatic late medieval period during which the eclipse of Mediterranean power was accompanied by the artistic and philosophical flowering of the Renaissance.

IDEOLOGY, CONQUEST AND SOCIAL STRUCTURE

Christianity and Islam

As noted briefly above, while the medieval Mediterranean world was an arena of ideological conflict between Christianity and Islam, it was also contested between factions within each of those principal belief systems. The Christian gospels post-date Christ by upwards of two hundred years and were written, revised and censored for a Roman audience (Chadwick 1967). The doctrine of the Church, including the Marian cult, is even later, dating originally from the third and fourth centuries and subjected subsequently to important medieval revisions. For example, much of the liturgy and dogma of the western Church, including the sanctity of marriage, are eleventh-century additions, reflecting the interaction of large-scale changes in the structuring of society and the Gregorian reforms of Latin Christendom (Duby 1984). These latter established the concept of the 'high-medieval papal monarchy', the ecclesiastical parallel to the system of absolute secular monarchy which was to be a principal characteristic of late-medieval Europe (Bartlett 1993). The Crusades against Islam in the Holy Land, which began in 1095, affirmed the pope's claim to supreme territorial as well as spiritual power.

Nevertheless, because only the very learned in medieval European society knew or understood Christian theology – although almost everyone comprehended elements of it – standardisation was difficult, one reason explaining the persistent outbreaks of heresy in the Mediterranean world. These reflected, too, the competing representations of Christian dogma, in particular the Gnostic concept that the material world is alien to the supreme God while goodness is a creation of inferior powers (Chadwick 1967). The dualist, or Manichean, heresy, in which the God of the material world is seen as the devil – the demiurge – while Christ cannot be the son of a good and living God in a world of evil, was the ultimate statement of such beliefs. Its most potent and enduring expressions were manifested in the Bogomil and Cathar heresies in the Balkans and Languedoc respectively (Moore 1985). The Cathars were subjected to the brutality of the Albigensian Crusade in the early thirteenth century, an event that well demonstrated the assertive absolutist nature of the medieval papacy and its temporal inclinations. Heresy, however, was but one manifestation of schism within the nominally Christian Mediterranean. By the eleventh century, the dichotomy between

Latin and Orthodox Christianity was deepening, leading to a north–south theological boundary between east and west Europe that stretched from the Baltic to the Adriatic and was fully recognisable by the fourteenth century (Bartlett 1993).

The Islamic religion was founded by Mohammed at Medina in 622 (the first year of the Islamic calendar). The key to Islam, as expressed in the *Koran*, lies in its interlinkage of the spiritual and the secular, expressed through the interdependence of religion, society and state and, by extension, militarism and the concept of *jihad* or holy war (Fossier 1989). However, its fundamental weakness lay in the failure to define the legitimacy of power and particularly laws of accession. The intrigue, conflict, and complex dynastic politics which followed the Prophet's death ultimately resulted in rival caliphates in Medina and Damascus (and later Baghdad) and the Sunni-Shi'ite fissure which split the Islamic world.

The spread of Islam

Despite this early history of schism, Islam underwent a dramatically rapid expansion, partly attributable to its emergence at the hub of trade routes linking the Middle East to sub-Saharan Africa and to central Asia and northern China (Park 1994) (Fig. 6.1). Before 750, it held sway in all of Arabia and Persia and had penetrated deeply into the central Asian steppes. To the west, Alexandria became an Islamic city before 650, the Moslem armies moving rapidly west across northern Africa to take Cyrenaica and Tripolitania (modern Libya) before subduing the Berber tribes of the Maghreb – contemporary Tunisia, Algeria and Morocco (Fig. 6.2). Tunis was founded at the end of the seventh century while, in 711, Muslim forces crossed the Straits of Gibraltar from Tangier to Tarifa to defeat the crumbling Visigoth kingdoms of Iberia. In less than a decade, they had crossed the Pyrenees where Merovingian France almost suffered the fate of its Visigothic neighbour, before Charles Martel eventually defeated the Islamic forces near Poitiers in 732. This reverse did not prevent the consolidation of Muslim control over Iberia, excepting only the mountain kingdoms of the far Atlantic north of the peninsula. The reconquest of Iberia for Christianity – the *Reconquista* – was to occupy the ensuing seven centuries and remained uncompleted until the kingdom of Granada fell to the forces of Catholic Spain in 1492 (Lomax 1978) (Fig. 6.3).

Between the eighth and tenth centuries, Moslem Spain – al-Andalus – was ruled by a Ummayad dynasty from its capital in Cordoba while, slowly, the northern marches from Navarra through Asturias into Galicia emerged as the heartland of Christian resistance. In reclaiming territory from Islam, Christianity was able to address one of its critical ideological problems. Although Christ had blessed the peacemakers, which made warfare between Christians difficult to justify, war against Islam could be presented as crusade or righteous war, waged under the symbol of the cross. Thus, the *Reconquista* can be envisaged as part of the spiritual and temporal construction and expansion of Latin Christendom. By 1085, it had reached the Tagus where the fall of Toledo proved a pivotal defeat for the Muslims. The reconquest of the subsequent four centuries was a spasmodic process, aided by the collapse of al-Andalus as a centralised power and its fragmentation into a number of smaller states (Jiménez 1989). Nevertheless, partly due to successive reinforcements from Morocco, to which southern Spain was increasingly a periphery, the iconic Islamic cities of Cordoba and Seville only fell to the Christian forces in 1236 and 1248 respectively, while Granada held on – through negotiation – until 1492.

Although the ultimate unification of Spain under the rule of the 'Catholic monarchs' Ferdinand (of Castile) and Isabella (of Aragon) – powerfully symbolised by their joint tomb in the triumphal surroundings of Granada cathedral's *Capilla Real* – has often been explained as a product of the linked processes of Christian reconquest and resettlement during centuries of war against the infidel, it must be emphasised that this representation is a fundamental

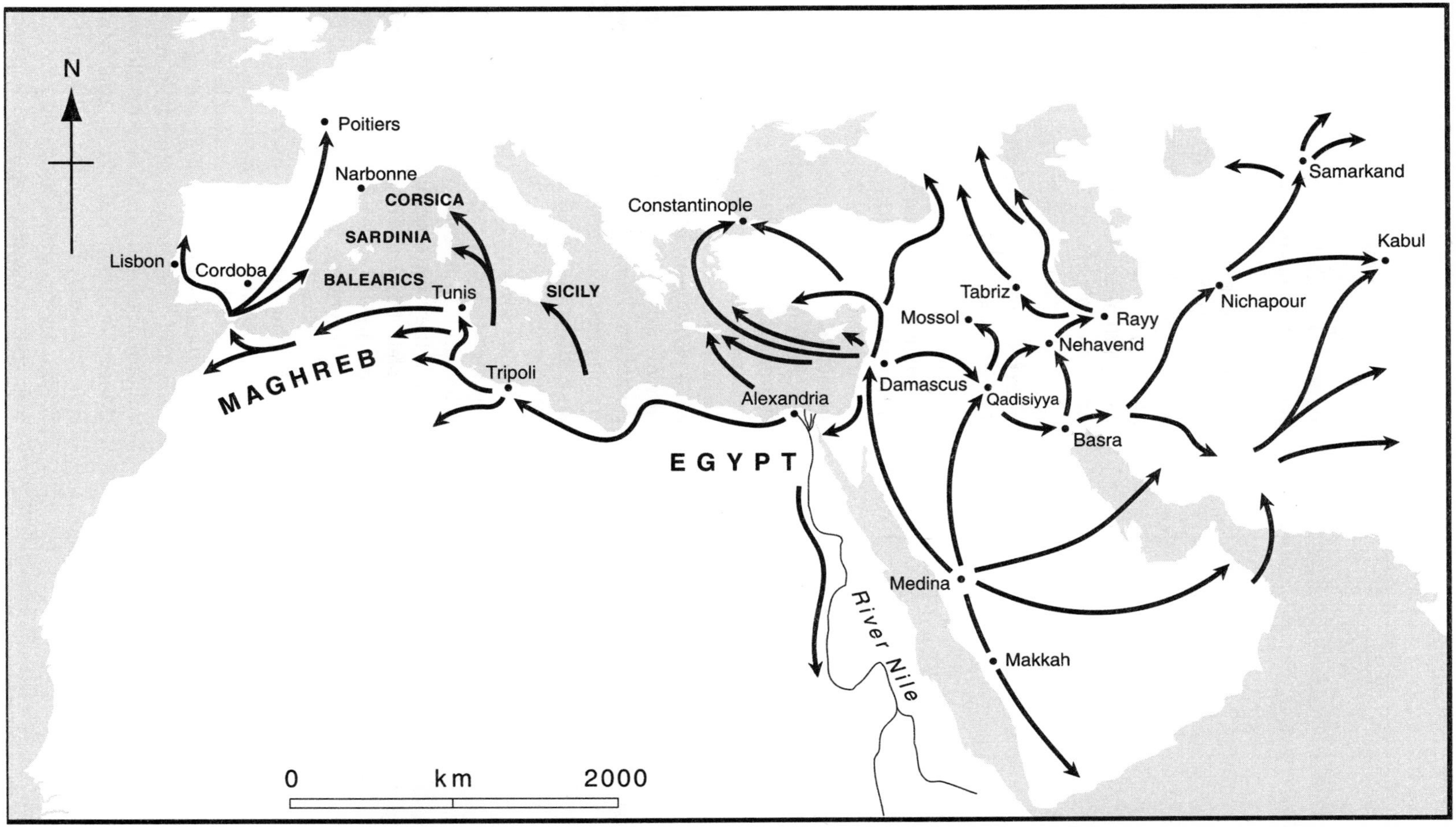

FIGURE 6.1 The spread of Islam 632 – *ca.* 730

Source: After Fossier (1989, p. 198)

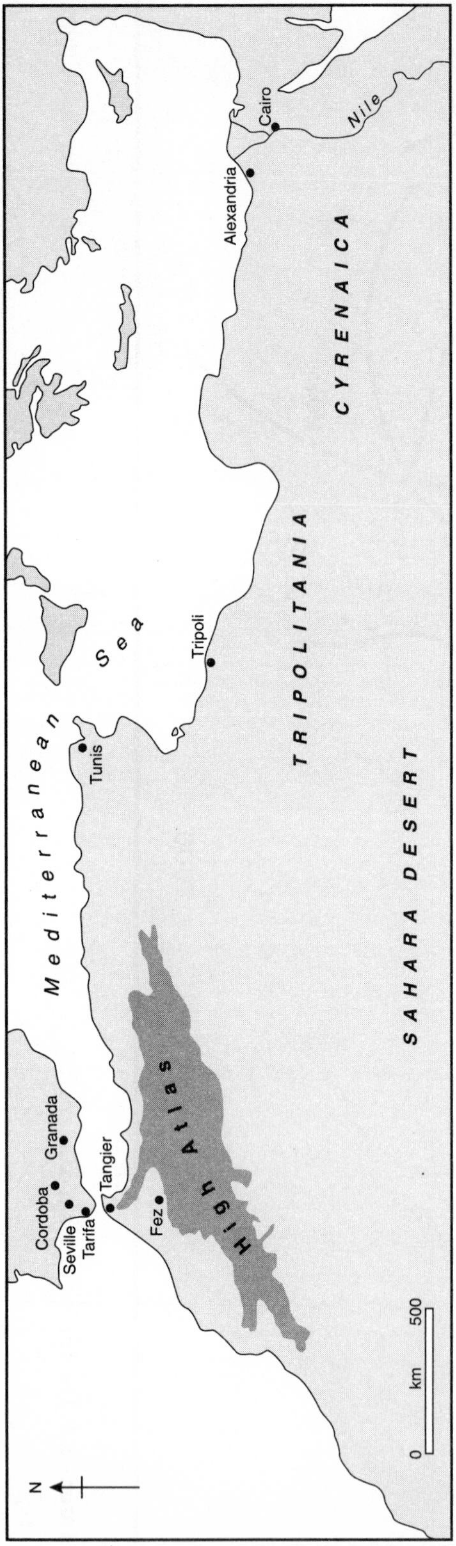

FIGURE 6.2 Mediterranean North Africa

element of orthodox Spanish nationalist history. Contemporary interpretations tend to depict a less Catholic and rather more pluralist medieval Spain in which the barrier between Islam and Christianity was a permeable one across which Christian, Muslim and Jewish influences intermingled (Collins 1983; MacKay 1977, 1989). Thus, the *Reconquista* can be visualised as part of the wider processes of internal European medieval colonisation, organisation and state formation.

State formation and the medieval Mediterranean

If events in Spain, particular though they were, are interpreted within such a context, we return again to the difficulties raised by the very construct of a medieval Mediterranean *world*. In particular, the *Reconquista* was but one facet of the slow emergence of centralised states, formed through the accretion of peripheries to cores such as Aragon–Castile in Spain and the Île-de-France. As Dodgshon (1987) argues, power in the early European Middle Ages was devolved downwards, forces of disintegration being dominant. From Catalonia to Italy, sovereignty fragmented into city–states, episcopal domains and other localised lordships (Tilly 1994). Polities such as the County of Toulouse, which exercised some degree of hegemony over Mediterranean France or the Midi, replicated the roles and functions of the royal centres to which they were nominally subject. The most complete breakdown of centralised power occurred in Italy, leading, particularly in the eleventh century, to the emergence of city-states such as Pisa, Lucca and Florence, which exercised complete control over their own affairs (Barber 1992) (Fig. 6.4). In the eastern Mediterranean, centralised power also became eroded, the Byzantine Empire being more of a commonwealth of small states, linked by religion and cultural tradition, than a consolidated polity (Morris 1988) (Fig. 6.5). The Islamic Middle East was similarly fragmented, although by the twelfth century the Abbasid caliphate in Baghdad was able to establish spiritual hegemony over

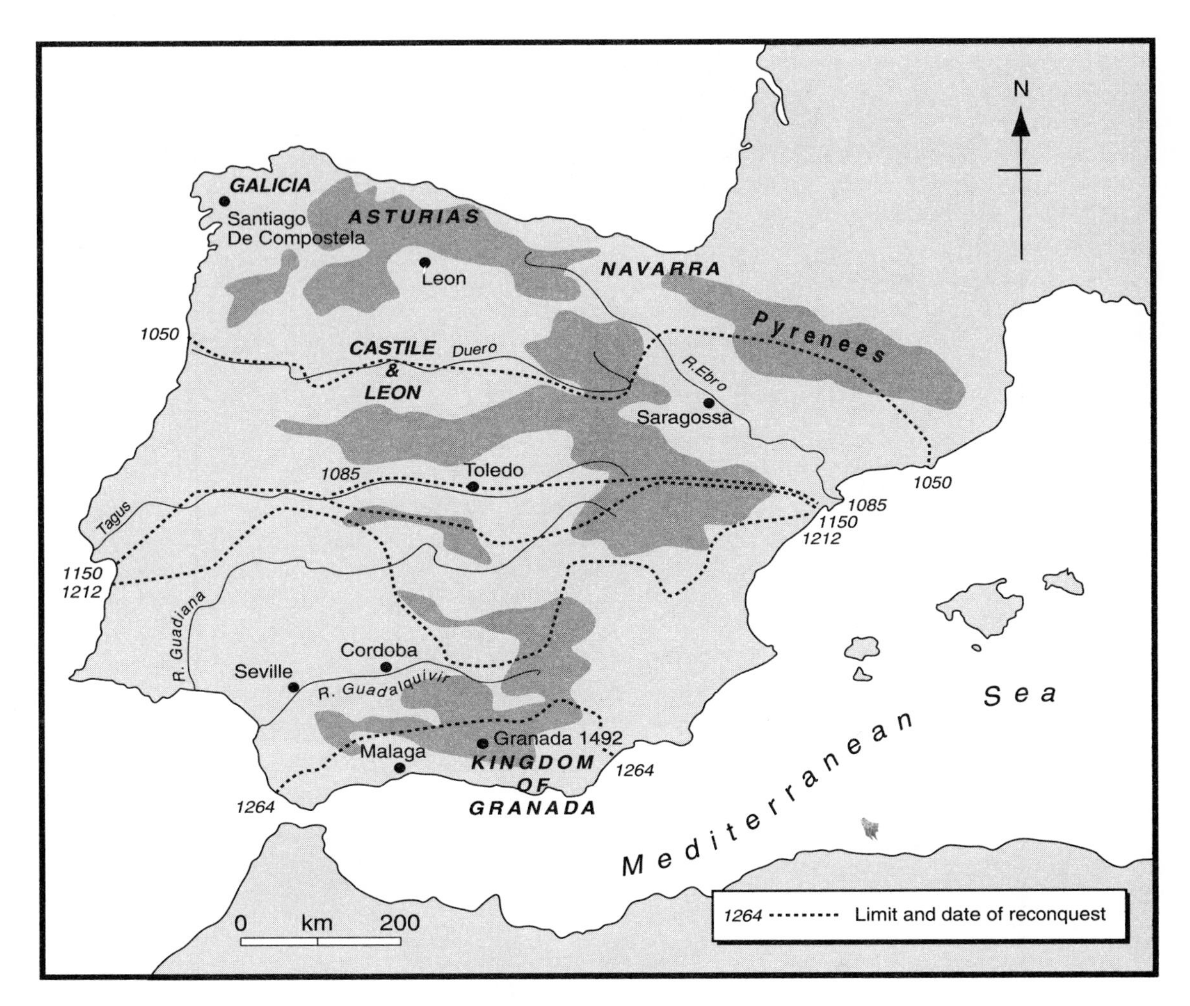

FIGURE 6.3 The Christian *Reconquista* of Iberia *Source:* Adapted from Holmes (1988, p. 204)

the Moslem world. *Ca.* 1000, the western Mediterranean was dominated by relatively large Islamic states, both in southern Spain and northern Africa.

However, between the eleventh and sixteenth centuries, a slow maturation of state systems in north-west Europe transformed the western Mediterranean from core to periphery. Political and commercial power eroded away from the city–states of the Mediterranean to the newly emergent states and relatively subordinated cites of the Atlantic. Baltic and Atlantic trade began to eclipse that of the Mediterranean, a potent factor undermining the power of the Italian city–states (Tilly 1994). Newly unified Spain and Portugal began to look south to Africa and west across the Atlantic, while the consolidation of France around its northern core reduced the Midi to little more than a backwater. In all this, northern Italy remained the exception, its commercial hegemony undermined by the survival of the increasingly anachronistic individual sovereignty of the city–state.

Christian and Islamic interaction

Between the eleventh and fourteenth centuries, however, in the era of political fragmentation and Christian–Islamic confrontation, these same city–

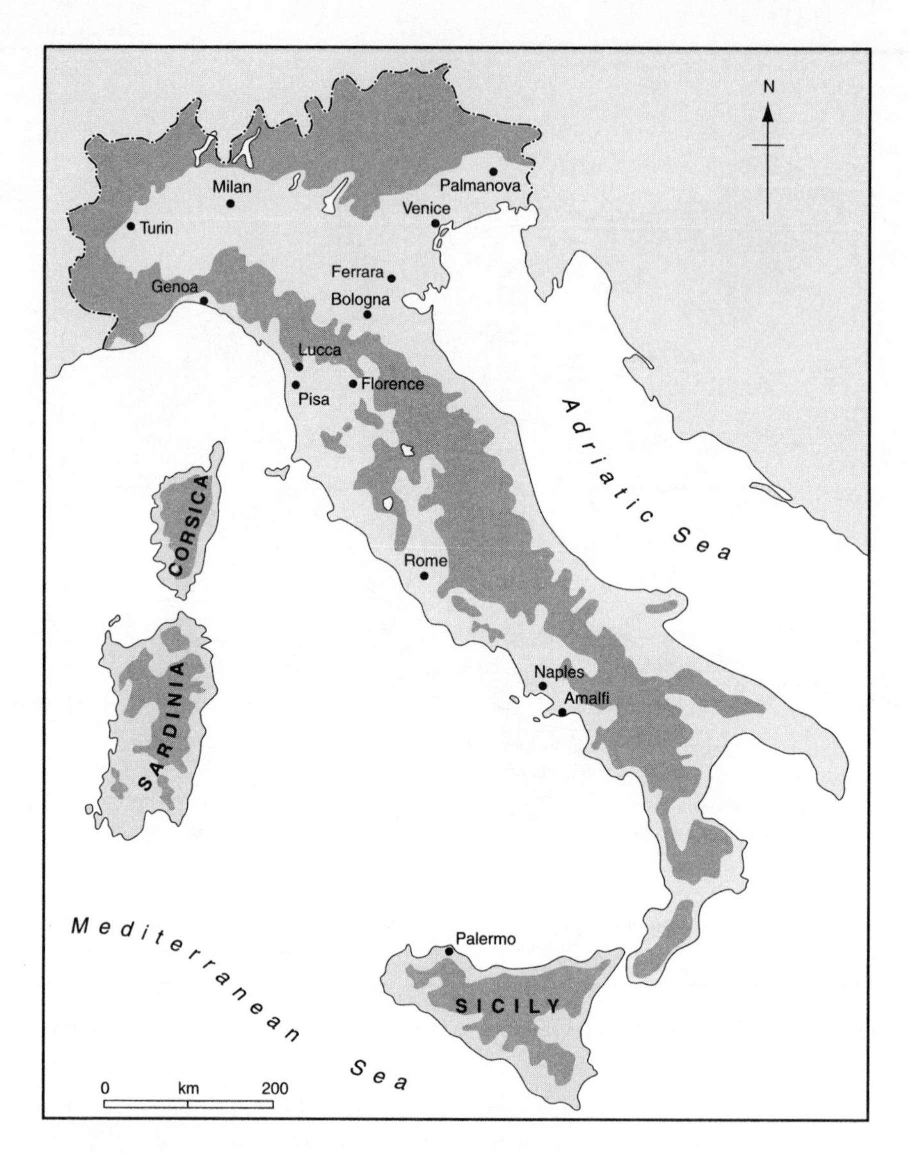

FIGURE 6.4 Medieval and Renaissance Italian cities

states dominated the Mediterranean's economic geography. They controlled trade and provided the fleets that carried the crusaders to the Holy Land. The Crusades, which began in 1096 and continued intermittently until the end of the thirteenth century, can only be interpreted within the context of the socio-economic and political changes occurring in Western Europe, in particular the emergence of assertive Latin Christianity under its centralising papacy. They also reflected the Church's adoption of the warrior cult of the west and the tensions created by the growing importance of primogeniture in feudal society (inheritance of land by the eldest son only). The Holy Land provided a safety valve, offering the lure of land to a surplus nobility.

Similar factors explain the internal colonisation that was occurring everywhere in Europe between the eleventh and thirteenth centuries, to which the Crusades formed an extension. The *Reconquista* was part of this frontier development, as was the expulsion of the Moors

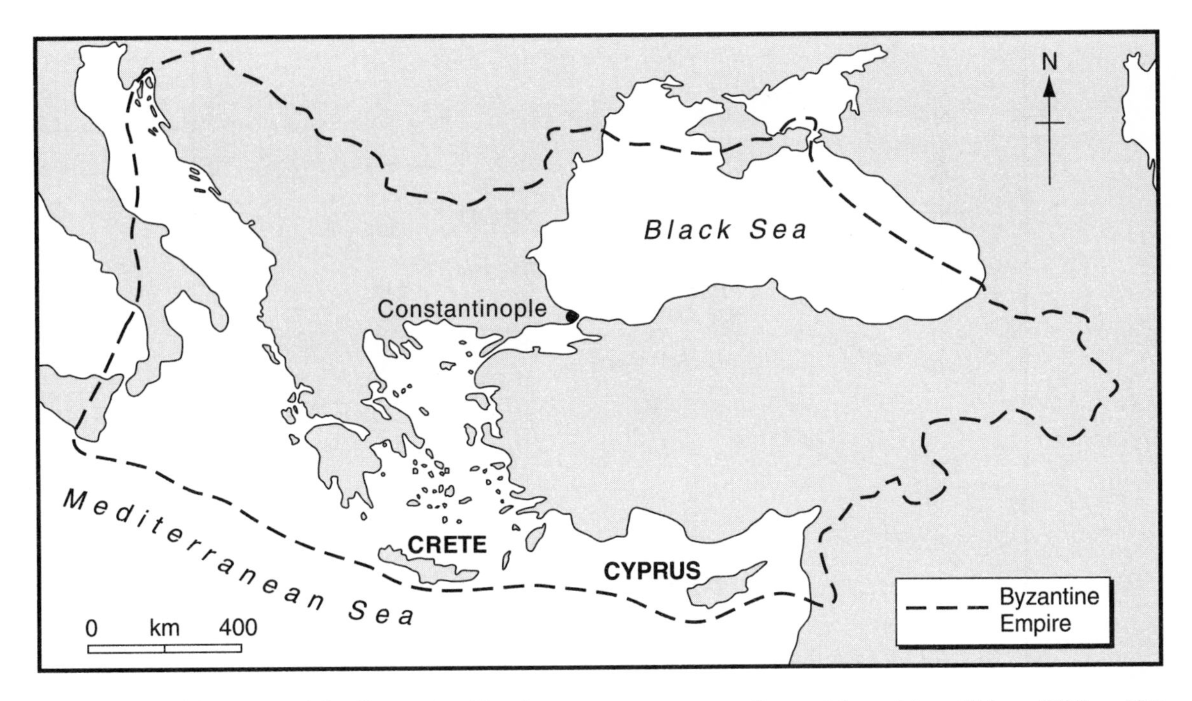

FIGURE 6.5 The extent of the Byzantine Empire *Source:* Adapted from Holmes (1988, p. 177)

from the Balearics, Sardinia and Corsica and the eleventh-century Norman subjugation of southern Italy and Sicily. The First Crusade led to the conquest of Jerusalem in 1099 and the establishment of Outremer – the Latin Christian states of Jerusalem itself, Edessa, Antioch and Tripoli, together with their fortresses, none more magnificent than Krak des Chevaliers, now in western Syria (Fig. 6.6). But these were to be short-lived successes. Edessa, hopelessly exposed beyond the Euphrates, fell as early as 1144, provoking the preaching of the Second Crusade, launched in 1147. This ended in political fiasco and, less than 90 years after its capture, the Christian cause was so weakened that Jerusalem fell to the Islamic armies in 1187. A Third Crusade stemmed the final collapse of Outremer although the recapture of Jerusalem proved impossible (Barber 1992; Runciman 1951–4), while the Fourth Crusade ended in ignominy when the Crusaders actually turned on Constantinople, sacking the city in 1204. The whole compromised Christian endeavour in the Holy Land ended with the fall of Acre in 1291.

None the less, the Crusades had significant repercussions. They helped unite the Islamic world while weakening the Byzantine, the capture of Constantinople by the forces of the Fourth Crusade being a tremendous blow to the latter. Byzantium's lands were divided between Venice and the Frankish nobility and, although the Byzantines were subsequently able to retake their capital in 1261, the emperors remained economically at 'the mercy of the powers they had ousted', most notably Venice and Genoa (Denley 1988, p. 256). Like the Crusader states of Outremer, Byzantium itself was doomed.

Economy, trade and urbanisation in the medieval Mediterranean

The nature of production

Traditionally, Mediterranean agricultural production revolved around the triad of wheat, vines and olives, supplemented by other fruit and tree crops. These latter demonstrate a further facet of

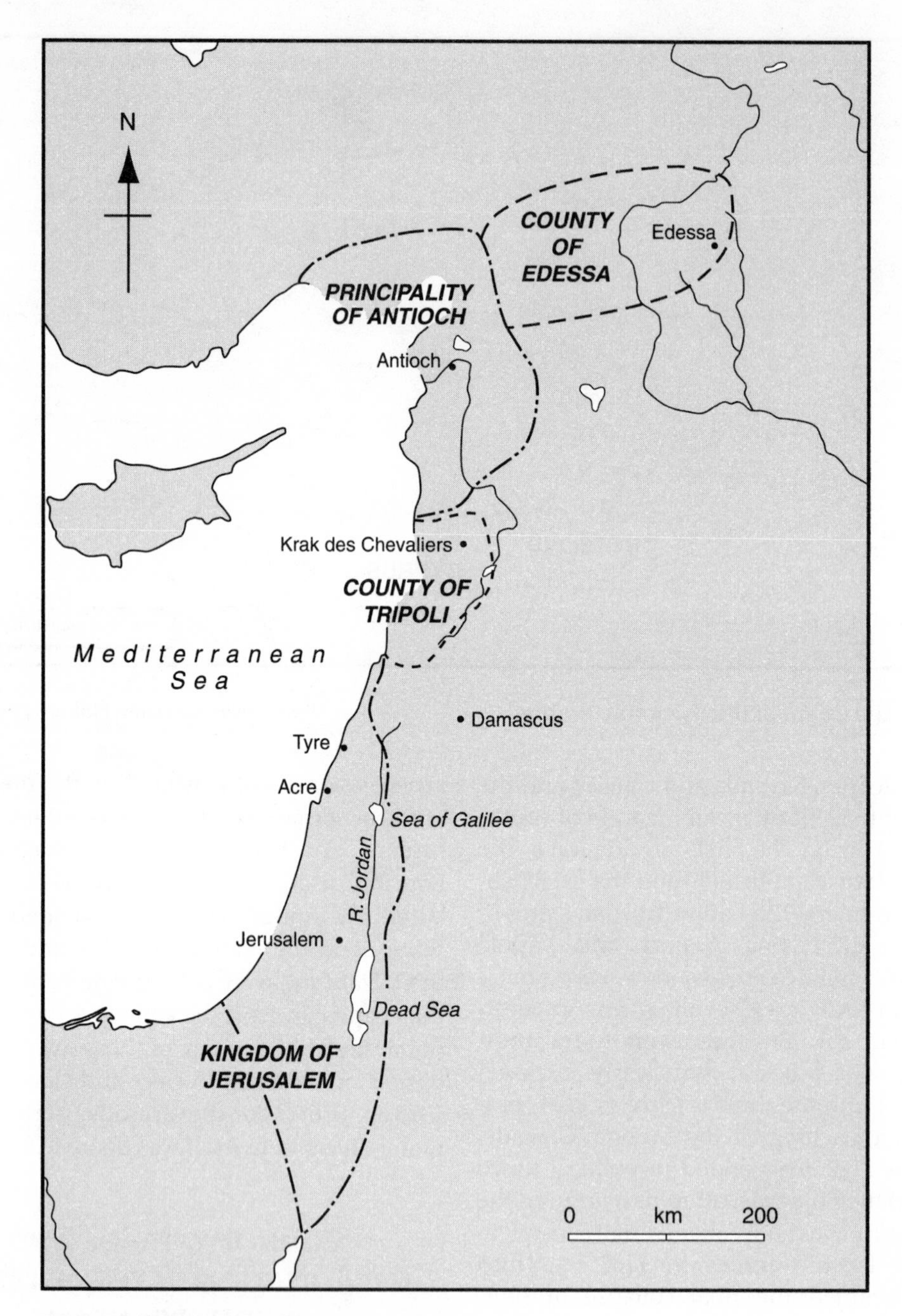

FIGURE 6.6 The Crusader states in Outremer

Source: Adapted from Holmes (1988, p. 220)

Christian–Islamic interaction, citrus fruits, sugar, strawberries and melons being Moorish introductions. Nevertheless, there was a considerable continuity between the classical and medieval worlds in terms of agricultural practices, landholding and rural settlement. Stock-rearing of sheep and goats and associated patterns of transhumance constituted one ancient mode of production that survived into the medieval world. Irrigation may well have been another, although

Islamic hydraulic technology and techniques were clearly more advanced than those likely to be encountered in early medieval Mediterranean Europe (Delano Smith 1979; Smith 1978).

It is also probable that medieval rural settlement forms and distributions reflected considerable continuity with the classical era. However, the evidence, drawn from place-names and morphological forms, is complex, diverse and contested. It appears that the recoil of rural settlement away from the littoral of southern Europe to mountain or hilltop locations postdated the classical era, being partly attributable to Arab piracy. The resulting tightly nucleated villages and agro-towns often survive – albeit in attenuated form – in the contemporary landscape. Although the commercialisation of agriculture in the later Middle Ages stimulated some relocation of farms, large-scale movements away from these settlements intensified only in the nineteenth century.

Nevertheless, such aggregate patterns disguise, for example, what Le Roy Ladurie (1974) has identified as a complex pulsation process of expansion and contraction defining the dual frontier of Mediterranean France – the brackish *étang* (lagoonal) environment of the littoral and the forested limestone Cévennes and desolate *garrigues* to the north-west of Nîmes (Fig. 6.7). Although this model was conceptualised within the specific context of the western Midi, it is of relevance elsewhere, particularly if aridity is substituted for topography as in northern Africa. In Languedoc itself, reclamation and forest clearance conquered the dual frontier by the eleventh century but, four centuries later, much of the coastal and upper hill-lands had reverted to waste. A complex array of forces – demographic expansion and contraction, economic crisis and war – combined with environmental degradation in accounting for this contraction. Slope erosion and deforestation led to declining soil fertility in the hills and encouraged the return of stagnant, malarial conditions to a littoral marked by creeping salinity. The sixteenth century, however, saw a reconquest of both frontiers, driven by arboriculture, particularly of olives, and viticulture.

Trade and the southern European medieval city

Despite such cyclical vicissitudes, the enduring peasant economy lay at the base of an economic system in which trans-Mediterranean trade provided the essential dynamic. The concept of the medieval Mediterranean as an arena of ideological conflict should not obscure its role as an economic artery. The trading system, linking north-west Europe to the Levant and Arabia, central and south Asia and sub-Saharan Africa, was by no means a closed one (Fig. 6.8). Around 1000, Europe can be visualised as lying on the periphery of a huge commercial system, controlled by the cities of the Mediterranean littoral. As markets became more functionally differentiated and hierarchical, these centres derived even greater power as self-regulating trading states (Dodgshon 1987).

The extent and dating of the early medieval Mediterranean crisis, in which the economic unity of the classical era was destroyed by the Islamic incursions, have caused intense debate. In the classic explanation, Henri Pirenne (1925, 1939) argued that Mediterranean commerce continued despite the Germanic invasions of Europe that succeeded the fall of the Roman Empire. As late as the eighth century, Marseilles, for example, was still the great port of Gaul, linked to it by the Rhône–Saône corridor (Loseby 1992). However, Pirenne maintained, the unity of the ancient world could not survive the advance of the Moslems who closed the sea to European trade and, in addition to Iberia, conquered the Balearics, Corsica, Sardinia and Sicily, where Palermo became their principal base. Islamic fleets subjected the coasts of Italy and Provence to constant pirate attacks, the results being that the economic development of the western Mediterranean reached its lowest ebb during the ninth century.

In a detailed consideration of more recent archaeological evidence, Hodges and Whitehouse agree that Mediterranean trade did indeed survive the Germanic invasions but argue that decline began as early as the fifth century. They see the Arab advance after 630 as 'the conse-

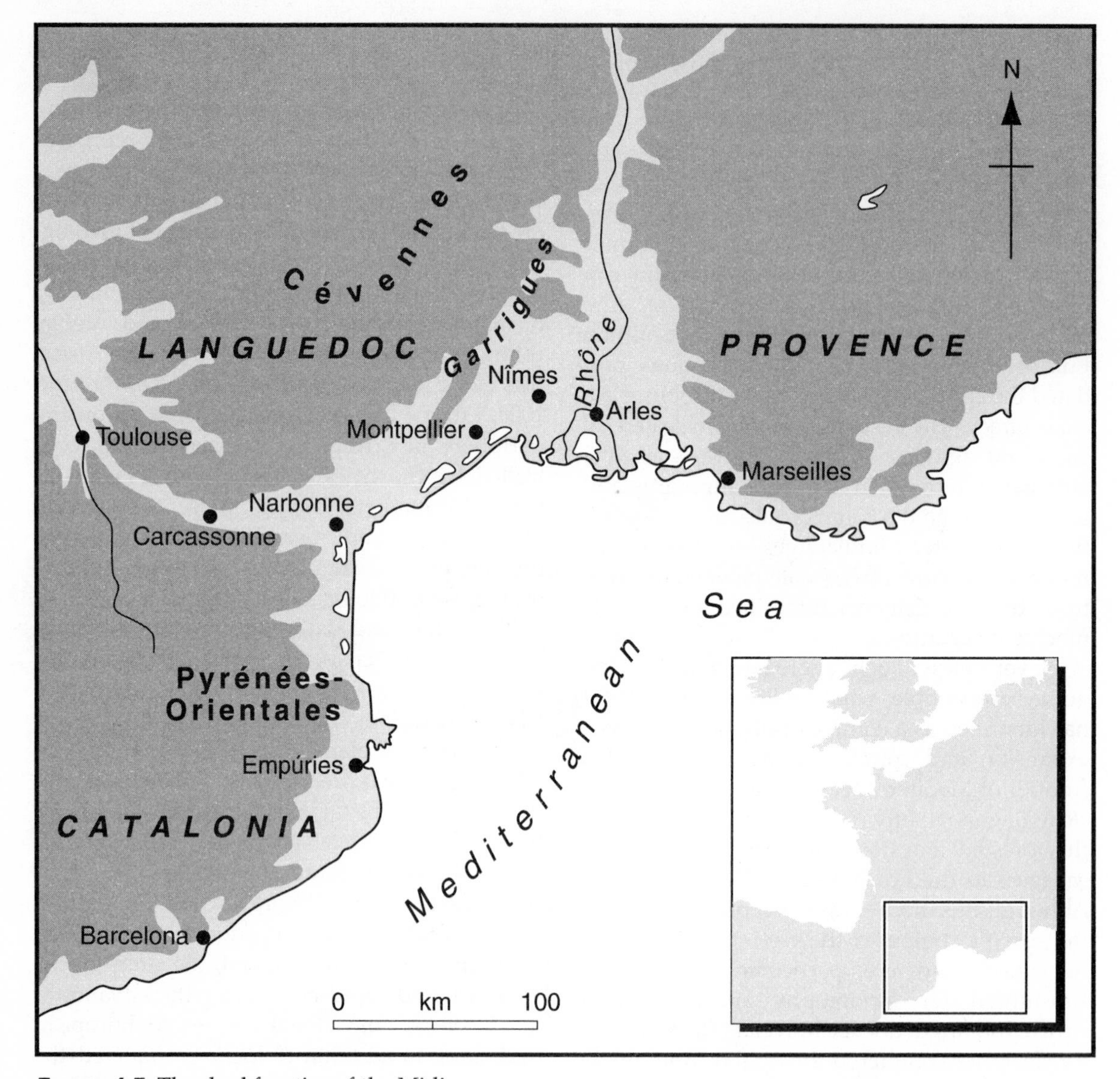

FIGURE 6.7 The dual frontier of the Midi

quence rather than the cause of the catastrophe' (1983, p. 52) and argue that the Mediterranean did not switch from being a Roman lake into a Moslem one as Pirenne had proposed. Rather, prior to the advance of Islam, it had already become divided into two major regions, one focused on Rome, the other on Constantinople. After their eclipse by the new Islamic order, there is very little evidence to show any substantial continuity in long-distance trade (Hodges 1982). Instead, as occurred elsewhere in early medieval Europe, economic exchange was increasingly organised at the local level. Thus, in the ninth and tenth centuries, Amalfi was the only European port carrying on an appreciable trade in the western Mediterranean. Nevertheless, there was trade by ship and camel train across northern Africa between the caliphate of Cordoba and Egypt, and also some between Western Europe and the Moslem world. Much of this was controlled by Jewish merchants who could pass freely between the two (Constable 1994).

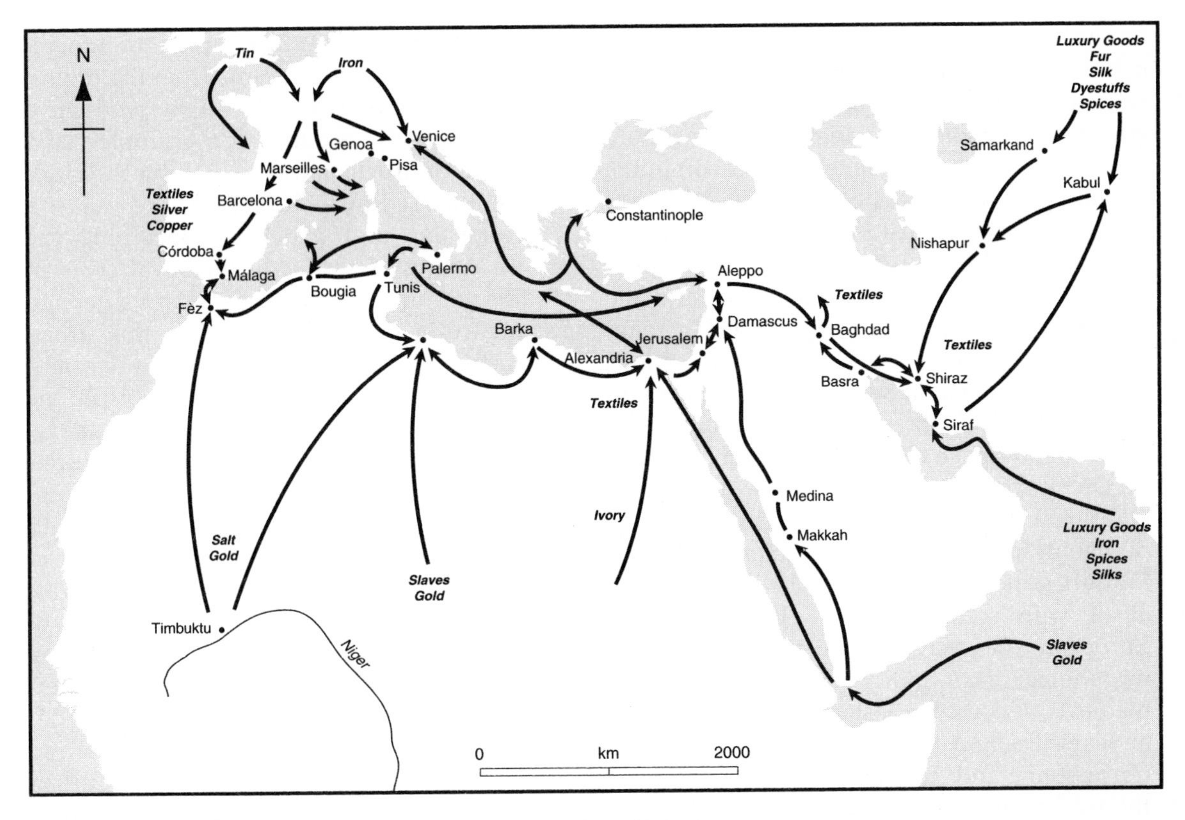

FIGURE 6.8 The principal trading routes of the medieval Mediterranean

Source: Adapted from Holmes (1988, p. 226)

There is considerable agreement, however, on an eleventh-century trade revival which held very significant repercussions for the southern European city. Under the aegis of the Italian city–states of Venice, Genoa and Pisa, trans-Mediterranean trade – later boosted by the Crusades – began to increase in volume. Venice and Genoa competed for primacy, the former holding sway after the conquest of Constantinople in 1204 and its purchase of Crete, the latter becoming ascendant after the Byzantine retaking of their capital in 1261. Genoa was notably dominant around the Black Sea where it established a succession of entrepôts on the trade routes leading to central Asia and China. Such long-distance trade was dominated by spices, sugar, silk, dyestuffs and precious stones (Pounds 1990). Bartlett (1993) envisages these Italian colonial empires as series of islands and headlands, located along the principal commercial axes and linking their respective metropolis to distant markets. During the twelfth and thirteenth centuries, the revival of commerce also extended to the western Mediterranean where Barcelona and Montpellier emerged as major commercial centres, competing with Genoa and Pisa for the trade in wheat, salt and wool (Reyerson 1994). By the thirteenth century, Majorca had a thriving trade with Granada and northern Africa (Abulafia 1994), part of a north–south commercial axis that had grown at the expense of the earlier east–west Muslim trading routes. Increasingly, this north–south trade passed into Christian hands (Constable 1994). The failure of the Crusades and the advances of the Ottoman Turks restricted, but did not terminate, the pan-Mediterranean trading activities of the Italian cities, which continued to be extensive throughout the fifteenth century (Pounds 1990).

The growth in trade and the evolution of urbanisation were thus symbiotically linked. Although epithets such as 'the Mediterranean city' disguise enormous variations in the nature and processes of urban development within the region (see Chapter 12), at the present level of generalisation we must be content with distinguishing southern European cities from their Moslem counterparts in North Africa and the Middle East, while recognising a hybrid variant that includes places such as Palermo, Cordoba and Seville which, at various periods, were in both Islamic and Christian hands. Many of the southern European cities were originally Roman or Greek foundations and, while there were occasional instances of partial or complete abandonment, for example, the Catalonian port of Empúries, most survived into the medieval era, albeit with some difficulty. Thus, southern European cities may be characterised by direct morphological continuity between classical and medieval periods. For instance, the Greek walls at Marseilles and their Roman counterparts at Carcassonne and Toulouse were still in use in the twelfth and thirteenth centuries, while the street layout of medieval Barcelona was one among a number that preserved the Roman grid-iron plan (Carter 1983). As at Bologna, Narbonne or Arles, the commercial zone of the medieval town often coincided with the Roman *forum*. A further important element of functional continuity was provided by the territorial organisation of the Latin Church into episcopal sees, marked in the townscape by the juxtaposition of cathedral (generally Romanesque in style) and civic monuments of Roman life including arenas and theatres.

Some commentators (for example, Delano Smith 1979) have mentioned the 'countrified' medieval cities of western Mediterranean Europe, referring to the inclusion of agricultural activities such as vineyards within their walls and also to the existence of urban peasantries. But there was also an urbanised countryside in the sense that agricultural production was controlled by, and directed towards, the towns. Agriculture was the first stage in a town's economic life, followed by industry and banking (Braudel 1972). The capital for agricultural improvement often came from the towns, the result being that virtually all the productive agricultural land around, for example, late-medieval Montpellier was owned by the city's bourgeoisie (Le Roy Ladurie 1974).

Irrespective of the debate concerning the precise dating of the early medieval crisis in Mediterranean Europe, there remains a general acceptance of Pirenne's conclusion that urban and economic life in the Christian western Mediterranean reached its lowest ebb in the ninth century. However, as observed above, the tenth and particularly eleventh centuries witnessed a marked revival of urban life, foremost perhaps in Italy where, first, Amalfi and then Venice, Genoa and Pisa capitalised on their locations as bridgeheads between east and west. Venice, effectively an outpost of Byzantium, was instrumental in maintaining Constantinople as the greatest of the Mediterranean cities. None the less, in areas under Byzantine rule – including the Balkans and coastal regions of southern Italy – urban growth was checked and the defensive contraction of earlier centuries remained a permanent characteristic (Benevolo 1993). Thus, even by the time of the Ottoman conquest at the end of the fourteenth century, the urban system of south-east Europe comprised only a few substantial cities, characterised by a diversified division of labour, developed trading structures and mercantile elites, and standing in contrast to a much larger number of functionally limited administrative, strategic and religious centres (Panova and Gavrilova 1992).

The Islamic Mediterranean city

In discussing the urban life of the medieval Mediterranean, it is important that a continuing reverence for the particular notion of a medieval southern European city, mediated through its continuity with the classical era, should not distort an understanding of urban life elsewhere in the region. As the Italian and Spanish cities recaptured from the Arabs readily demonstrate – even today – the Islamic city was (and remains, as we shall see in Chapter 12) organised on very different principles, its morphology dominated by a configuration of contiguous enclosures in which the buildings faced inwards to central courtyards. Architectural detailing was confined within the interior structures of the houses, little care being given to exterior appearances (Benevolo 1993). Once the Muslims had been displaced, the hybrid cities of southern Europe were gradually transformed through the consolidation of their internal enclosures and the construction of enclosing walls. Nevertheless, in Granada, for example, separated by the Rio Darro from the Alhambra Palace – perhaps the most evocative surviving signification of Islamic culture in the contemporary Andalusian landscape – the old Moorish Albaicín quarter, with its low, courtyard houses and formless alleys, remains markedly distinct from the rectilinear morphology of the sixteenth- and seventeenth-century Spanish Renaissance town located around the cathedral (Fig. 6.9).

Thus, the plan of the Islamic Mediterranean cities reflects a different set of priorities to those apparent in most of southern Europe. The Roman grid city was designed to serve unrelated family houses whereas its Islamic counterpart was divided into separate neighbourhoods, based on kinship, tribe and ethnicity and often separately walled. Roman grid cities such as Damascus were transformed into the characteristic dense urban fabric, served by cul-de-sac alleys (Kostof 1991). These opened off marginally wider through-routes – mostly roofed over to provide shade – that linked the city gates to a centrally located mosque. Such axes, often no wider than was necessary to admit pack-animals, were lined by the small individual shops of the *suqs*, the city's markets (Morris 1994).

This variant of medieval Mediterranean urban morphology is still readily identifiable today,

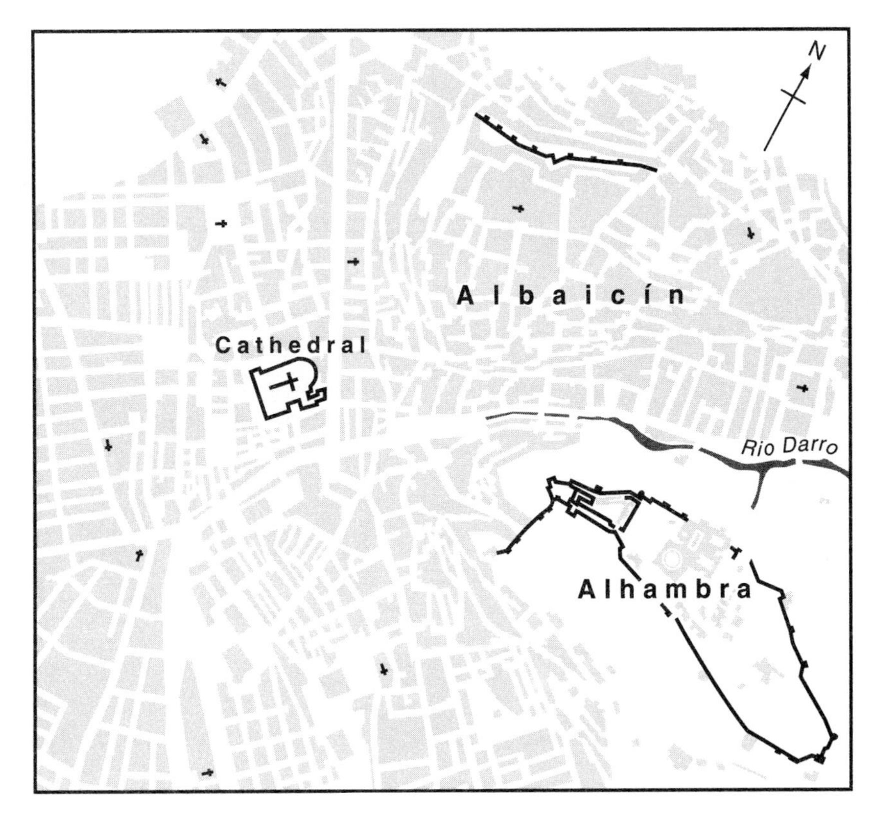

FIGURE 6.9 The street-plan of contemporary Granada, Spain

nowhere more so than in the Moroccan city of Fez (Fig. 6.10). Founded in 808 and possessing no classical antecedents, the oldest part of the city – the *medina* of Fez El Bali – comprises a number of separately fortified quarters and *suqs*, each dedicated to a distinct economic activity such as tanning, metal-working of various sorts or particular forms of retailing. Dominated by an incoherent maze of cul-de-sac alleys and straggling through-routes, the city's plan lacks the discernible patterning and morphological aggrandisement of its European contemporaries. Even the great medieval Kairaouine Mosque – one of the holiest places in western Islam – is difficult to see, given the absence of any axial perspective in the street-plan, or a place of public assembly corresponding to the European cathedral square. As Morris (1994) observes, however, although most medieval Islamic Mediterranean cities conformed to the 'contiguous cellular essence' exemplified by Fez El Bali, a few, most notably Cairo, were characterised by comparatively 'high-rise' houses facing onto the city streets and ventilated by screened windows.

CRISIS AND RENAISSANCE

To conclude, we must return to the enigma of the late medieval Mediterranean crisis. Despite the ideological conflict between Islam and Christianity, an internal coherence and dynamic to the region can be identified throughout much of the medieval period, although, as observed above, the fragmentation and concomitant decline of the Mediterranean are apparent very much earlier than the sixteenth century. In

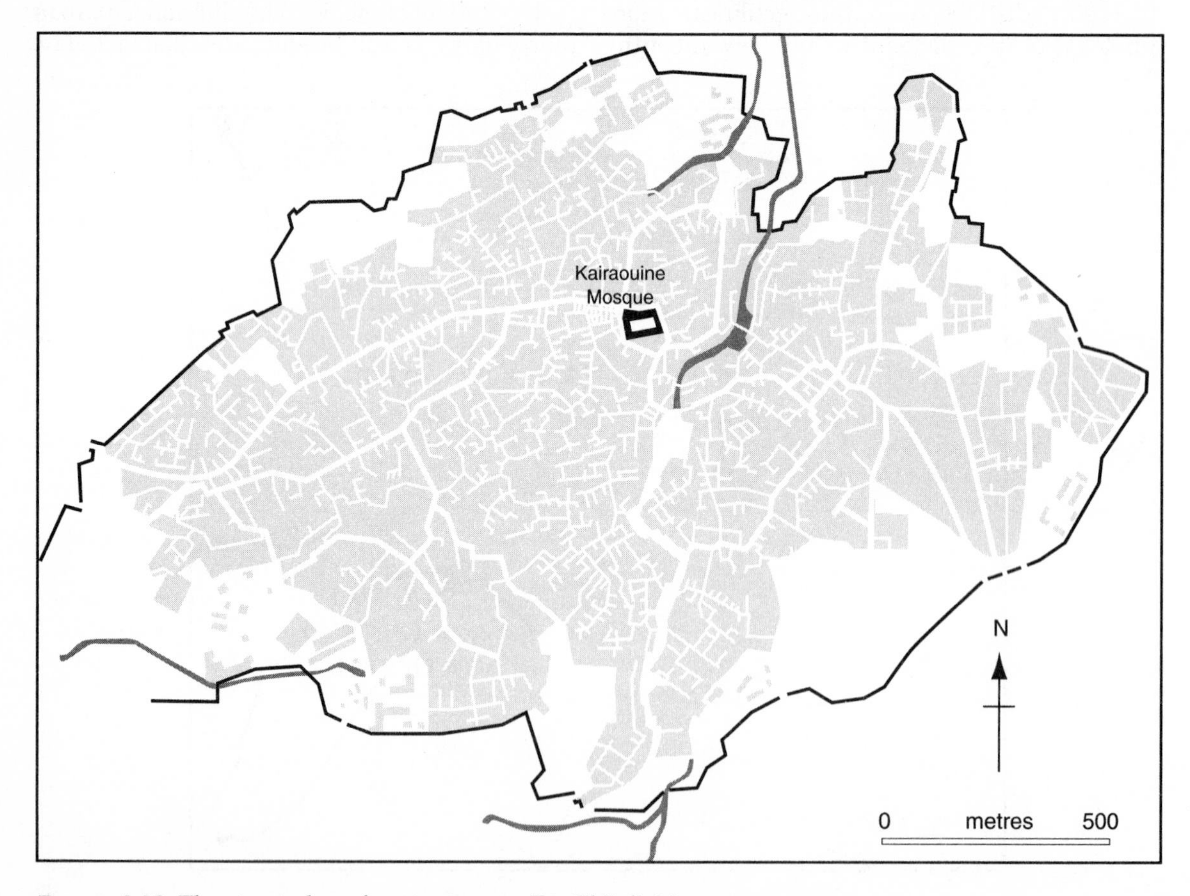

FIGURE 6.10 The street-plan of contemporary Fez El Bali, Morocco

general terms, Europe as a whole suffered from economic and demographic contraction for the second half of the fourteenth and much of the fifteenth centuries. While the Black Death was not alone in accounting for this, the virulence of the plague – which apparently arrived in Europe in 1347 on the ships of Italian merchants – was a crucial factor in decline. However, as Denley (1988) observes, that contraction was not uniform throughout the Mediterranean, the fourteenth and fifteenth centuries seeing shifts in emphasis rather than uniform crisis, as regions and cities responded to new patterns of production (Epstein 1991, 1993) and the relocation of Europe's power centre away from the classical world to the north-west. The emergence of centralised states and absolute monarchies was ultimately to dissolve the political independence of northern Italy, the most dynamic region in the European Mediterranean, while the creation of unified states in Spain and Portugal began the overseas territorial expansion of Europe and, eventually, the transition to a capitalist world-economy. In the fifteenth century, for example, the Portuguese fleets were able to capture much of the trans-Saharan economic traffic, ending the gold bullion trade and destroying the economic position of North Africa (Braudel 1972; Wallerstein 1974).

Enigmatically, however, this period of decline and shifting relations of economic dominance and power was also that of the Renaissance. Literally meaning rebirth, this movement, which originated in early fifteenth-century Florence under the patronage of the city's wealthy merchants, encapsulated a revival of interest in the classical art forms of Rome and Greece. As Epstein (1991) observes, the Florentine Renaissance seems to have flowered at the expense of an impoverished and subdued regional economy, one expression of the changing spatial fortunes of the late medieval crisis. None the less, northern Italy continued 'for a while to be thought of as the cross-roads of European culture' (Denley 1988, p. 296). Matching its commercial hegemony of the high Middle Ages, by the fifteenth and sixteenth centuries, the region had emerged as the 'centre of production and exchange' in European culture. We must accept, however, the ethnocentric nature of the term, 'Renaissance', and be wary of employing it to define and conceptualise the entirety of the late medieval Mediterranean. Just as the southern European medieval city with its classical origins did not define Mediterranean urbanisation, Renaissance ideas had little relevance to – or impact upon – the Islamic Mediterranean world, which continued to reflect its very different principles of social organisation.

First in Italy, and then throughout much of Europe, the Renaissance witnessed the application of particular principles to the landscape. As Cosgrove (1993) argues, these represented a belief in the unification of nature and a perfect and harmonious order of creation. Renaissance artists, sculptors and architects were striving to create a new visionary landscape. In more prosaic terms, the most obvious manifestation of such ideas was the incorporation of linear perspective into urban plans. The essence of Renaissance urban structure revolved around four design components: the primary straight street, grid-iron districts, formal enclosed spaces, piazzas or *places* and symmetrical enclosing defences. Several thinkers formulated ideal schemes of urban morphology although, as Morris (1994) observes, these were generally flawed by the need to reconcile circular or faceted polygonal defensive perimeters with grid-iron street patterns (Fig. 6.11). Only very occasionally, as at Palmanova near Venice, was such ambitious theory ever translated into reality.

In most cities, therefore, Renaissance ideas were adapted to extant structures. Despite its pivotal role, Florence remained essentially medieval in morphological terms while in many other larger cities, including Rome, Venice, where the largely fifteenth- and sixteenth-century Piazza San Marco is perhaps the single most expressive urban rendition of Renaissance principles, Milan or Turin, the application of these ideas was largely confined to extensions beyond the medieval walls. In some contrast,

FIGURE 6.11 An example of a mid-sixteenth-century ideal Renaissance city: plan by Pietro Cataneo

Source: Adapted from Morris (1994, p. 171)

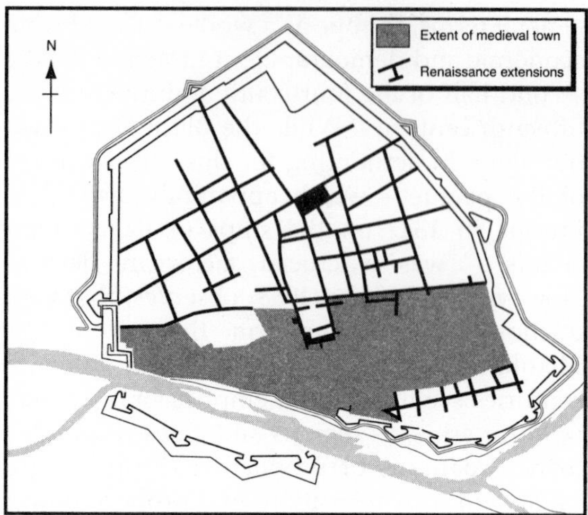

FIGURE 6.12 The fifteenth-century Renaissance extensions to Ferrara, Italy

Source: Adapted from Benevolo (1993, p. 95)

the plans of several smaller cities were more substantially altered (Benevolo 1993). At Ferrara, for example, the loosely linear medieval layout was supplemented by an ambitious grid-iron of large streets and piazzas during the later fifteenth century (Fig. 6.12).

Nevertheless, the flowering of the Renaissance served to conceal a further dimension of crisis. Political power had slipped away from the southern European Mediterranean. Even if capital from the Italian city–states was helping to fund the overseas expansion of Europe, the concept of the Mediterranean world was losing any residual significance by the sixteenth century. The integration of European power, forged through the late medieval decline in feudal lordship and the concomitant growth of absolute monarchy, was reshaping the political map of Europe. Commercial hegemony gradually passed to metamorphosing, centralised states in north-west Europe, their wealth vested in industrial production and trade along the Atlantic–Baltic axis. These power shifts relegated the trading cities of northern Italy and, indeed, Spain – despite its imperial expansion – to the 'semi-periphery' within the international economic hierarchy (Wallerstein 1974). If the medieval Mediterranean had been an arena of ideological conflict, the sea also provided the coherence for an economic region that linked Europe, Africa and Asia. Increasingly, however, it was now to become the southern boundary of Europe, a physical barrier separating that continent from its neighbours, a role neatly symbolised by the irrelevance of the Renaissance to the Islamic world beyond.

REFERENCES

ABULAFIA, D. 1994: *A Mediterranean Emporium: The Catalan Kingdom of Majorca*. Cambridge: Cambridge University Press.

BARBER, M. 1992: *The Two Cities: Medieval Europe, 1050–1320*. London: Routledge.

BARTLETT, R. 1993: *The Making of Europe Conquest, Colonization and Cultural Change, 950–1350*. London: Allen Lane.

BENEVOLO, L. 1993: *The European City*. Oxford: Blackwell.

BRAUDEL, F. 1972: *The Mediterranean and the Mediterranean World in the Age of Philip II*. 2 vols. London: Collins.

BUTLIN, R.A. 1993: *Historical Geography: Through the Gates of Space and Time*. London: Arnold.

CARTER, H. 1983: *An Introduction to Urban Historical Geography*. London: Edward Arnold.

CHADWICK, H. 1967: *The Early Church*. London: Pelican.

COLLINS, R. 1983: *Early Medieval Spain: Unity in Diversity, 400–1000*. Basingstoke: Macmillan.

CONSTABLE, A.R. 1994: *Trade and Traders in Muslim Spain: The Commercial Realignment of the Iberian Peninsula, 900–1500*. Cambridge: Cambridge University Press.

COSGROVE, D. 1993: *The Palladian Landscape: Geographical Change and its Cultural Representations in Sixteenth-Century Italy*. Leicester: Leicester University Press.

DELANO SMITH, C. 1979: *Western Mediterranean Europe*. London: Academic Press.

DENLEY, P. 1988: The Mediterranean in the age of the Renaissance, 1200–1500. In Holmes, G. (ed.), *The Oxford Illustrated History of Medieval Europe*. Oxford: Oxford University Press, 235–96.

DE PLANHOL, X. 1994: *An Historical Geography of France*. Cambridge: Cambridge University Press.

DODGSHON, R.A. 1987: *The European Past: Social Evolution and Spatial Order*. Basingstoke: Macmillan.

DUBY, G. 1984: *The Knight, the Lady and the Priest*. London: Allen Lane.

EPSTEIN, S.R. 1991: Cities, regions and the late medieval crisis: Sicily and Tuscany compared. *Past and Present* **130**, 3–50.

EPSTEIN, S.R. 1993: Town and country: economy and institutions in late medieval Italy. *Economic History Review* **46**, 453–77.

FENECH, D. 1993: East–west to north–south in the Mediterranean. *GeoJournal* **31**, 129–40.

FOSSIER, R. (ed.) 1989: *The Cambridge Illustrated History of the Middle Ages, 350–950*. Cambridge: Cambridge University Press.

HODGES, R. 1982: *Dark Age Economics: The Origin of Towns and Trade, AD 600–1000*. London: Duckworth.

HODGES, R. and WHITEHOUSE, D. 1983: *Mohammed, Charlemagne and the Origins of Europe*. London: Duckworth.

HOLMES, G. (ed.) 1988: *The Oxford Illustrated History of Medieval Europe*. Oxford: Oxford University Press.

JIMÉNEZ, M.G. 1989: Frontier and settlement in the Kingdom of Castile (1085–1350). In Bartlett, R. and MacKay, A. (eds), *Medieval Frontier Societies*. Oxford: Clarendon Press, 49–74.

KOSTOF, S. 1991: *The City Shaped: Urban Patterns and Meanings through History*. London: Thames and Hudson.

LE ROY LADURIE, E. 1974: *The Peasants of Languedoc*. Urbana. University of Illinois Press.

LOMAX, D.W. 1978: *The Reconquest of Spain*. London: Longman.

LOSEBY, S.T. 1992: Marseille: a late antique success story? *Journal of Roman Studies* **82**, 165–85.

MACKAY, A. 1977: *Spain in the Middle Ages: From Frontier to Empire, 1000–1500*. Basingstoke: Macmillan.

MACKAY, A. 1989: Religion, culture and ideology on the late medieval Castilian-Granadan frontier. In Bartlett, R. and MacKay, A. (eds), *Medieval Frontier Societies*. Oxford: Clarendon Press, 217–43.

MOORE, R.I. 1985: *The Origins of European Dissent*. Oxford: Basil Blackwell.

MORRIS, A.E.J. 1994: *History of Urban Form before the Industrial Revolutions*. London: Longman.

MORRIS, R. 1988: Northern Europe invades the Mediterranean, 900–1200. In Holmes, G. (ed.), *The Oxford Illustrated History of Medieval Europe*. Oxford: Oxford University Press, 175–234.

PANOVA, R. and GAVRILOVA, R. 1992: Bulgarian urban historiography. *Urban History* **19**, 257–68.

PARK, C.C. 1994: *Sacred Worlds: An Introduction to Geography and Religion*. London: Routledge.

PIRENNE, H. 1925: *Medieval Cities: their Origins and the Revival of Trade*. Princeton, NJ: Princeton University Press.

PIRENNE, H. 1939: *Mohammed and Charlemagne*. London: Allen and Unwin.

POUNDS, N.J.G. 1990: *An Historical Geography of Europe*. Cambridge: Cambridge University Press.

PRINCE, H. 1975: Fernand Braudel and total history. *Journal of Historical Geography* **1**, 103–6.

REYERSON, K.L. 1994: Montpellier and Genoa: the dilemma of dominance. *Journal of Medieval History* **20**, 359–72.

RUNCIMAN, S. 1951–4: *A History of the Crusades*. 3 vols. Cambridge: Cambridge University Press.

SMITH, C.T. 1978: *An Historical Geography of Western Europe before 1800*. London: Longman.

TILLY, C. 1994: Entanglements of European cities and states. In Tilly, C. and Blockmans, W.P. (eds), *Cities and the Rise of States in Europe, AD 1000 to 1800*. Boulder, CO: Westview Press, 1–27.

WALLERSTEIN, I. 1974: *The Modern World System I: Capitalist Agriculture and the Origins of the European World-Economy in the Sixteenth Century*. London: Academic Press.

7

THE OTTOMAN MEDITERRANEAN AND ITS TRANSFORMATION, *CA.* 1800–1920

LINDSAY PROUDFOOT

Although Fernand Braudel (1972, pp. 14, 134–8) could still insist on the 'unity and coherence of the Mediterranean regions' in the sixteenth century, that unity was already in a process of dissolution. By 1500 AD, the emergent hegemonic Ottoman state already controlled mainland Greece, Macedonia and Bulgaria, and over the next 100 years it succeeded in incorporating much of the eastern and southern shores of the Mediterranean into an imperial Islamic polity centred on Turkey. In so doing, the Ottomans created a power bloc which eclipsed the earlier Genoese and Venetian Empires and was only partly rivalled by the Spanish Hapsburgs in Iberia and southern Italy to the west (Fenech 1993). The tottering Mamluk sultanate which had dominated Syria, Egypt and western Arabia was overthrown by 1517, while Hungary and Wallachia were seized after the decisive battle of Mohacs in 1529. Rhodes was captured by the Turks in 1522, Chios in 1566 and Cyprus – after a prolonged siege of Famagusta and Nicosia – in 1573 (Braudel 1972, p. 76; Lewis 1995, pp. 114–29). In North Africa, Cyrenaica, Tripolitania, Tunisia and Algeria had all been absorbed by the Ottomans before 1566 (Fig. 7.1).

The Mediterranean thus emerged as a boundary zone between the Islamic and Christian worlds, but it was a boundary which was to prove to be far from static. Over the next 200 years, the initial seemingly irresistible expansion of Ottoman authority gave way: first, to an uneasy stalemate with the European powers in the seventeenth century; and then, in the eighteenth century, to the increasingly frequent defeat of the Ottoman armies at the hands of their Russian and European Christian adversaries. These military reverses signalled the beginning of a fundamental change in the balance of power between European and Islamic interests in eastern Europe and the Mediterranean, and this was reflected in the successive treaties signed between the Ottomans and their opponents. In 1606 at Sitvatorok, they negotiated with the Hapsburgs as equals for the first time. In 1699 at Carlowitz, it was the Hapsburgs who dictated the terms to the Ottomans after the latter's resounding defeat following the failure of their siege of Vienna in 1683. At Küçük Kaynarca in 1774, the Turks were forced to accept even more humiliating terms from the Russians. These allowed the Russians unprecedented commercial access

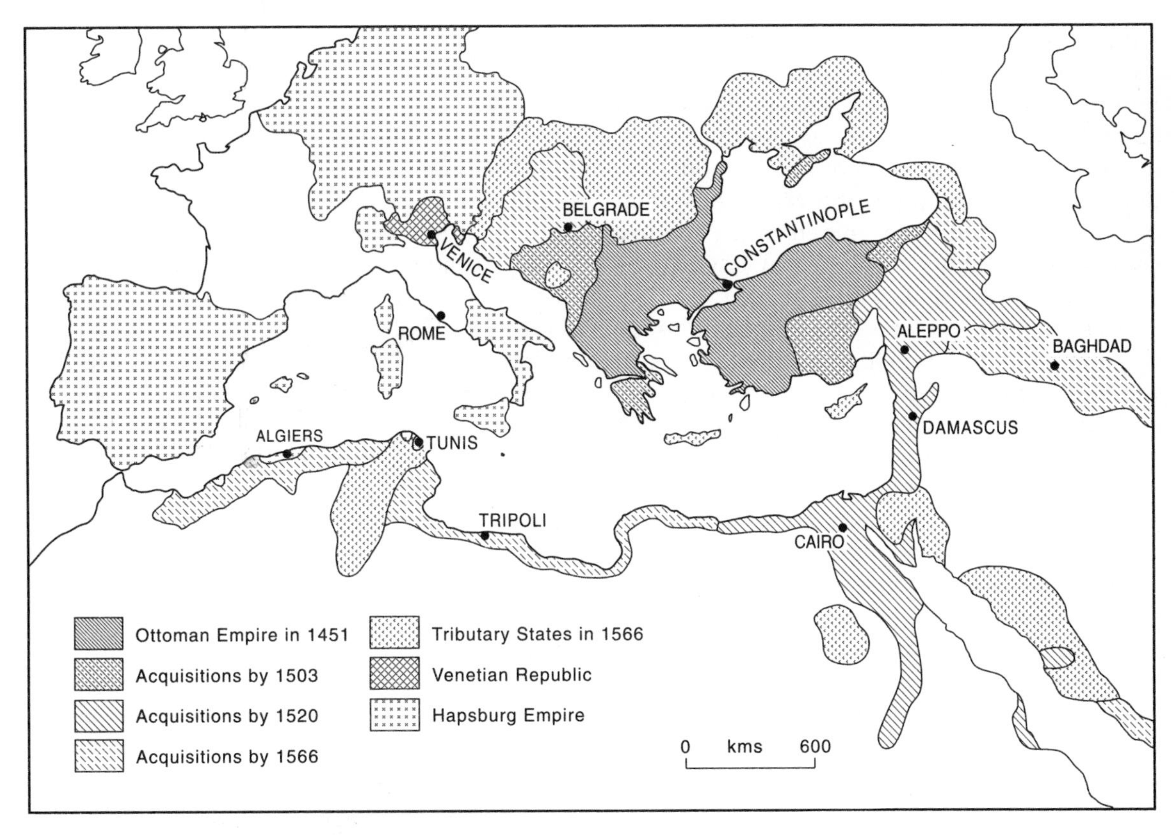

FIGURE 7.1 The expansion of the Ottoman Empire in the Mediterranean

Source: After Moore (1981, p. 86)

and cultural influence within Turkey itself, and paved the way for their rapid territorial annexation of the Crimea and other lands around the Black Sea over the next 20 years (Lewis 1995, pp. 276–9).

In Lewis's view, Küçük Kaynarca marked a turning point in relations between Europe and the Middle East. The treaty ushered in a period of accelerating westernisation, territorial fragmentation and internal reaction and reform which first transformed and then extinguished the Ottoman Empire by the end of the First World War. Previous tendencies towards political decentralisation had been encouraged by the empire's structure, but had been more or less accommodated by it (Hourani 1991, pp. 228–9). In the nineteenth century, by contrast, such changes were frequently externally driven, and arose in large part from the growth of European power and influence in the Mediterranean, either through trade, the encouragement of nationalism or outright colonisation. France and Britain were pre-eminent among these imperial protagonists, but Germany, Spain and Italy were also involved. Their ability to extend their Mediterranean interests in this way reflected the Eurocentric nature of the nineteenth-century capitalist world system, the technological advances which had allowed this to develop, and the economic, demographic, political and military power which northern European states in particular derived from their role as the system's 'core' (Andersen *et al.* 1993, pp. 54–5; Fenech 1993; Wallerstein 1974). On the other hand, the Ottoman Empire's own relative structural weakness, its cultural introspection, ecological fragility, consumerist attitudes, fragmentary forms of government and failure to develop an early coherent state identity, ensured that it was unable to keep pace with the West and thus unable to combat these encroachments effectively. Accordingly, the Empire moved inexorably from a position as an autarchic world power in the sixteenth century, to one of increasing peripheralisation and dependency within the world capitalist system by the end of the nineteenth century (Andersen *et al.* 1993, p. 42; Karpat 1988; Lewis 1994, pp. 28–9; McGowan 1994, p. 651).

This chapter explores the implications of these changes for the Ottoman possessions in the Mediterranean. The perspective is selective, and deliberately concentrates on events and issues which highlight the extent of the ideological and cultural discontinuities which existed at this time between the Islamic and Western worlds. Two themes predominate: first, the fragmentary and regionally diverse nature of the structures of empire within the Ottoman world at the beginning of the nineteenth century; and second, the ways in which these were modified by the processes of territorial and institutional adjustment during the nineteenth century.

THE OTTOMAN MEDITERRANEAN, *CA.* 1800

Government and society

By 1800, the Ottoman Empire was already showing signs of the territorial fragmentation and shrinkage that were to characterise it over the ensuing hundred years. In the north, around the shores of the Black Sea, Russia had made good its annexation of the Crimean Khanate, while to the west the neighbouring province of Podolya had been ceded to Poland. Further west again, Transylvania, comprising much of modern Hungary, had been lost to the Hapsburgs, giving them control of a zone running from Split on the Adriatic coast, across the upper Danube Basin to the upper Dniester in the east. In the far south, the Empire's control over its North African provinces had long been exercised at a distance through a series of increasingly autonomous local leaders centred at Cairo, Tripoli, Tunis and Algiers (Hourani 1991, pp. 220–9).

Everywhere, prior to the emergence of Turkish nationalism in the late nineteenth century, the Empire was socially rather than territorially constructed. It may be interpreted as a 'horizontal' social mosaic rather than as a 'vertically' structured class system (Yapp 1987, pp. 2–3). Various elements contributed to this mosaic. The ecological diversity of the physical environment determined in large measure the location, nature and importance of agriculture and pastoralism, and the sedentary or nomadic nature of the societies each supported (Owen 1981, pp. 25–32). These, in turn, were further differentiated on ethnic, linguistic and religious grounds, and stratified, particularly in the Arab provinces, on the basis of kinship, tribe and occupation (Andersen *et al.* 1993, pp. 3–11; Valensi 1977, pp. 13–24).

The form of Ottoman government mirrored this fragmentation. Consequently, it was characterised from the start by a tendency towards devolution and the growth of local administrative autonomy. Imperial revenues, for example, had from the earliest times been organised around a system of tax-farms, the *mukataat*, in which responsibility for tax collection was devolved either to financial *consortia* (in the case of major customs or bullion taxes) or – in the case of land taxes – to private landholders, *sipahis*, *ayans*, *pashas* and others (Inälcik 1994, pp. 64–6; Owen 1981, pp. 10–21). Nevertheless, in 1800 the Empire still retained its distinguishing characteristics as an agrarian state ruled by an absolute monarch (the sultan), who was supported by an elite class of military and bureaucratic slaves, the *janissaries*. These dominated the machinery of government until their abolition as a corps in 1826, and until then effectively prevented the formation of other privileged classes or loyalities opposed to Ottoman interests. Of equal importance in maintaining Ottoman authority was the sultans' success in allying themselves with the *ulema* or community of Islamic scholars. Traditionally, the *ulema* had frequently found themselves in the role of an informal religious opposition to the ruling dynasties, as they attempted to assert *Qu'ranic* values as a counterweight to the power of the political leadership, even though the latter upheld the *Shari'a* or Islamic law (Lewis 1995, pp. 186–91). By allying themselves with the *ulema* and acknowledging – in theory at least – their right to depose a sultan for religious reasons, the Ottoman sultans ensured the religious and therefore social legitimacy of their political authority within Islam.

Within the Empire the most important administrative unit for non-Muslims was the ethno-religious community or *millet*. This was non-territorial in conception, and consisted of all the members of a given confessional group throughout the Empire, who were allowed military protection, religious freedom and self-government under their respective religious leaders. Each *millet* was held collectively accountable for the crimes and debts of its members, including public debts arising out of any tax farms such individuals might have held (Inälcik 1994, pp. 190–1). The *millet* system had developed during the later fifteenth century as the Empire expanded to incorporate progressively larger non-Muslim and non-Turkic minorities. The most important were the Christian Orthodox *millet* (under the Greek Orthodox patriarch), the Armenian *millet* (which included all other non-Orthodox eastern Christians), and the Jewish *millet*, led by the chief rabbi. Unlike the Jewish *millet*, in which ethnicity, religion and a sense of community were more or less synonymous, the Christian *millets* were further divided along ethnic and linguistic lines which reflected the spatial organisation of their respective churches (Karpat 1988, pp. 39–44).

Although the *millet* system was extended to include the communities of European merchants who became established within the Empire during the later eighteenth century, the Muslim community itself was never formally included within it. The leadership functions of the chief *mufti* were performed by the Ottoman government, which although it did not identify itself specifically with the Muslim community prior to the mid-nineteenth century, nevertheless ruled according to the *Shari'a* (Karpat 1988, pp. 44–5). One consequence was that the *millet* system was much more visible in the northern parts of the Empire, in the Balkans and Anatolia for example, where relatively large non-Muslim minorities existed, than it was in Egypt or the Maghreb. Here, the important ethnic differences between Arabs, Berbers, Mamluks and other descendants of the *janissaries* who constituted the quasi-independent ruling dynasties, were subsumed beneath their shared and predominantly *Sunni* or *Sufist* Islamic faith.

Population, agriculture and rural settlement

The population served by these faiths was very unevenly distributed, and in parts of the Empire at least, may still have been in a spiral of plague and famine-induced decline which had begun during the seventeenth century (Faroqhi 1994, pp. 438–47). The lack of uniformly reliable and comparable population statistics for the period makes any attempt to estimate total figures either for the Empire or its provinces a matter of guesswork. Recent estimates for the period between 1800 and 1820 have put the Empire's population at between 26 and 30 million. Of this, Egypt may have accounted for 3.5 million, the three Maghreb provinces 4.5 million, Anatolia 6 million, and the European Balkan provinces, 9 million (Owen 1981, pp. 24–5; Quataert 1994, pp. 778–9; Yapp 1987, pp. 10–11).

In fact in 1800 the most heavily populated and productive region was the Balkans. Their subsequent loss during the 1820s and after not only imposed a severe strain on the Ottoman economy, but also accelerated existing migration flows of Muslim refugees from erstwhile Ottoman provinces into Anatolia and the Syrian borderlands. By 1913, an estimated 5 to 7 million refugees had moved in this way, and they contributed significantly to the decisive shift towards Islam which characterised confessional allegiances in the Empire during the nineteenth century. Indeed, the Empire's continuing contraction from an estimated 3 million square kilometres in 1800 to 1.3 million in 1914 masked the fundamental demographic change which occurred during the same period. As a result of gradually improving nutrition and health care and the progressive sectoral modernisation of the economy, the overall population grew for the first time since the sixteenth century, and achieved the same level, *ca.* 26 million, that the much larger Empire had supported in 1800 (Quataert 1994, pp. 777–95).

But in the early nineteenth century the determinants of population distribution remained very much what they had traditionally been. At least 80 per cent of the population was rural, and the demands made by sedentary agriculture for fertile soils and an assured water supply ensured that beyond the European provinces, the densest populations were found on coastal plains such as the Tunisian Sahel, along river valleys such as the Syrian Orontes or the Egyptian Nile, and, in drier regions, around oases. Although the agriculture, landholding and settlement patterns which supported these populations reflected regional variations in environmental and social conditions, certain structural characteristics recurred with sufficient frequency to suggest that they represented a 'norm' from which more localised variations might be measured. Thus dry farming, especially for winter cereals, predominated in all regions, but was supplemented wherever water supplies were adequate by more intensive and productive irrigated agriculture. These areas specialised in a wider variety of crops for both the export and domestic markets, including tobacco, grapes, sugar, rice and cotton.

Irrigation offered considerable productivity gains: yields might be eight times higher than those produced by dry farming, but irrigated areas such as those in the Nile Delta remained the exception rather than the rule. Generally speaking, agricultural productivity remained low, hindered by inefficient technology, intensive labour inputs yielding low *per capita* gains, the extreme fragmentation of peasant holdings in many areas, and the disincentive afforded by the relatively high levels of taxation demanded from the peasantry (Owen 1981, pp. 25–44; Quataert 1994, pp. 843–87; Valensi, 1977, pp. 27–9).

Although large estates, frequently worked by share-croppers, existed everywhere within the Empire, in Anatolia, the Balkans and parts of the Levant the most widespread unit of landholding was the peasant family farm or *çift-hane*. This comprised three elements: a unit of land workable by a team of oxen, the *çiftlik*; the family household which provided the labour; and the team of oxen. The settlement structures and field patterns associated with these *çift-hanes* reflected the cumulative effects of previous changes in the balance between population pressure and resources. Normally the land was designated as state or *miri* land, and the peasant enjoyed a status as a perpetual tenant with a hereditary right of possession in the male line. Each *çift-hane* also constituted the basic unit of land-tax assessment, and thus the peasantry were locked into the imperial fiscal system and linked to the structures of landownership by virtue of their payments to the local tax farmers, who were frequently also the local landowners (Inälcik 1994, pp. 155–78).

In the Arab provinces of the Maghreb the *çift-hane* system was less pervasive. The structures of land occupation were more diverse and reflected the varied cultural traditions and inherent tribalism of Maghreb society, as well as the constraints imposed by the equally varied agricultural environment. In some Berber districts, including parts of the Kybilia and Aurès mountain ranges in Algeria, systems similar to the *çift-hane* did develop, but on freehold or *milk* land. These agrarian landscapes were characteristically terraced, intensively cultivated and irrigated, and, in common with *milk* land elsewhere in the Empire, were devoted to horticultural and orchard crops. In other parts of the Algerian *tell* and Tunisian *sahel*, where Arab colonisation had been more extensive and where periodic water deficiencies were a serious problem, more extensive agricultural enterprises developed with a greater emphasis on herding, particularly of sheep, and irregular cereal rotations. In these areas, agrarian landscapes were less enclosed, and in the more sedentary districts of the *sahel*, supported settlement patterns based on *ksour* or fortified villages. These were essentially defended granaries which acted as low-order central places within each tribe's territory. Tribal membership gave each peasant access to plots of land which might amount to 10 or 15 hectares, and which he retained by virtue of his customary cultivation of them. Pasture, meadow and forest provided additional, communally owned resources, and this communality also found expression in the informal co-operative strategies frequently used by the peasantry to make good the deficiencies in their technological base (Valensi 1977, pp. 25–34).

Urbanism

Nowhere in the Empire were these rural economies completely autarkic. All were linked via trade or taxation to complex networks of inter-regional commodity flows which in 1800 still vastly outweighed the volume and value of Ottoman exports to Europe (McGowan 1994, pp. 723–39). The urban network articulated these flows. Throughout the Empire, the major towns and cities were either located on important caravan routes or on maritime trade routes. Consequently, the pattern and functions of Ottoman urbanism reflected both the consumerist and agrarian nature of the state. It also provided an arena in which the conflicting ideologies of Islam and the West could find material expression. In the Arab cities of the Maghreb in particular, the irregular traditional morphologies of the central *medinas* stand in eloquent contrast to the orthogonal European suburbs laid out during the colonial period of the later nineteenth century.

In 1800, however, these developments still lay in the future. The urban population accounted for around 15 to 20 per cent of the total, and thus by the standards of pre-industrial Europe, the Empire was relatively highly urbanised (Roberts 1979, p. 43). Moreover, it also showed signs of the urban primacy which Carter (1983) argues typified advanced pre-industrial economies in general. With a population of perhaps 750 000, Istanbul was unrivalled in size, and as the imperial capital, performed the sort of overarching governmental and political functions which Carter identifies as a necessary precondition for such primate status and which attracted a diverse range of ancillary functions and labour (Carter 1983, pp. 96–112). The other major urban centres of the period were considerably smaller. Cairo, the next largest, had a population of about 260 000, while the various other 'port cities' which were to come to prominence during the nineteenth century under the impact of intensifying European trade were smaller still, varying in size from 100 000 (Smyrna) to less than 15 000 (Alexandria and Beirut). On the North African coast, Algiers, Tripoli and Tunis varied between 50 000 and 120 000.

Despite this clear size differential between Istanbul and the other major cities, it is by no means clear whether they were fully integrated within a single functional imperial hierarchy by 1800. The near-autonomy of Egypt and the Maghreb provinces generated a level of regional primacy in cities such as Cairo and Tunis which was analogous to that seen on an imperial scale at Istanbul. Moreover, even within the core of the Empire, it is debatable whether conditions were conducive to the early and complete functional integration of the urban network as a rank-order hierarchy. Indeed, Tekeil (1972) has argued that while an integrated urban hierarchy existed in Anatolia during the sixteenth century, this disintegrated under the pressures of decentralisation over the following two centuries. Thus, given the social rather than territorial basis of the administration and its tendencies towards decentralisation, and the existence of the 'mosaic' of different ethnic and cultural groups which seemingly still lacked a cohesive sense of national identity, it may be more appropriate to view provincial Ottoman urbanism at this stage as a dynamic and loosely connected network of regional systems, rather than as a closely integrated imperial hierarchy.

The traditional forms of Islamic urbanism which existed within the Empire at this time are understood more clearly. While it is doubtful whether we may speak of 'the Islamic city' as a single entity, given the diverse cultural influences which modified Islamic urbanism in the Maghreb as compared, say, with India or Afghanistan, certain functional and morphological characteristics appear to have typified Arab cities at least (Costello 1977, p. 8; Hourani 1970, pp. 9–11). Central to our understanding of the Ottoman city, whether Arab or Turkish, is the precept that whatever its role as a market place or religious, administrative or defensive centre, it expressed these through an exclusively Islamic conceptualisation of society (Roberts 1979, pp. 36–42). Consequently, the same emphasis on social, ethnic and religious fragmentation which characterised the administration of the Empire found expression in the division of Islamic towns into residential quarters, each of which formed the basis of urban life for its inhabitants. The degree of functional separation between quarters could be striking, and it is clear that the concept of a corporate urban identity was alien to the Islamic city, despite the presence of a number of shared facilities, such as the main ('Friday') mosque and associated bazaar, baths and *madrasa* (religious school), and *caravanserai*. Each quarter was administered separately by its own *kadi* or *shaik*, who together with the *imam* of the local mosque, was responsible for the maintenance of law and order and the collection of taxes. The social cohesion provided by such shared responsibilities was reinforced in other ways, by the allegiance of the quarter's residents to a particular political faction, for example, but above all by the likelihood that they were members of the same ethnic, religious or occupational group (Lapidus 1969, pp. 49–50) Indeed, so strong were these internal identities, that the primary functional linkages for a particular quarter might be with similar communities in other cities or their own hinterland, rather than with neighbouring quarters in the same city.

This fragmented, cellular social ecology found clear expression – as was also noted in Chapter 6 – in the irregular morphologies which characterised the structuration of traditional Islamic urban space (Fig. 6.10). Each quarter usually contained at least one mosque, as well as an area devoted to craft production. Many were physically divided from their neighbours by walls and gates, a development which seems to have become more common in the period after 1500 (Wagstaff 1980, p. 23). Within each quarter, domestic residential space was given pre-eminence, reflecting the importance placed on the family in Islamic ideology. Houses were inward-looking and normally built around a central courtyard, and in their internal design reflected both the modesty of the Muslim domestic tradition and the need to ameliorate the extremes of the summer climate. Public space was correspondingly limited, its secondary importance symbolised by the characteristic narrowness and irregularity of the maze of communicating lanes and alleyways.

TRAJECTORIES OF CHANGE, *CA.* 1800–1920

The subsequent transformation of the Ottoman Mediterranean during the nineteenth and early twentieth centuries thus acted upon economic, social and political structures which were profoundly diverse. For purposes of discussion, two broad categories of change may be identified: those arising out of the Empire's own attempts at internal reform in the face of the growth of European influence; and those consequent upon the outright loss of territory, either as a result of nationalist secession, as in Greece and the Balkans, or European colonisation, as in the Maghreb. In reality, of course, both were expressions of the same fundamental peripheralisation of the Ottoman Empire within the world capitalist system, and the willingness and ability of the European 'core' states to exploit this to their own advantage.

Internal reform

The origins of the internal reforms may be traced to the late eighteenth century, and the growing penetration of European, particularly

French Revolutionary, ideas on social order, justice and equality. The mechanisms of penetration varied. The urgent need for reform of the Ottoman army in the face of the growing threat posed by the operations of European armies during the Napoleonic Wars, led to its reorganisation as a conscript and professional body. Trained on European lines by European, particularly French, advisers, its officer *cadre* was exposed to western ideologies which were as much philosophical as military. At the same time, Revolutionary France had not been above proselytising its ideals concerning *égalité* and *fraternité* among the Ottomans' European subjects, especially the Greeks, while Napoleon's brief occupation of Egypt between 1798 and 1801 also resulted in the penetration of modernising social, economic and political ideologies (Lewis 1994, pp. 35–8, 1995, pp. 305–21).

In all of these ways, the administrative structures of the Ottoman military–bureaucratic complex were exposed to reforming ideas, but for the first twenty or thirty years of the nineteenth century, the conservative vested interests within it, particularly the *janissaries*, were sufficiently strong to impede change. Owen stresses that at this stage, the immediate perception was of the need for military reform. Its economic consequences, the reform of government and taxation, and the encouragement of industrial and agricultural improvement, were secondary, and resulted from the need to enhance imperial revenues in order to achieve the desired military objectives, particularly following the loss of the relatively wealthy Balkan and Greek provinces (Owen 1981, pp. 57–62).

The effects of these early attempts at military and fiscal efficiency were far-reaching. They have been argued to have been partly responsible for the successionist movements in the Balkans. They were certainly a proximate cause of the revolt of the *janissaries* in 1826, following which the corps was destroyed by the sultan, Mahmoud II, and the way opened up for further military bureaucratic reforms and economic liberalisation. The *Tanzimat* or westernising reforms which followed were ushered in by the 1838 Anglo-Turkish Convention, which was signed under duress to gain British support against Egyptian expansionism, and given formal expression in the Imperial Rescript of Gülhane, promulgated by Abdülmecid I a year later. In their initial form they lasted until 1876, and the accession of the absolutist Abdülhamid II. He continued the programme of technological modernisation, but attempted to stem the tide of cultural Westernisation associated with it, and thus came to symbolise the resurgent Islamist–traditional consciousness which constituted one of the most potent socio-political developments of the period.

The *Tanzimat* reforms constituted a radical programme for the modernisation of the whole government project. Their primary objective was the reassertion of central governmental authority over the provinces and the modernisation of the economy and society through the encouragement of European trade and technology. The associated cultural changes involved everything from the promotion of Western literature to demands for gender equality. As the state took over many of the functions which had previously been delegated to semi-autonomous provincial administrations, there was an inevitable rise in the size of the bureaucracy, which may have numbered half a million by 1914. It is important to stress, however, that many of the long-term shifts in production encouraged by the *Tanzimat* had already begun in the late eighteenth century, as the Empire's increasing economic peripheralisation within the world system forced a switch from the export of uncompetitive manufactured goods to raw materials, and a consequent rise in the import of European manufactures (Quataert 1994, pp. 761–6).

Inevitably, the *Tanzimat* reforms led to various adjustments to the pattern of settlement and land-use. These were regionally uneven but important in both their symbolic and substantive content. Agriculture made significant gains, and continued to employ over 80 per cent of the Empire's population. By 1914 it still provided over 57 per cent of the 'national' income, but in many regions occupied as little as 5 per cent of the land area. Generally speaking, however,

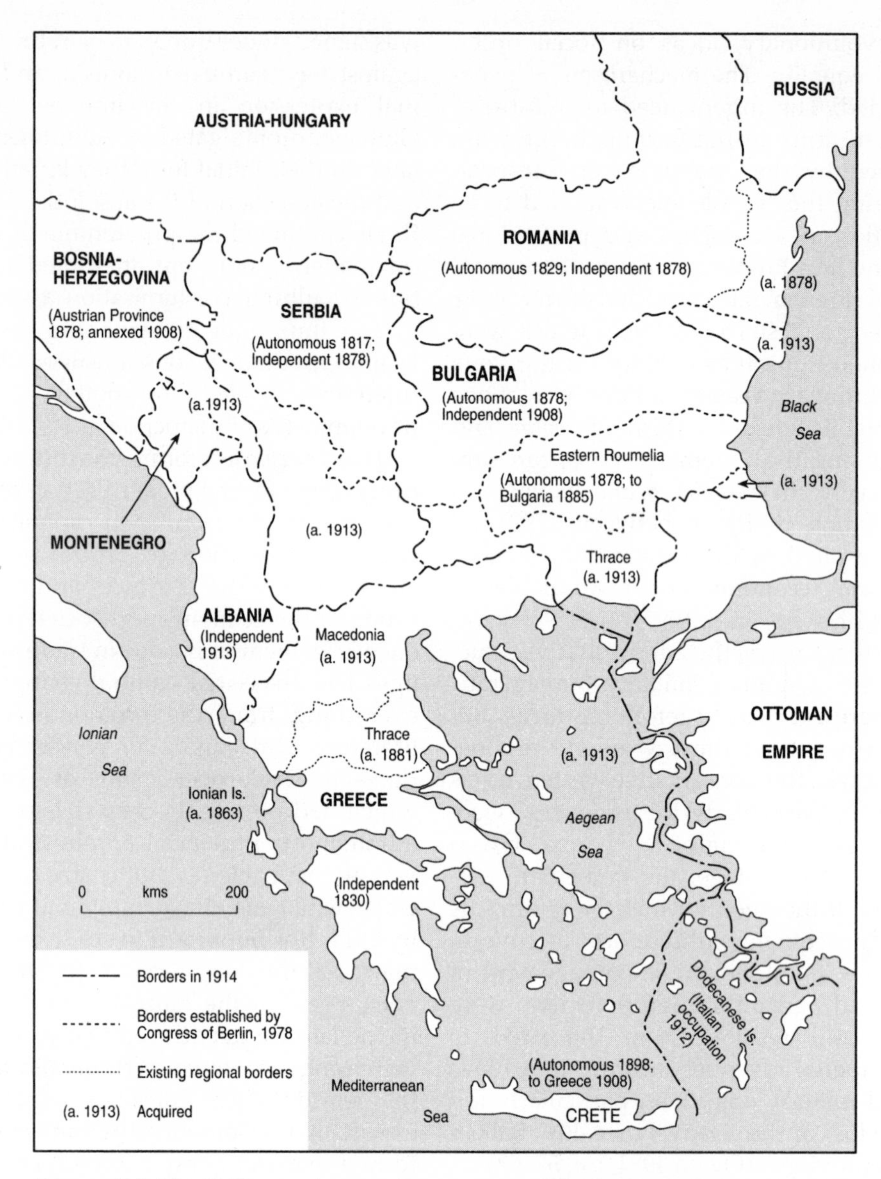

FIGURE 7.2 The Balkans: nineteenth-century political geography
Source: After Moore (1981, p. 121)

under the impact of the reforms the area under cultivation expanded to reach around 6.3 million hectares by 1909. The crop mix remained much as before, with the dry farming of cereals remaining the single most important sector. In the 1880s, over 82 per cent of all cultivated land was used in this way. Such statistics as do exist suggest that output was already increasing prior to 1860, largely in response to the rise in European demand and the prevalence of more peaceable domestic conditions. Thereafter, agricultural output may have doubled or even trebled, but this time in response to growing tribal sedentarisation, the continuing influx of Balkan refugees, and the growth in the domestic urban market.

These expansionist trends imposed their own regionally diverse patterns on the agrarian landscape. In Anatolia and northern Syria, for example, large numbers of Balkan refugees were settled in new *çift-hane* villages on the Anatolian plateau and in the upper Euphrates valley, although in some districts opposition from nomadic tribes anxious to preserve their traditional pastures was sufficient to forestall this. Similarly, the final pacification of Palestine after 1850 following local revolts against the higher taxes imposed under the reform programme led to similar settlements being established in the Jordan valley. Elsewhere, traditional structural constraints on agriculture continued to survive, but could not mask the increasing commercialisation of peasant agriculture, as growing numbers of peasants began to take advantage of improvements in communications to produce cash crops for market. Moreover, while levels of technology remained low, there were signs of improvement, as for example in the gradual increase in the use of fertilisers, iron ploughs and steam-powered threshing technology on some of the larger estates in Anatolia in the 1880s. Arguably, the most significant constraint on agricultural productivity remained one which the *Tanzimat* reforms had been most clearly directed against: the partial survival despite the 1858 Land Law of the earlier system of tax-farming landowners, whose local exactions impeded both peasant productivity and the government's tax receipts alike (Quataert 1994, pp. 843–87; Yapp 1987, pp. 20–9).

The reforms also affected the urban network. In a fundamental restructuring, port cities such as Alexandria, Smyrna (Izmir), Salonica, Trabzon and Beirut grew rapidly in response to the broader shifts in the Ottoman economy as it sought to meet the demands of European trade, and in doing so replaced older, inland cities such as Aleppo, Damascus or Homs as the major regional centres (Keyder *et al.* 1993). In many ways, Smyrna typified both the causes and limitations of this trade-led urban growth. Second only in size in Anatolia to Istanbul, Smyrna's population more than doubled during the nineteenth century to reach *ca.* 200 000, and included some 50 000 foreigners involved in the city's commerce. The basis of the city's wealth was its role as an entrepôt for western Anatolia, but the nature of its trade reflected the inherent weaknesses in the Ottoman economy. The major exports were all primary agricultural products: cotton, figs, raisins and dyes; the imports, all goods of European manufacture. Consequently, despite developing complex mercantilist social and commercial structures, neither Smyrna nor its fellow port cities offered an opportunity for the sort of economic growth which could break the dependencies which characterised the Empire's trading relationships with the west (Kasaba 1993).

At Istanbul, the changes were appropriately symbolic. Under the *Tanzimat* charter, the traditional Islamic pattern of local self-government by the city's districts was replaced by a western-style municipal corporation vested with city-wide authority. At the same time, in a pattern determined by Istanbul's numerous and frequently devastating fires, large areas on both sides of the Golden Horn were formally replanned on 'scientific' (i.e. regular) principles under European advice and direction (Çelik 1993, pp. 43–145). The symbolism was clear enough: the irregular morphologies encapsulating the traditional values of Istanbul's past were to be shed as the city reinvented itself as a modern, western capital. In the event, the incremental nature of the opportunities for reconstruction, as well as local opposition to the unilateral nature of the intended changes, ensured that none of the rebuilding projects achieved the overall coherence their protagonists intended.

Territorial fragmentation

Prior to its final dismantling in the years immediately following the First World War, the major territorial losses suffered by the Ottoman Empire occurred in the Balkans and the Maghreb. The significance of the territorial losses in the Balkans can hardly be overstated. Between 1812 and 1913, the Empire lost control successively of Bessarabia, Serbia (1817), Greece (in 1828, following the War of Independence which had begun seven years earlier), Moldavia and Wallachia (1856), and in 1878, Bosnia, Herzegovina and Bulgaria, as well as Cyprus and Crete. Finally, in 1913, following the Balkan War between the Ottomans and Greece, Serbia and Bulgaria, Albania declared its independence and Macedonia was ceded to Greece (Fig. 7.2).

These provinces had formed the economic core of the Empire, but their cultural, ethnic and religious complexity had always constituted a potentially destabilising factor. Not surprisingly, therefore, the prime motivation for these successionist movements appears to have been the assertion of regional nationalist identities, which if they shared nothing else, were united in their non-Islamic and non-Turkic character. The growth of Turkish nationalism in the late nineteenth century can only have encouraged these trends, but they are likely also to have been exacerbated by the growing influence of Western liberal ideas and the increasingly onerous financial exactions which the more centralised reformed Ottoman government attempted to impose.

The economic consequences of this 'Balkanisation' varied. They have already been noted to have led to major refugee movements and loss of revenues for the Empire, each of which have been shown to have had longer-term consequences for agriculture and the structure of government. In the newly autonomous or independent states the pattern of subsequent events varied. In Bulgaria, for example, the declaration of autonomy in 1878 coincided with a cereals- and tobacco-led boom in agricultural exports which funded the extensive peasant purchase of lands formerly owned by the departing Islamic *sipahis*. In Serbia, by contrast, the traditional highly fragmented landownership pattern survived unchanged. In other respects, the pace of Balkan modernisation exceeded that of the Empire. By 1914, for example, Romania, Bulgaria, Serbia and Greece had between them built over 8 000 kilometres of railway, more than in the entire Empire itself. In Greece, these infrastructural improvements formed part of a remarkable programme of national reconstruction which had been necessitated by the devastation caused during the War of Independence, and which included the planning of 174 new towns and cities between 1828 and 1912 (Hastaoglou-Martinidis *et al.* 1993; Quataert 1994, pp. 805, 852, 864).

In Egypt and the Maghreb, the process of fragmentation was in some ways more straightforward, and the loss of these provinces had fewer economic consequences of any significance for the Ottoman Empire. The relatively high degree of autonomy they enjoyed within the Empire at the beginning of the nineteenth century has already been noted. Egypt, especially, experienced an economic boom under its early nineteenth-century governor, Muhammad Ali Pasha (1805–49), who sought to turn the province into an independent principality. Following the end of the French occupation of the country in 1801, Egypt was ruled by a succession of unstable Mamluk dynasties until Muhammad Ali's seizure of power. In a development which can only be described as an attempt to create a military–industrial complex, Muhammad Ali modernised the Egyptian army and created a long-staple cotton monoculture and manufacturing industry to generate the wealth this required (Batou 1993). In the short term, this modernisation succeeded sufficiently to allow Muhammad Ali to engage in a programme of territorial expansionism in the Levant and Arabia which posed a real threat to the survival of the Ottoman Empire. It was only the intervention of the European powers, particularly the British, who in the face of French territorial ambitions had no interest in unnecessarily weakening the Ottoman Empire, that prevented Muhammad Ali from succeeding. Nevertheless, he did create a dynasty which continued to rule Egypt until the 1920s, first as a virtually independent state, and then, following the British invasion in 1882, as a protectorate and colony (Hourani 1991, pp. 282–4).

The transition to British rule in Egypt followed a pattern which had already been set by the French in Algeria in 1830 and Tunisia in 1881. In each case, continued attempts at modernisation, for example in improved communications such as the Suez Canal (opened in 1869), or in agricultural restructuring, proved to be insupportable in the long term by the characteristically underdeveloped local economies. Increasing European investment coupled with the growing indebtedness of the local administrations led inexorably to further fiscal involvement by the sponsoring European powers and, ultimately, outright annexation. A similar sequence of events obtained in Morocco (Abun-Nasr 1971, pp. 235–50).

The geographical transformations which followed these colonial interventions were particularly profound in the French dominions. In Algeria they arose out of the creation of a settler or *colon* state, and its gradual and grudging incorporation into metropolitan France. Following their initial invasion of Algiers in 1830, the French were drawn increasingly and unwillingly into Algerian affairs, and it was not until the late 1840s that the decision was taken to proceed with complete colonisation. This was based on mass immigration by peasant colonists, generally French but also including some Spanish and Italians, many of whom were officially sponsored by the French government. Demographically speaking, the policy was successful. By 1901 there were over 364 000 French citizens as well as 184 000 non-French Europeans in Algeria; between them, they represented approximately one-eighth of the total population of around 4 million.

The new colonists required land, and despite the misgivings of the initial military administration, tribal lands were sequestered and made available for colonial settlement, destroying the ecological and social balance of traditional agriculture. Neither the French civilian authorities in the colony nor the metropolitan government in Paris showed much awareness of the complex nature of Muslim land tenure. The most sought-after land lay in the northern Mediterranean zone, or *tell*, where annual rainfall of around 400–600 mm permitted sedentary agriculture. Of the 14 million hectares in this zone, 4.5 million were freehold or *milk* land; 5 million were tribal lands for which village elders paid taxes and tithes; 1.5 million were in state control but given to tribes in return for military service; and 3 million hectares consisted of forests and grazing land owned by the state but used for tribal pastoralism.

From the outset, French policy was designed to facilitate the transfer of this land for private and official colonisation as speedily as possible. By 1851, over 250 000 hectares of state and tribal land had been sequestered, and was used to provide a fund of free land for official colonists, who by this time numbered over 10 000. Despite the temporary introduction of a more conciliatory land code in 1863 designed to preserve Muslim interests, the weakness of the metropolitan government following the Franco-Prussian War and the growth in the agricultural export market, particularly for wine, ensured that the demand for land, and its acquisition, continued. By the 1920s, official colonisation amounted to 1.5 million hectares and private colonisation to 1 million, while 3.1 million hectares of taxable tribal land had been acquired as 'state domains' (Abun-Nasr 1971, pp. 249–58; Issawi 1982, pp. 138–40). In Tunisia, the pattern was fundamentally different. After the French Protectorate was established in 1881, would-be colonists were required to purchase their own land, leading to the creation of a smaller but better capitalised settler class, within an overall population that probably did not exceed 1.9 million (Abun-Nasr 1971, pp. 280–5; Amin 1970, p. 33). European settlement was concentrated in the north-west around Bizerte, Sfax and in the Mejerdha valley, and most land was left in native hands. Accordingly, while both countries saw the 'Europeanisation' of the agrarian landscapes in the colonial districts by the construction of new farms, villages and communications infrastructures, this imprint was most extensive, if no more deeply rooted, in Algeria (Karabenick 1991).

The urban network was similarly transformed. As in Morocco and, following the Italian invasion of 1911, Tripolitania and Cyrenaica (Segrè 1974), this involved both the extension of existing Islamic centres such as Algiers or Tunis, and the foundation or refoundation of new towns such as Bône, Oran and Philippeville in Algeria (Mièrge 1985). The level of urbanisation varied. By 1910, 16 per cent of Algeria's population were town dwellers compared to 10 per cent in Tunisia, but everywhere urban growth was concentrated on the coast. This locational shift paralleled the similar changes that had begun rather earlier in the Ottoman provinces of the eastern Mediterranean and occurred for the same reasons. As colonies, the Maghreb states served as a source of readily available land and cheap agricultural exports for Europe, and the location, functions and structure of the newly founded or extended towns and cities were designed to facilitate this. As administrative and commercial centres they oversaw the progressive subordination of the colonial economies to the requirements of the colonising power. Every centre supported a large European community, who in the case of the Algerian cities, formed a majority of the population. European values imbued city life, and found material expression in their formal replanning and new architecture. In Tunis, for example, orthogonally planned suburbs were laid out to the east of the old *medina*, partly on the site of the old Christian ghetto and partly on land reclaimed from the city's lagoon. Suitably equipped with the symbols of imperialism, triumphal avenues and dominant cultural icons (for example, the cathedral of St Vincent de Paul, built in 1882), the area aptly signalled the creation of European rather than Islamic cultural space, and the assertion of European, rather than Ottoman, legitimacy.

THE END OF EMPIRE

By the eve of the First World War, the transformation of the Ottoman Mediterranean was virtually complete. The territorial losses of the past century had proved permanent, while the processes of modernisation, although temporarily hindered by the Empire's *de facto* bankruptcy in 1875 and the accession of the culturally hostile sultan Abdülhamd II a year later, were asserted with renewed vigour following the Revolution of 1908 and the accession to power of the reformist Young Turk administration. Where their policies might eventually have led is unclear. In the event, the First World War intervened, and the Ottomans, having allied themselves with Germany, were defeated. Their remaining possessions in Arabia and the Levant were mandated by the victorious Great Powers, who also encouraged the formation of the Kemälist secular Turkish state. European imperialism had replaced Ottoman imperialism, but the challenges posed by the seemingly irresolvable conflict between nation, state and identity remained. The stage was set for the creation of the modern geography of the eastern Mediterranean.

REFERENCES

ABUN-NASR, J.M. 1971: *A History of the Maghrib*. Cambridge: Cambridge University Press.

AMIN, S. 1970: *The Maghreb in the Modern World: Algeria, Tunisia, Morocco*. London: Penguin.

ANDERSEN, R.R., SEIBERT, R.F. and WAGNER, J.G. 1993: *Politics and Change in the Middle East: Sources of Conflict and Accommodation*. Englewood Cliffs, NJ: Prentice Hall.

BATOU, J. 1993: Nineteenth-century attempted escapes from the periphery: the cases of Egypt and Paraguay. *Review* **16**, 279–318.

BRAUDEL, F. 1972: *The Mediterranean and the Mediterranean World in the Age of Philip II*. 2 vols. London: Collins.

CARTER, H. 1983: *An Introduction to Urban Historical Geography*. London: Edward Arnold.

ÇELIK, Z. 1993: *The Remaking of Istanbul: Portrait of an Ottoman City in the Nineteenth Century*. Seattle: University of Washington Press.

COSTELLO, V.F. 1977: *Urbanization in the Middle East*. Cambridge: Cambridge University Press.

FAROQUI, S. 1994: Crisis and change, 1590–1666. In Inälcik, H. with Quataert, D. (eds), *An Economic and Social History of the Ottoman Empire, 1300–1914*. Cambridge: Cambridge University Press, 411–636.

FENECH, D. 1993: East–west to north–south in the Mediterranean. *GeoJournal* **31**, 129–40.

HASTAOGLOU-MARTINIDIS, V., KAFKOULA, K. and PAPAMICHOS, N. 1993: Urban modernization and national renaissance: town planning in 19th century Greece. *Planning Perspectives* **8**, 427–69.

HOURANI, A.H. 1970: The Islamic city in the light of recent research. In Hourani, A.H. and Stern, S.M. (eds), *The Islamic City: A Colloquium*. Oxford: Bruno Cassirer, 9–24.

HOURANI, A.H. 1991: *A History of the Arab Peoples*. London: Faber and Faber.

INÄLCIK, H. 1994: The Ottoman state: economy and society, 1300–1600. In Inälcik, H. with Quataert, D. (eds), *An Economic and Social History of the Ottoman Empire, 1300–1914*. Cambridge: Cambridge University Press, 9–379.

ISSAWI, C. 1982: *An Economic History of the Middle East and North Africa*. London: Methuen.

KARABENICK, E. 1991: A postcolonial rural landscape: the Algiers Sahel. *Association of Pacific Coast Geographers Yearbook* **50**, 87–108.

KARPAT, K. 1988: The Ottoman ethnic and confessional legacy in the Middle East. In Esman, M.J. and Rabinovich, I. (eds), *Ethnicity, Pluralism and the State in the Middle East*. Ithaca, NY: Cornell University Press, 35–53.

KASABA, R. 1993: Izmir. In Keyder, C., Özveren, Y.E. and Quataert, D. (eds), Port-cities of the Eastern Mediterranean 1800–1914. *Review* **16**, 387–409.

KEYDER, C., ÖZVEREN, Y.E. and QUATAERT, D. 1993: Port-cities in the Ottoman Empire: some theoretical and historical perspectives. In Keyder, C., Özveren, Y.E. and Quataert, D. (eds), Port-cities of the Eastern Mediterranean 1800–1914. *Review* **16**, 519–57.

LAPIDUS, I. 1969: Muslim cities and Islamic societies. In Lapidus, I. (ed.), *Middle Eastern Cities*. Berkeley: University of California Press, 47–74.

LEWIS, B. 1994: *The Shaping of the Modern Middle East*. Oxford: Oxford University Press.

LEWIS, B. 1995: *The Middle East: 2000 Years of History from the Rise of Christianity to the Present Day*. London: Weidenfeld and Nicolson.

MCGOWAN, B. 1994: The age of the *Ayans*, 1699–1812. In Inälcik, H. with Quataert, D. (eds), *An Economic and Social History of the Ottoman Empire, 1300–1914*. Cambridge: Cambridge University Press, 637–742.

MIÈRGE, J.L. 1985: Algiers: colonial metropolis (1830–1961). In Ross, R.J. and Telkamp, G.J. (eds.), *Colonial Cities*. Boston, MA: Martinus Nijhoff, 171–9.

MOORE, R.I. (ed.) 1981: *The Hamlyn Historical Atlas*. London: Hamlyn.

OWEN, R. 1981: *The Middle East in the World Economy 1800–1914*. London: Methuen.

QUATAERT, D. 1994: The Age of Reforms, 1812–1914. In Inälcik, H. with Quataert, D. (eds), *An Economic and Social History of the Ottoman Empire, 1300–1914*. Cambridge: Cambridge University Press, 759–933.

ROBERTS, M.H.P. 1979: *An Urban Profile of the Middle East*. London: Croom Helm.

SEGRÈ, C.G. 1974: *Fourth Shore. The Italian Colonization of Libya*. Chicago, IL: University of Chicago Press.

TEKEIL, I. 1972: On institutionalised external relations of cities in the Ottoman Empire – a settlement model approach. *Études Balkaniques* **8**, 49–72.

VALENSI, L. 1977: *On the Eve of Colonialism: North Africa Before the French Conquest*. London: Africana.

WAGSTAFF, J.M. 1980: The origin and evolution of towns, 4000 BC to AD 1900. In Blake, G.H. and Lawless, R.I. (eds), *The Changing Middle Eastern City*. London: Croom Helm, 11–33.

WALLERSTEIN, I. 1974: *The Modern World System I: Capitalist Agriculture and the Origins of the European World-Economy in the Sixteenth Century*. London: Academic Press.

YAPP, M.E. 1987: *The Making of the Modern Near East 1792–1923*. London: Longman.

8

POLITICS AND SOCIETY IN THE MEDITERRANEAN BASIN

NURIT KLIOT

INTRODUCTION

As Chapter 1 pointed out, the Mediterranean has been variously portrayed as a region of homogeneity or one of heterogeneity. Braudel (1972, p. 14) characterised the Mediterranean as 'the sum of its routes in which the essence of the region is the product of intellectual and commercial intercourse' – a functional definition. Branigan and Jarrett (1975, pp. 3–4) emphasised the unifying influence of 'the landlocked sea ... on the countries which border it or come within its scope'. The homogeneous features which unify the various parts of the region were then, and still are now, the Mediterranean Sea and climate, the Mediterranean coastal landforms and vegetation and, perhaps, the Mediterranean agricultural and rural way of life. Although the agrarian way of life is vanishing and in places has been replaced by tourism, agriculture still plays an important role in the Mediterranean, as we shall see in Chapter 13. On the other hand, the area is characterised by ethnic, cultural and religious diversity; it is very heterogeneous in its political regimes and political alignments and in its economic and social features.

In 1981 the region had 19 countries which were divided into four main subgroups: countries which were part of the European core area; countries of the European periphery; the Arab states; and the socialist states. Little similarity was to be found in the economies, organisational networks, societies and political systems of these Mediterranean lands (Kliot 1981, pp. 347–8). In 1997, the number of Mediterranean littoral countries is 21 as a result of the disintegration of Yugoslavia, and the Mediterranean area is facing new challenges. This chapter will review the various factors and processes – social, political and economic – that shape the Mediterranean region at the present time. The chapter will start with a short historical review and will then survey, first, the homogeneous and functional factors which exercise a unifying effect on the area as a whole and, second, the disintegrating factors, which tend to split the region. Finally, four cases of unresolved conflicts which endanger Mediterranean stability at large will be presented.

THE HISTORICAL DEVELOPMENT OF THE MEDITERRANEAN

Let us start by picking out a few key features from the rich historical geographies presented in the last three chapters. The Mediterranean region is known as the cradle of human civilisation, being the birthplace of the ancient Egyptian, Greek and Roman cultures. The last of these was extremely important in the development of Mediterranean trade and commerce and the spread of its civilisation all over the Mediterranean (see Chapter 5). After the break-up of the Roman Empire, a period of decline and political, social and economic regression began; from the fourth century onwards there were continuous incursions into the Western Mediterranean lands by barbarian tribes, followed, in the seventh and eighth centuries, by the armies of Islam which conquered most of the Levant, North Africa and Iberia. In the fourteenth century the Ottoman Empire was established and its impact continued to be felt in the eastern Mediterranean until the early twentieth century, as we saw in Chapter 7. Of great significance in the history of the Mediterranean lands is the fact that none of these invading peoples, with the exception of the Moorish Arabs, had a well-developed culture of their own to replace that which had been destroyed. Only within the past two centuries have the Mediterranean lands begun to re-attain the dignified position they held in ancient times.

From the earliest times of antiquity until the late Middle Ages and the Renaissance, the principal determining factor in the evolution of the Mediterranean Basin was the strengthening of intra-Mediterranean relations, which despite their conflicting nature, emphasised the region's unity (Amin 1989, p. 2). Then, during the following four centuries, the main movement was in the other direction: shattering the Mediterranean region through centrifugal attractions. The Mediterranean became the frontier zone between the new centre of Europe and the new Afro-Asian periphery. The era of colonisation in the Mediterranean brought new external influences to the region and established the foundations for the North–South confrontation. From the destruction of Napoleon's fleet at Trafalgar until 1945, Britain ruled the Mediterranean and occupied Egypt, Cyprus, Malta, Gibraltar and Palestine. France dominated the Maghreb, Syria and Lebanon, and Italy controlled Libya for a while. By the 1950s almost all the former colonies had become independent states, mostly preserving special economic and political relations with the metropolitan countries. The years 1945–70 were years of accelerated economic growth in the European Mediterranean countries, especially Spain, Portugal, Greece, Italy, and France (Williams 1984, p. 4). Economic growth was accompanied by political changes: dictatorship ended in Greece and Portugal in 1974, in Spain in 1977. The effect of their membership of the European Community then became enormously significant for these three countries.

The North African states became independent during the post-war decades: first Libya in 1951, then Morocco and Tunisia in 1956 and finally Algeria, after a long struggle, in 1962 (Drysdale and Blake 1985). All three former French colonies retained extensive commercial ties with France and hundreds of thousands of Maghrebi migrant workers found employment there. The other facet of the North African national identity is Muslim and Arabic, and the North African states keep close economic and political ties with other Arab countries. Lebanon and Syria are also more connected to the Arab Muslim world than to the European core, though the two countries retained special relations with France, and Lebanon continues to do so. The Middle Eastern countries within the Mediterranean were particularly exposed to the East–West superpower competition for dominance in the region. USSR and NATO competition in the Mediterranean manifested itself in the military bases for both naval and air forces that the two antagonists acquired for themselves in the region. Algeria, Libya and Syria served the Soviet navy, and Soviet army personnel were stationed in these countries. NATO had bases in Spain, France, Italy, Greece and Turkey, and the British kept military bases in Cyprus. Egypt did not provide military bases for the USSR but became an ally of the Soviet Union in the 1960s when President Nasser was seeking Soviet aid to develop the Egyptian economy and for the construction of the Aswan Dam.

In addition, Albania and Yugoslavia, each with its own version of socialism, were either non-aligned or pro-Eastern bloc in their policies. Thus, in the 1970s and 1980s differences in historical evolution, variation in social systems, and the differing political affiliations turned the Mediterranean Basin into a region with very high tension levels and extremely diverse and often contradictory forces which were pulling all the parts in opposing directions. The most recent set of developments, namely the disintegration of the USSR, the fall of socialism as a state religion in the Eastern bloc, the weakening of Russia as a world power, and the emergence of the USA as a single superpower, are again changing the fortunes of the Mediterranean. However, the disappearance of East–West rivalry, followed by less chance of military confrontation between East and West, was accompanied by greater North–South cleavages and by exacerbation of regional conflicts such as the chaos in former Yugoslavia. Some of these recent historical developments will be reviewed in the next two sections of this chapter as they almost certainly point to the future of the Mediterranean in the next century.

MEDITERRANEAN UNITY

A variety of social, political and economic processes still coalesces the Mediterranean Basin countries and reinforces its homogeneous and functional nature as one region. We shall present unifying factors in agriculture and tourism, in commerce and the environment and, most importantly, the unifying effect of the EU.

Agriculture or, more accurately, Mediterranean farming and the rural way of life, was considered in the past as a common denominator in Mediterranean societies. Table 8.1 shows that the Mediterranean is still fairly agricultural and rural in character, with an impressive contribution to the GDP and to employment. The rural population proportion is very high, even in the more industrialised societies such as Italy, Greece, Cyprus and Spain. The adherence to Mediterranean crops such as olives, grapes and wine, citrus and grains, is reflected in the high position enjoyed by Mediterranean countries in the world production of these crops. It seems appropriate to conclude that Mediterranean societies are still unified by their allegiance to Mediterranean agriculture and the rural way of life.

On the other hand, Mediterranean patterns of trade and tourism are less supportive of Mediterranean regional cohesiveness. Table 8.2 shows that most Mediterranean countries do have, among their most important trade partners (for both export and import) at least one other Mediterranean country, but all these partners are EU members, and it is clear that it is their EU membership, rather than their Mediterranean location, which makes them a desired trade partner. Also significant in trade relations are former colonial ties: France is extremely important in the trade of Morocco, Tunisia, Syria and Lebanon – all former colonies. Similar trade ties exist between Libya and Italy and between Malta and the UK. Cultural and political ties within the Arab Muslim Middle East shape trade relations between Mediterranean Arab and Islamic countries: Lebanon trades with Syria and Saudi Arabia, Syria with Turkey and Turkey with Saudi Arabia. The bulk of the trade of the Mediterranean members of the EU is with other countries of the EU. Qualitative differences are also to be noted in trade flows moving in different directions within and across the Mediterranean Basin. Thus, the commercial exchanges from south to north assure the latter of the major part of its energy supplies. The flow from north to south is comprised mainly of foodstuffs and producer goods (Amin 1989, p. 10).

Another area which reflects probable Mediterranean internal (but also external) connectedness is tourism, considered in more detail in Chapter 14. The top earners amongst Middle Eastern Mediterranean countries are Egypt and Syria. Egypt, and to a lesser extent, Syria, were countries which were most visited by European tourists, but more recently this trend has changed because of terrorist attacks on European tourists in Egypt. Arab Middle Eastern tourists tend to

TABLE 8.1 Rurality and agriculture as indicators of Mediterranean unity, *ca.* 1990

Country	Rural population (%)	Employment in agriculture (%)	Agriculture's contribution to total GDP (%)	World ranking in agricultural production
Albania	65	48	36	
Algeria	48	24	13	Olives 9
Bosnia	No data	4	No data	
Croatia	No data	5	No data	
Cyprus	47	13	6	Grapefruit 8
Egypt	53	40	17	Cotton 9
				Lemons 6
				Oranges 8
France	26	5	4	Apples 5
				Barley 4
				Beef 6
				Grapes 2
				Peaches 6
				Wheat 5
				Wine 1
Greece	38	24	17	Nectarines 3
				Olives 3
Israel	8	4	3	Grapefruit 7
Italy	31	9	4	Beef 10
				Grapes 1
				Lemons 1
				Olives 1
				Oranges 6
				Wine 2
Lebanon	16	8	9	
Libya	30	13	6	
Macedonia	No data	8	No data	
Malta	13	2	3	
Morocco	52	36	19	Olives 5
Slovenia	No data	3	6	
Spain	22	10	4	Barley 6
				Grapes 4
				Lemons 5
				Olives 2
				Oranges 4
				Wine 3
Tunisia	46	23	18	Olives 6
Turkey	39	47	11	Apples 4
				Barley 7
				Cotton 7
				Grapes 6
				Olives 4
				Wheat 7
Yugoslavia-Serbia	(Former Yugoslavia 44)	5	No data	

Sources: Europa World Yearbook. London: Europa Publications, 1993
Philips Geographical Digest. London: Heinemann, 1994
World Resources 1992–1993. New York: Oxford University Press, 1992

TABLE 8.2 Trade and tourism as indicators of unity in the Mediterranean countries

Country	Main trading partners: Imports	Main trading partners: Exports	Main sources of tourists	Membership and association with EU
Albania	Italy, East Germany, Czechoslovakia	Italy, Czechoslovakia, Bulgaria		Commercial agreement with EC
Algeria	France, USA, Italy	Italy, USA, France	Morocco, Tunisia, France, Italy	Commercial agreement
Croatia	Germany, Italy, USSR	Germany, Italy, USSR	Germany, Italy, UK, Austria	
Cyprus	France, Germany, Greece	Egypt, Germany, Greece, Italy	UK, Scandinavia, Germany, Lebanon, Greece	Association agreement
Egypt	USA, Germany, France, Italy	Italy, USSR, Japan, USA	OECD countries, Arab countries	Commercial agreement
France	Germany, Italy, Belgium	Germany, Italy, Belgium, UK	Germany, Belgium, Italy, Switzerland, EC countries, Yugoslavia, USA	EC member
Greece	Germany, Italy, France	Germany, Italy, Egypt, UK	EC countries, Yugoslavia, USA	EC member
Israel	USA, Belgium, Germany	USA, Japan, UK, Germany		Commercial agreement
Italy	Germany, France, Netherlands, UK	Germany, France, UK, USA	Germany, Switzerland, France, Yugoslavia	EC member
Lebanon	Italy, France, USA, Germany	Saudi Arabia, Syria, Kuwait		Commercial agreement
Libya	Italy, Japan, Germany, UK	Italy, France, Greece, Spain		
Malta	Italy, UK, Germany, France	Italy, Germany, UK, Libya	UK	Association agreement
Morocco	France, Spain, UK	France, Spain, Italy	Arabs, France, Spain, Germany	Commercial agreement
Spain	Germany, France, Italy, USA, UK	France, Germany, Italy, UK	France, Portugal, Germany, UK	EU member
Syria	France, Germany, USA, Turkey	Italy, France, USSR, Saudi Arabia	Lebanon, Jordan, Iran	Commercial agreement
Slovenia	Germany, Italy, France	Germany, Italy, France	Germany, Austria, UK	Trade agreement
Tunisia	France, Italy, Germany	France, Italy, Germany	Libya, Germany, France, Algeria	Commercial agreement
Turkey	Germany, USA, Italy, Saudi Arabia	Germany, USA, UK, Italy	Germany	Association agreement
Yugoslavia-Serbia	Germany, USSR, Italy	Germany, USSR, Italy	Germany, USSR, Italy	

Note: For trade and tourism partners, countries are listed in ranked importance.
Sources: Hunter (1993); *Europa World Yearbook*. London: Europa Publications, 1993

visit, first of all, other Arab states, then the UK, Germany, Italy and France. The top destinations for European–Mediterranean tourists are other European countries. Tourists from France visit Spain, the UK, and Italy in that order. Only 0.8 per cent and 8.1 per cent visit Middle Eastern and North African countries, respectively. Italian tourists (87 per cent of them) visit other European countries – the most important being France and Spain. Spanish tourists visit Portugal, France, Italy and the UK (World Tourism Organization 1994).

Two more elements draw the Mediterranean together. The Conference on Security and Cooperation in the Mediterranean (CSCM) was proposed in 1990 by ten countries on opposite shores of the Mediterranean: France, Spain, Italy, Egypt, Cyprus, Malta, Libya, Algeria, Morocco and Tunisia (Farley 1994). The CSCM is building bridges of a political, economic and strategic nature between the north and south within the Mediterranean. Second, the Mediterranean region has been at the forefront of international environmental cooperation for nearly 20 years. The Mediterranean Action Plan or 'Med Plan' is the framework for coordination, research and monitoring programmes concerning environmental pollution of the Mediterranean Sea (Lempert and Farnsworth 1994). By 1990 18 Mediterranean states had signed the Barcelona Convention, the foundation for the 'Med Plan'. The Med Plan succeeded in bridging differences between countries of the north and the south, and between traditional enemies such as Israel and her neighbours. Haas (1990) has argued that the Med Plan was largely achieved by cooperation between the scientific elites of the member countries but that accomplishments in pollution reduction are modest. The most important factor in the success of the Mediterranean Action Plans are EU regulations which compel France and Italy, the major polluters, to adhere to EU environmental regulations and reduce their waste disposal into the Mediterranean Sea. For a more detailed discussion see Chapter 17.

In conclusion, the evidence for the thesis of Mediterranean unity is mixed. Trade, tourism and the continuing importance of agriculture provide patchy support for the basic contention. The EU's role as a unifying factor, expressed, for example, through trade and tourism movements, has less to do with intrinsic qualities of Mediterraneanism than with economics and geopolitics. A shared concern for the environment and security cooperation does bind the region together, but at a rather low level of intensity.

DIVERSITY AND DIVISION IN THE MEDITERRANEAN REGION

Beyond the unifying factors lie deep divisions between the Mediterranean countries – gaps and social cleavages which are difficult to bridge. As has been pointed out, the region's strategic importance and its religious, cultural and ethnic plurality have always served as destabilising factors and at least one war and three unresolved conflicts continue to afflict the region. At present, the most important geopolitical division is the North–South divide which has replaced the East–West struggle for power. Other cleavages are socio-economic and religious–ethnic cultural divides. Each of these will now be discussed separately.

Geopolitical features of the Mediterranean

In a world of bi-polarity, Cohen (1973, 1982) defined two major geostrategic realms that geopolitical regions either lie within or are caught between. The two realms were the Trade Dependent Maritime World which included Anglo-America and the Caribbean, Western Europe, and Africa south of the Sahara; and the second was the Eurasian continental power based on the USSR and Eastern Europe. The Mediterranean European countries and the Maghreb were included in the Trade Dependent Maritime World, but the Middle East, Libya and Egypt fell within shatterbelts – areas caught up in major power conflict (Cohen 1973). Also, the Mediterranean region contained several nations which either aspire to or are perceived as having second-order power status – power rank second only to the superpowers. Among such countries were Israel, Yugoslavia, Turkey, Egypt, Algeria, Spain and Morocco (Cohen 1982, p. 233). Using the Correlates of War Dataset compiled by Small and Singer (1982), van der Wusten and Nierop (1990) found that shatterbelts have the highest incidence of both civil and inter-state wars as well as militarised disputes. Between 1960 and 1990 Turkey and Egypt were engaged in war for four years, Algeria five, Cyprus one, Syria more than a year, Lebanon eleven, and Israel at least a year while shorter periods of conflict and terror were found in other shatterbelt states. More recently, the war between Iran and Iraq, the Iraqi invasion of Kuwait, and Operation Desert Storm which followed, and the Balkan crisis have all had an immediate effect on the Mediterranean region.

The most important global development was the disintegration of the Soviet Union and its disappearance as superpower and leader of the communist bloc. The implication for the Arab client states (Libya, Algeria, Syria) was a strong sense of desertion (Gazit *et al.* 1993, p. 8). The Arab countries in the Middle East discovered that they were no longer as important as they had believed themselves to be in the past and found that the Western world was no longer as influenced by their activities as it once was. The threat of an oil embargo, for example, could not dictate major strategic decisions in the 1990s. Global strategy was now directed from one single centre, Washington, even if the USA did, at times, prefer to work through the UN in order to lend legitimisation to its policies (Operation Desert Storm, sanctions against Libya, the recent war in Bosnia). The USA exercised its power in the Mediterranean Basin through the NATO Alliance, in which other Mediterranean countries such as Spain, Italy, France, Greece, and Turkey are members, though France and Spain are not fully integrated into the military structure of the Alliance (Aliboni 1992, p. 3).

NATO military forces took part in Operation Desert Storm and were involved in the war in Bosnia. In addition to NATO involvement in Mediterranean security, UN Peace-Keeping Forces are set up in southern Lebanon and Kurdistan; they separate Israel from Syria, and Turkey from Iraq. Until the crisis in the Balkans, many of the threats to Mediterranean security and stability came from the *South*: Iran–Iraq, the invasion of Kuwait, Israeli Palestinian–Arab conflict, and the old conflict between Turkey and Greece which culminated in the Turkish invasion of Cyprus. Since the end of the Cold War the Europeans (NATO, for example) have shifted their attention from the East (or central front) towards the southern and eastern Mediterranean.

Implicit in this shift is the belief that future threats to Western security will emanate from different quarters in the 1990s (Farley 1994). One of the many consequences of this change has been Southern Europe's increased willingness to step up its initiatives towards its southern neighbours, especially in the Maghreb (de Vasconcelos 1992, p. 21). Spain and France tend to concentrate on Morocco, Italy on Libya (and Malta) and France also on Algeria and Lebanon.

On the negative side, because of the collapse of the USSR, and its replacement by 15 independent republics, huge stocks of weapons and military hardware became available for sale (Gazit *et al.* 1993, p. 9). Middle Eastern and North African countries (including Iran) dedicated enormous budgets to purchasing military hardware. In 1988, Algeria was importing 10.6 per cent of the world's total arms import, Egypt 3.1 per cent, Israel 12.6 per cent, Libya 11.5 per cent, Morocco 1.9 per cent and Turkey 5.4 per cent. More recently, the Middle Eastern countries purchased further military hardware for very large sums of money. In 1991 and 1992 Algeria invested $875 million in weapons, Egypt $4.1 billion, Israel $6.3 billion, Lebanon $230 million, Libya $1.4 billion, Morocco $1.4 billion and Tunisia $331 million (Gazit *et al.* 1993). Defence spending in the late 1980s ranged around 10–20 per cent of total GNP in most of these countries.

A final and important point is Bouchat's

(1994) observation that the main cause for Mediterranean instability and insecurity is that North–South battles are being fought *within* sovereign states – examples being the conflicts in Lebanon, Cyprus, Yugoslavia and the Israel/Palestinian confrontation. These four cases will be debated later.

The social and economic gap within the Mediterranean

Table 8.3 shows that the divide between north and south across the basin is very real. This must be regarded as a preliminary table; the next chapter contains a more detailed discussion of socio-economic data and the evolution of uneven development. In synthesis, Table 8.3 shows that the north has modest population growth, life expectancy ranging 5–10 years longer than nations of the south, GNP per capita that is 5–10 times higher and much lower infant mortality. Some countries like Albania and the new emerging states of former Yugoslavia are north-oriented in their social features, but their incomes are lower than typical northern incomes.

The most critical problems of the south are its continuing population explosion and high rates of unemployment. These give rise to the population problems and migration pressures discussed in Chapter 11. Already by 1982 France had nearly half a million North African migrant workers (322 000 Algerians, 170 000 Moroccans and 77 000 Tunisians) and ten years later it hosted a total Islamic population of 2.6 million (Peach and Glebe 1995). Less immigration, both legal and illegal, is clearly the objective of European policy, especially in the Mediterranean EU countries. However, there is growing understanding in the EU that aid to develop the economies of the South in order to expand local employment opportunities is the only way to try to solve the problem of illegal migration. This developing political and economic relationship between the EU and the southern Mediterranean is the special theme of Chapter 10.

Political and ethnic diversity

Mediterranean countries are affiliated to many different organisational and political frameworks (Table 8.4). The southern European nations and Turkey are affiliated to European core organisations such as the Council of Europe, OECD and NATO. All the Muslim countries are affiliated to Arab organisations such as the Arab League, the Arab Monetary Fund or Arab Cooperation Council. It is particularly characteristic of the region that the countries of the Mediterranean are affiliated to three (continental) UN organs: the Economic Commission of Europe, the Economic and Social Commission of Western Asia and the Economic Commission of Africa. Outside of the UN-type organisations in which almost all are members, only two frameworks exist in which North and South cooperate; these are the Mediterranean Action Plan and the Conference on Security and Cooperation of Europe, as already noted.

Part of the explanation for this organisational fragmentation is provided by the data in Table 8.5 which points to the major cultural and ethnic divisions in the Mediterranean countries. Except for Libya, Tunisia and Malta, all the Mediterranean states have various religious and linguistic minorities. Some of these groups were politically strong enough to start a secession often accompanied by terror and war. The Turkish Cypriot minority, with the assistance of Turkey, seceded into its small state in north Cyprus; in Lebanon, decades of civil war and ethnic rivalry ended with *de facto* occupation by Syria and Israel; Yugoslavia has been torn by war in Bosnia; and the Basque separatists still launch sporadic attacks in Spain. Some of the ethnic minorities are relatively large: at least one-quarter of Algeria's population is comprised of Berbers, 18 per cent of the population of Israel is Arab, 40 per cent of the population of Morocco is Berber, one-fifth of the population of Spain is Catalan and 8 per cent are Basques. Some of the minorities, even when they are relatively small in their proportion to the population, are politically mobilised and participate in what is viewed by the central government as

TABLE 8.3 The Mediterranean North–South gap: some socio-economic indicators

	Average annual population growth 1990–95 (%)	Life expectancy (years)	Infant mortality (per 1000 live births)	GNP per capita $US (1991)
North				
Cyprus	0.37	79	10	8640
France	0.25	73	7	20600
Greece	0.76	74	13	6530
Israel	0.09	74	10	12293
Italy	negative	73	9	18580
Malta	0.16	72	9	6850
Spain	negative	74	9	12460
South				
Albania	0.84	75	32	1000
Libya	3.47	63	68	5500
Turkey	2.05	65	62	1793
Algeria	2.71	66	67	1991
Egypt	2.20	62	57	611
Morocco	2.40	63	68	1033
Tunisia	2.06	56	43	1504
Syria	3.58	57	39	1141
Lebanon	2.00	65	34	2500
Bosnia	No data	No data	15	1800
Croatia	0.70	No data	11	No data
Slovenia	No data	No data	8	7150
Yugoslavia (Serbia-Montenegro)	No data	69	23	No data

Sources: Hunter (1993); *World Resources 1994–95*. New York: Oxford University Press and the World Resources Institute, 1994

subversive activity. The Kurds in Turkey, Basques in Spain and Corsicans in France are just a few examples (Kliot 1989, pp. 154–5). However, most of the current research points to the rise of fundamentalist Islam as the major threat to Mediterranean security.

Extreme Muslim regimes in the Sudan and Iran support extremist Islamic groups in the Middle East and even in Europe. Iran and Sudan promote clandestine and terrorist activities in Egypt and Lebanon. In Algeria, the victory of the Islamic Salvation Front led the Algerian government to cancel the second round of elections and to impose a general state of emergency. More recently, the Algerian extremists have exported their terrorist activities into France. Islamic groups are gaining power also in Turkey and Jordan, and the Islamic movement poses the strongest opposition to the Tunisian regime (Kam 1993). In Gaza and the West Bank fundamental Palestinian Muslim groups objecting to the peace negotiations between Israel and the Palestinians launch frequent attacks on Israeli targets – and sometimes also on moderate Palestinian targets. Nor is opposition to peace talks lacking on the other side, as Rabin's assassination by a fanatical Jewish student illustrate. The four case studies presented in the following section will point to the destructive effect of ethnicity in its various forms on the stability of the Middle East.

UNRESOLVED CONFLICTS

War and ethnic conflicts in the Balkans

In addition to recent and current ethnic strife,

TABLE 8.4 Mediterranean countries: organisational frameworks

Country	ACC	AL	AMF	ECE	ESCWA	ECA	CE	CSCE	MAP	OECD	AMU	BC	NATO
Albania				+				+	+				
Algeria		+	+			+			+		+		
Bosnia				+				+					
Croatia				+				+					
Cyprus				+		+	+	+	+			+	
Egypt	+	+	+		+		+		+				
France				+				+	+	+			+
Greece				+			+	+	+	+			+
Israel									+				
Italy				+			+	+	+	+			+
Lebanon		+	+		+				+				
Libya		+	+			+			+		+		
Malta				+			+	+	+			+	
Morocco		+	+			+			+		+		
Slovenia				+				+					
Spain							+	+		+			+
Syria		+	+		+				+				
Tunisia		+	+			+			+		+		
Turkey				+			+	+	+	+	+		+
Yugoslavia			+				+	+	+				
Serbia													
Montenegro								(Former Yugo-slavia)	(Special status – former Yugo-slavia				

Notes: ACC=Arab Cooperation Council; AL=Arab League; AMF=Arab Monetary Fund; ECE=Economic Commission of Europe; ESCWA=Economic and Social Commission of Western Asia; ECA=Economic Commission for Africa; CE=Council of Europe; CSCE=Conference on Security and Cooperation of Europe; MAP=Mediterranean Action Plan; OECD=Organization for Economic Cooperation and Development; AMU=Arab Maghreb Union; BC=The British Commonwealth; NATO=North Atlantic Treaty Organization

Sources: Banks (1991); Hunter (1993); *Europa World Yearbook*. London: Europa Publications, 1993

the Balkans have many possible bases for future conflicts. First in line is Albania and its relations with its neighbours. Albania is a small country but every second Albanian lives outside the country – in Kosovo, Montenegro, Macedonia or Greece (Lendvai 1994). This diaspora shapes Albania's foreign relations with Serbia, Greece and especially Macedonia, where around one-third of the population is Albanian. To complicate matters, Greek politicians launched a nationalist campaign against the Albanian treatment of the estimated 300 000 ethnic Greeks who live in 'Northern Epirus'. Greece raised the spectre of a territorial claim on southern Albania by insisting that Albania should be willing to grant ethnic Greeks the same rights it demands for the 2 million ethnic Albanians in Kosovo. The Albanian response was a call to halt the 'Hellenisation' of southern Albania. The Albanian media has alleged a 'secret deal' between Greece and Serbia on southern Albania, Kosovo and Macedonia (Biberaj 1993).

Macedonia, which declared its independence in the early 1990s, is another source of tension in the Balkans. Greece's veto prevented EU recognition of Macedonia until 1993 when

TABLE 8.5 Ethnic and cultural diversity in the Mediterranean

Country	Population
Albania	3.39 million. Albanians 82%; Ethnic Greeks 9% (1990)
Algeria	28.58 million. Arabs 78.5%; Berbers 19.4%; Europeans 1%
Bosnia	4.3 million. Moslems 45%; Serbs 30%; Croats 17%; Others 8%
Croatia	4.2 million. Croats 80%; Serbs 11–12%
Cyprus	0.71 million. Turkish Cypriots 80%; Greek-Cypriots 19%
Egypt	58.5 million. Sunni Muslims 94%; Christian Copts 6%
France	57.77 million. French 62.5%; Occitans 20%; Bretons 7%; Alsatians 3%; Catalans 0.5%; Basques 0.4%; Corsicans 0.4%
Greece	10.25 million. Greeks 98%; Turks and Albanians less than 1%
Israel	5.88 million. Jews 81.9%; Arabs 18.1% (Muslim 14.0%, Christian 2.1%, Druze 1.7%)
Italy	57.91 million. Italians 84%; Sards 3%; Friulians 1.5%; German-speaking Tyrolese 1%; Slovenians 0.2%
Lebanon	3.03 million. Shi'ite 32%; Sunni 21%; Druze 6%; Alawis 1%; Maronite 21%; Greek Orthodox 8%; Greek Catholic 5%; Armenians 4%
Libya	5.41 million. Arabs 92.3%; Berbers 4%; Africans 2.5%; Europeans 1.2%
Macedonia	2.03 million. Macedonians 64%; Albanians 25%; Turks 4%; Gypsies 3%; Others 4%
Malta	0.36 million. Maltese 99%
Morocco	28.26 million. Arabs 59.6%; Berbers 39.5%; European 0.9%
Slovenia	1.98 million. Slovenes 90%
Spain	39.28 million. Castilians 60%; Catalans 21%; Basques 8%; Galicians 11%
Syria	14.78 million. Sunni 78%; Alawaite 12%; Druze less than 1%; Kurds 8%; Armenians 0.6%
Tunisia	8.93 million. 95% Arabs (Sunni)
Turkey	62.03 million. Turks 76%; Kurds 22%; Armenians less than 1%; Greeks 0.1%
Yugoslavia, Serbia, Montenegro	9.71 million. Serbs 68%; Albanians 18%; Montenegrins 5%; Hungarians 4%; Moslems 4%; Croats 1%

Sources: Fischer (1980); Ulmert (1986); Hunter (1993)

it was admitted to the UN (Fenske 1993). Greece was opposed to the existence of a country on her northern border bearing the name of an ancient Greek province, because that newcomer (Macedonia) might dispute the border between the two countries. Albania strongly supported the independence of the former Yugoslav Republic of Macedonia (FYROM) but relations continue to be marred by disagreement regarding the status of ethnic Albanians in Macedonia. Macedonian extremists fear an 'Albanian Action' against their country (Biberaj 1993).

Other sources of tension in the region are the large Hungarian and Muslim minorities in Yugoslavia–Serbia, the Turkish minority in Greece, and Greek Orthodox Church property in Istanbul. The major point of conflict between these two NATO allies is Cyprus, which will be discussed separately.

There is no doubt, however, that Mediterranean and European peace has been, and perhaps remains, in danger because of the bloody disintegration of Yugoslavia. Ex-Yugoslavia is a good example of the contradiction between respect for borders and national self-determination. The first factor suggests that Yugoslavia should have stayed whole, while the second implies that Slovenia, Croatia, Bosnia and Macedonia all had a perfect right to secede (Fenske 1993). The major reasons for the falling-apart of Yugoslavia are: (a) political conflicts and weakening of the regional communist parties; (b) the deteriorating economic conditions and the widening economic gap between the members of the federation; (c) the erosion in the strategic importance of Yugoslavia at the end of the Cold War; (d) ethnic strife and lack of national cohesion; and (e) external political factors, most notably the encouragement of Slovene and Croatian secession by Germany (Burg 1993; Hall 1994; Thomas 1994).

We may put different weight on some of the above-mentioned factors, but the most important one, no doubt, is the lack of Yugoslav cohesion and *raison d'être*. The proportion of population that declared itself to be 'Yugoslav' rather than an ethnic identity (Serb, Albanian, Slovene, etc.) in the national censuses was always small: 1.3 per cent in 1971; 5.4 per cent in 1981 (Burg 1993). The various ethnic groups, especially the Serbs and Croats, developed their national identities as nations without a state under the Hapsburg Empire and the Bosnians considered themselves to be an elite under the Ottoman Empire. Some researchers question the haste with which the principle of self-determination was applied to former Yugoslavia and point also to the territorial scale in the application of the principle of self-determination. They ask, why can the principle of national self-determination be granted to Slovenes, Croats, Bosnian Muslims and Macedonians, but not to the Serbs in Bosnia and Croatia (Thomas 1994)?

There is also no doubt that ethnic–national identity was reinforced by the economic disparity in the level of development among Yugoslavia's republics. The more developed areas, Croatia and Slovenia, were in the north; Bosnia, Montenegro, Macedonia and Kosovo, the less-developed areas, were in the south. The best indicators for this economic disparity are the unemployment rates which by the mid-1980s reached 56 per cent in Kosovo, 27 per cent in Macedonia and 24.5 per cent in Montenegro (Hall 1994, p. 132). Slovenia and Croatia refused to carry the burden of the less developed areas and complained that most of the income that was generated in their areas was siphoned through Belgrade for the benefit of the South (Bouchat 1994, p. 130).

Finally, the haste with which the EU rushed to recognise Slovenian and Croatian secession arguably also affected the rapid disintegration of Yugoslavia. In retrospect, the internationalisation of the conflict in former Yugoslavia appears to have consistently accelerated the violence rather than contained it (Remington 1993). The secession of Slovenia was relatively smooth but Croatia's birth was accompanied by a seven-month war in 1991 and 1992 in which Croatian forces fought against the combined forces of the Yugoslav army and Serb paramilitia. The war left Croatia with considerable civilian and military casualties, and damage to at least 10 per cent of the economic capacity or an estimated $123 billion (Cviic, 1993). The war in Bosnia claimed 200 000 lives, left 2 million homeless and more than 1 million Bosnian refugees in former Yugoslavia and outside the country (Remington 1993). This war, for which the ugly term 'ethnic cleansing' was coined, has been one of the most vicious and cruel conflicts of modern times. Ethnic cleansing (be it in ex-Yugoslavia, the former Soviet Union, Cyprus or Lebanon) is the sustained suppression by all means possible of an ethnically or religiously different group with the ultimate aim to expel or eliminate it altogether (Ahmed 1995, p. 7). And, indeed, Bosnia has now been divided up according to principles of ethnic control: some areas remain under Bosnian rule whereas others are integrated into Serbia and Croatia.

The partition of Cyprus

The current problems of Cyprus originate in the Ottoman conquest of the island. With the Ottoman Empire came Turkish immigrants who settled on an island which was Greek and Hellenic. For several hundred years Greek and Turkish Cypriots lived in peace, but when Greece (under Ottoman rule) began its struggle for independence from the Ottoman Empire, Greek Cypriots identified with Greek nationalism. In 1878 Turkey handed the island over to British administration which continued for 80 years. Until the 1950s only minor incidents of inter-communal violence were reported (Crawshaw 1978), but between 1955 and 1959 inter-communal conflict erupted in Cyprus for the first time. It arose, on the one hand, from the Greek Cypriot struggle for independence from British rule, and on the other, from Turkish and Greek Cypriot ethnic clashes. The two communities collided on their basic notions on the future of Cyprus. Many Greek Cypriots wanted *Enosis*, unification with Greece, whereas the Turkish Cypriots at first wanted to remain a British colony but later advocated *Taksim*, partition of the island between Greek and Turkish entities (Attalides 1979, p. 77).

The compromise which emerged from the conflict of 1955–59 was that Cyprus would become an independent state with a consociational constitution. Consociationalism, which is also practised in Lebanon, is a political regime in which each ethnic group is assigned positions in the parliament, judiciary and government according to its proportion in the population. This type of constitution is intended to secure minority rights in government, but it worked neither in Lebanon nor in Cyprus. In 1963, three years after Cyprus gained its independence, the first constitutional crisis took place, followed by civil intercommunal war which ended in 1964 with the intervention of the UN and the stationing of UN forces in the island to separate the warring communities. The 1963–4 war led many Turkish Cypriots to desert mixed villages in which they had been settled for many generations and move to larger villages and towns where Turkish Cypriot militias could defend them. Altogether, 42 Turkish Cypriot-held enclaves were created in the early 1960s and no government forces were allowed into them. The Turkish Cypriots developed an alternative administration which acted as government for the enclaved Turkish Cypriots (Kolodny 1971; Patrick 1976). Turkish and Greek Cypriots never lived together after this period.

Examination of the Turkish and Greek Cypriot communities before independence and afterwards shows that they failed to develop a common base of cooperation or communal *raison d'être*. Turkish and Greek Cypriots differ in every characteristic: religion, language, culture, ethnic origin and allegiance to different mother countries – Turkey and Greece. Both communities celebrated the national holidays of their own mother country by raising Greek or Turkish flags – but only rarely the Cypriot flag (Loizos 1988; Volkan 1979). Their failure to evolve a sense of community was evidenced by their failure to bridge their differences in the 1960s and 1970s. The intervention of Greece and Turkey on behalf of their respective communities did not help to moderate the hostility between the two communities.

Responding to the action of the Greek junta which deposed the moderate Archbishop Makarios as Cypriot leader, in 1974 Turkish forces invaded the northern side of the island and occupied 38 per cent of its territory, turning 180 000 Greek Cypriots into refugees in their own country. Since then Cyprus – including the capital Nicosia – has remained partitioned and UN forces (UNFICYP) have been monitoring the 180-km-long buffer zone between north and south Cyprus. In the last 20 years the north has passed through a process of 'Turkification'. Turkish settlers were brought to the Turkish occupied area and they allegedly constitute up to half of the North Cyprus population. All signs of Greek or Hellenic culture in the north, from place names and signs to churches, were eradicated or ruined (King and Ladbury 1981), replaced by Turkish monuments, slogans and icons (see Fig. 8.1).

The north was proclaimed a 'state' but the Turkish Republic of North Cyprus is recognised only by Turkey. All efforts at confidence-building and finding some kind of compromise in the conflict have failed. The prospect of confederation between the two entities is anathema to the Greek Cypriots since it conjures up fears that they will be swamped by mainland Turks as a first step in a Turkish take-over of the whole island. Turkey's strategic interests are best served by the Turkish zone already existing in the north (Crawshaw 1994). Negotiations have failed in the past because the underlying objectives of the Cypriot leaders are diametrically opposed to each other and at present it does not look as if this is going to change.

FIGURE 8.1 Since the partition of Cyprus in 1974, slogans in the Turkish north of the island stress the Turkish identity of this area. The text on this militaristic decoration on the Turkish side of the dividing wall that runs through Nicosia reads 'Let my blood colour my shroud and let my shroud be equal to my flag'.

Lebanon: war, disintegration and occupation

Lebanon is not dissimilar to Cyprus. In the 1990s Lebanon is not a truly sovereign state since most of her territory is either occupied by Syria or controlled by Israel. Lebanon became a pluralistic society as a result of the various waves of occupation and settlement which transformed it over successive generations. It was occupied by the Arabs, the Crusaders, the Mamluks, the Ottomans and finally, in the twentieth century, by the French. The first wave of sectarian conflicts took place between 1840 and 1860, when the Druze fought against Christians and the Druze and Maronites fought each other and engaged in 'ethnic cleansing' long before this expression was coined for the war in Yugoslavia. These sectarian conflicts resulted from the invasion by ethnic groups of territories which were held by other ethnic groups. The pattern of mingling, mainly of Christian Maronites and Druze who tended to settle together (similar to Turkish and Greek Cypriots), made them potential hostages for each other in times of trouble. The ethnic pattern became more complex when France, as the Mandatory Government, extended the territory of Ottoman Lebanon to incorporate the coast, which was settled by Sunni Muslims who were tied to the Syrian Arabs. In 1943 a 'National Pact' or unwritten agreement to define the balance between Christians and Muslims was made between the Maronite President and the Sunni Prime Minister. This formalised the practice of allocating the top political offices on a sectarian basis (Hitti 1957; Hudson 1968). Lebanon became independent in the late 1940s and in 1948 received a new addition to its already complex population structure: 150 000 Palestinians fled from Palestine and entered Lebanon. In 1958 there was a civil war between Lebanese Sunnis and the Maronites, which ended in the same manner as the 1858–60 conflict, with foreign intervention. The 1958 conflict contained sectarian, socio-economic, and external elements, and these features are common denominators in all four case studies of ethnic conflicts presented in this chapter. The third sectarian Lebanese civil war took place in 1975–6 and brought with it the virtual collapse of the Lebanese state. This was reflected in the disintegration of the government and the army as well as in the intervention of the Syrian army. The elements which led to the war were the sectarian issue and the creation of a Palestinian mini-state in southern Lebanon by the PLO, based on the Palestinian refugee camps there.

The central government of Lebanon lost control over its national territory and was replaced by Maronite and Druze mini-states, in addition to the Palestinian mini-statelets (Rabinovich 1984). Another issue which led to the 1975–6 war was the demand from Muslim groups for political reforms in which the consociational constitution would be changed and more power allotted to the Muslims who had grown considerably in number and felt themselves to be underrepresented in all branches of government. This was particularly true for the Shi'ites in southern Lebanon who felt underprivileged economically, socially and politically. The war ended with the occupation of most of the Lebanese territory by the Syrian army. In southern Lebanon, Israel actively intervened in order to suppress Palestinian attacks on its territory. In 1978 Israel invaded Lebanon in the 'Litani Operation' which was aimed at eliminating all hostile populations in an area extending from the Israeli border to the Litani River. An Israeli-supported militia was established in the south in order to prevent terrorist attacks on Israel. Also, the UN Interim Force (UNIFIL) was deployed in order to police the south. In 1982 Israel again invaded Lebanon in an operation aimed at destruction of the PLO infrastructure in south Lebanon (Rabinovich 1984). As a result, Palestinian military forces were evacuated from Lebanon, and Israel withdrew its forces from almost all the occupied areas. Shi'ite forces replaced Palestinian forces in launching terror attacks on Israel.

It is important to note that the 1975–6, 1978 and 1982 wars were accompanied by massive internal migration involving, perhaps, one-third of the population of Lebanon. Both Christians and Muslims fled from isolated and besieged regions to areas where their own communities predominated. As a result, the areas of sectarian exclusivity expanded significantly and continued to do so even after the cessation of general hostilities (Khalidi 1979, p. 107). Israeli invasion brought with it direct and indirect population movement also: Palestinians, Druze and Maronites were the main groups who moved to safe areas. In the last decade sectarian conflict has almost disappeared and the political system has returned, more or less, to its fragile consociational structure, this time with a formula offering more representation to the Muslim communities. But because Lebanon 'is a chessboard upon which divided international players vie for position and advantage, offering an open invitation to superpowers to engage in ambitious diplomacy', the victim in this struggle is Lebanon itself (Windsor 1989, p. 77). Windsor is also right in his observation that the international agenda which provided the cover for a *rapprochement* between Jordan and Israel would almost certainly make Lebanon more important in Syrian eyes and could lead to an intensified confrontation in that country between Israel and Syria. Lebanon has continued to be the main base for terrorist units headed for Israel, especially fundamental Shi'ite Muslim groups supported by Iran. In other spheres, Lebanon has been more successful in reconstructing and rehabilitating its economy and infrastructure; major efforts have been invested in Beirut. But the situation remains fragile, as recent events have shown.

The Israeli–Palestinian conflict

Israel and the Arab countries have been involved in a 'war in rounds'. First came the 1948–9 war when Israel became an independent state. As a result of that war many Palestinian Arabs became refugees in nearby Arab countries and Jordan annexed the West Bank. The 1950s were characterised by terror attacks on Israel originating in the Arab countries and by a major war between Israel and Egypt in 1956. In 1967 the third major Israeli–Arab war ended with Israeli victory and occupation of vast Arab territories: the Sinai Peninsula, Gaza, Jordan's West Bank and the Syrian Golan Heights. In 1973 another war broke out when the Arab countries tried to regain their occupied territories. The war ended with a 'no win' situation which led to Israeli–Egyptian negotiations. These culminated in 1979 in an Egyptian–Israeli Peace Treaty, and in 1982 Israel returned the whole of Sinai to Egypt. The next breakthrough in the long Israeli–Arab conflict followed the break-up of the Soviet Union and the Gulf War which reshaped the basic political order of the Middle East. The change in the political structure of the region has led the Arab world to reassess its attitude towards Israel.

The Arab nations have now reconsidered Israel's offer to enter into bilateral peace negotiations and to cooperate in building a new future for the Middle East. In October 1991, a conference was convened in Madrid to inaugurate these direct talks. Under the Madrid framework two tracks of negotiations have been started: the bilateral track and the multilateral track. Since the Madrid Conference direct bilateral negotiations are being held between Israel and Jordan, the Palestinians, Syria and

Lebanon. The talks led to two major agreements: in September 1993 Israel and the PLO (representing the Palestinians) signed the Israeli–PLO Declaration of Principles; and the subsequent Gaza–Jericho Agreement (May 1994). Israel and Jordan signed a common agenda followed by the Washington declaration of July 1994. The two documents which were signed constitute blueprints for peace. The Peace Treaty between Israel and Jordan was signed on 26 October 1994 and in summer 1995 the two countries normalised their relations; the border between them is now open. This peace is nicknamed the 'warm peace' in contrast to the peace with Egypt which is termed the 'cold peace'. Egypt was not anxious for peace between the countries to develop beyond a minimal level of transactions involving oil, telecommunications and unilateral tourism, most of which were stipulated by the peace treaty.

However, Israel and Jordan, in their peace treaty and the negotiations which followed, are promoting extensive economic and social cooperation and intend to transform their common border into an axis of cooperation in projects involving energy, transport and tourism. But the most dramatic event in the region was the Israel–Palestinian Accord in which the two adversaries recognised each other and decided on the phases in which Palestinian self-rule will be established in Gaza and the West Bank. Self-rule was established in the Gaza and Jericho areas and five specific spheres – education and culture, health, social welfare, direct taxation and tourism – were transferred to Palestinian control. At the time of writing, Israel and the Palestinians are in the process of negotiating the agreement of the next stage in which Israeli forces will withdraw from many towns and rural areas and the Palestinians will be empowered in other spheres of administration. However, there is strong opposition in both Israel and among the Palestinians to the peace process. Among the Palestinians the opposition is registered by a massive wave of terror attacks by Islamic fundamentalists who have killed more than one hundred Israelis since the process began. A more elaborate criticism comes from Palestinian scientists and academics (cf. Khalidi 1994). In Israel, opposition to the process comes from the settlers in Gaza and the West Bank and from the religious and right-wing parties who claim that the Accord with the Palestinians does not improve security for Israelis (Heller 1994). In recent months the assassination of peace-maker Rabin by an extreme-right Jew and the election of a right-wing government have posed further major question marks for the on-going peace process.

On the other fronts of the bilateral talks with Syria and Lebanon, progress has been very slow. Currently the talks are frozen as the two sides are still far from agreeing on fundamental issues such as security arrangements, the future of the civilian Israeli settlements on the Golan Heights, the future of southern Lebanon, and other crucial issues (Alpher 1993).

More progress was registered in the multilateral talks in which representatives from all Middle Eastern countries and the Maghreb are taking part together with the US, EU countries, Russia, Japan and others. These talks concern issues such as water scarcity, environment, arms control and regional security, refugees and regional economic development. There has been significant progress in these areas and a special Economic Summit was convened at Casablanca in October 1994 to present and promote economic development projects for the Middle East. The improvement in relations between Israel and Arab countries is felt in the evolution of low-level diplomatic relations which were initiated between Israel and Morocco, Tunisia and some of the countries in the Arabian Peninsula such as Oman. The Arab economic boycott against Israel which for years seriously damaged the Israeli economy is in a process of erosion and disintegration. However, the process is not yet completed and countries such as Libya, Iran, Algeria, the Sudan and even Saudi Arabia, remain firmly opposed to it.

CONCLUSIONS

The Mediterranean is a complex region in which relatively few features, mostly natural physical ones, exercise centripetal force towards a certain level of unity. However, this chapter has argued that the centrifugal forces which tear the nations of the region apart are stronger. Until the 1980s the East–West geopolitical rivalry combined with great ethnic and cultural variations within the region to keep its countries aligned towards external centres. Since the collapse of the USSR, the major division is along a north–south cleavage, and this has been reflected in the recent conflicts in the Middle East and by the difficulties in overcoming the large economic and social gaps between the countries of the northern and southern Mediterranean littorals.

The Mediterranean is also an arena of ethnic conflict in which former Yugoslavia poses serious dangers to world peace. Moreover, the Balkan region has other nodes of ethnic tensions which may turn sour at any time. The conflicts in Cyprus and Lebanon remain unresolved. The Israeli–Arab–Palestinian conflict has made an enormous leap in the right direction but is still active; shooting skirmishes and terror activities still upset all the residents of the region. It is unfortunate that the very basic feature of the region as a cradle of civilisation has also turned it into a focus of war and conquest, and a jigsaw of ethnic, religious and cultural groups who have not found a way to coexist.

References

AHMED, A.S. 1995: Ethnic cleansing, a metaphor for our time? *Ethnic and Racial Studies* **18**, 1–25.

ALIBONI, R. (ed.) 1992: *Southern European Security in the 1990s*. London: Pinter.

ALPHER, J. 1993: The Arab–Israeli peace process. In Gazit, S., Zeev, E. and Gilboa, A. (eds), *The Middle East Military Balance 1992–3*. Tel Aviv University: Jaffe Center for Strategic Studies, 45–65.

AMIN, S. 1989: Conditions for autonomy in the Mediterranean region. In Yachir, F. (ed.), *The Mediterranean between Autonomy and Dependency*. Tokyo: UN University, pp. 1–24.

ATTALIDES, M. 1979: *Cyprus: Nationalism and International Politics*. Edinburgh: Q Press.

BANKS, A. 1991: *Political Handbook of the World*. Binghamton, NY: CSA Publications.

BIBERAJ, E. 1993: Albania's road to democracy. *Current History* **92**, 381–5.

BOUCHAT, C.J. 1994: The north–south divide within the Mediterranean countries. *Mediterranean Quarterly* **5**(4), 125–41.

BRANIGAN, J.J. and JARRETT, H.R. 1975: *The Mediterranean Lands*. London: MacDonald and Evans.

BRAUDEL, F. 1972: *The Mediterranean and the Mediterranean World in the Age of Philip II*. Vol. 1. London: Collins.

BURG, S.L. 1993: Why Yugoslavia fell apart. *Current History* **92**, 357–63.

COHEN, S.B. 1973: *Geography and Politics in a World Divided*. New York: Oxford University Press.

COHEN, S.B. 1982: A new map of global geopolitical equilibrium: a developmental approach. *Political Geography Quarterly* **1**, 223–42.

CRAWSHAW, N. 1978: *The Cypriot Revolt*. London: George Allen and Unwin.

CRAWSHAW, N. 1994: Cyprus: a crisis of confidence. *World Today* **50**(4), 70–3.

CVIIC, C. 1993: Croatia's violent birth. *Current History* **92**, 370–5.

DRYSDALE, A. and BLAKE, G. 1985: *The Middle East and North Africa: A Political Geography*. New York: Oxford University Press.

FARLEY, J. 1994: The Mediterranean: southern threats to northern shores. *World Today* **50**(2), 33–5.

FENSKE, J. 1993: The West and the problem from Hell. *Current History* **92**, 353–6.

FISCHER, E. 1980: *Minorities and Minority Problems*. New York: Vantage Press.

GAZIT, S., ZEEV, E. and GILBOA, A. (eds) 1993: *The Middle East Military Balance 1992–3*. Tel-Aviv University: Jaffe Center for Strategic Studies.

HAAS, P.M. 1990: *Saving the Mediterranean*. New York: Columbia University Press.

HALL, G.O. 1994: Ethnic conflict, economics and the fall of Yugoslavia. *Mediterranean Quarterly* **5**(3), 123–43.

HELLER, M. 1994: The Israeli Palestinian Accord: an Israeli view. *Current History* **93**, 56–61.

HITTI, P. 1957: *Lebanon in History*. New York: St Martin's Press.

HUDSON, M. 1968: *The Precarious Republic: Political Modernization of Lebanon*. New York: Random House.

HUNTER, B. (ed.) 1993: *The Statesman's Yearbook 1993–1994*. London: Macmillan.

KAM, E. 1993: The regional arena – key issues. In Gazit, S., Zeev, E. and Gilboa A. (eds), *The Middle East Military Balance 1992–3*. Tel-Aviv University: Jaffe Center for Strategic Studies, 26–44.

KHALIDI, R. 1994: A Palestinian view of the Accord with Israel. *Current History* **93**, 62–6.

KHALIDI, W. 1979: *Conflict and Violence in Lebanon*. Cambridge, MA: Harvard University Press.

KING, R. and LADBURY, S. 1981: The cultural reconstruction of political reality: Greek and Turkish Cyprus since 1974. *Anthropological Quarterly* **55**, 1–16.

KLIOT, N. 1981: The unity of semi-landlocked seas. *Ekistics* **48**, 345–58.

KLIOT, N. 1989: Mediterranean potential for ethnic conflict: some generalizations. *Tijdschrift voor Economische en Sociale Geografie* **80**, 147–63.

KOLODNY, E.Y. 1971: Une communauté insulaire en Méditerranée Orientale: les Turcs de Chypre. *Revue de Géographie de Lyon* **46**, 5–56.

LEMPERT, R.J. and FARNSWORTH, G. 1994: The Mediterranean environment: prospects for cooperation to solve the problems of the 1990s. *Mediterranean Quarterly* **5** (4), 110–24.

LENDVAI, P. 1994: Albania: the rebirth of a nation. *World Today* **50** (1), 2–3.

LOIZOS, P. 1988: Intercommunal killing in Cyprus. *Man* **23**, 639–53.

PATRICK, R.A. 1976: *Political Geography and the Cyprus Conflict 1963–1971*. University of Waterloo: Department of Geography Publications Series No. 4.

PEACH, C. and GLEBE, G. 1995: Muslim minorities in Western Europe. *Ethnic and Racial Studies* **18**, 1–25.

RABINOVICH, I. 1984: *The Warfare in Lebanon 1976–1983*. Ithaca, NY: Cornell University Press.

REMINGTON, R.A. 1993: Bosnia: the tangled web. *Current History* **92**, 364–9.

SMALL, M. and SINGER, J.D. 1982: *Resort to Arms: International and Civil Wars 1816–1980*. Beverly Hills, CA: Sage.

THOMAS, R. 1994: Nations, states, and secessions: lessons from the former Yugoslavia. *Mediterranean Quarterly* **5** (4), 40–65.

ULMERT, Y. 1986: *Minorities in the Middle East*. Tel-Aviv: Ministry of Defence (in Hebrew).

VAN DER WUSTEN, H. and NIEROP, T. 1990: Functions, roles and form in international politics. *Political Geography Quarterly* **9**, 213–31.

VASCONCELOS de, A. 1992: The shaping of a subregional identity. In Aliboni, R. (ed.), *Southern European Security in the 1990s*. London: Pinter, 15–27.

VOLKAN, V.D. 1979: *Cyprus: War and Adaptation. A Psychoanalytic History of Ethnic Groups in Conflict*. Charlottesville, VA: University Press of Virginia.

WILLIAMS, A.M. 1984: Introduction. In Williams, A.M. (ed.), *Southern Europe Transformed*. London: Harper and Row, 1–8.

WINDSOR, P. 1989: Lebanon: an international perspective. In Shehadi, N. and Harney, B. (eds), *Politics and the Economy in Lebanon*. Oxford: The Centre for Lebanese Studies, 75–9.

WORLD TOURISM ORGANIZATION 1994: *Tourism Market Trends: Middle East 1980–1993*. Madrid: World Tourism Organization.

9

MEDITERRANEAN ECONOMIES: THE DYNAMICS OF UNEVEN DEVELOPMENT

MICHAEL DUNFORD

INTRODUCTION

The aim of this chapter is to explore the differential transformation of the Mediterranean world and the different degrees of success and failure of strategies of national and regional modernisation. Given the broad scope of this objective, the analysis deals with general processes and comparisons rather than detailed case studies. To set the scene, the chapter starts with a brief but essential background economic history of the Mediterranean and then focuses on the measurement and scale of differences in human development. To explain these development differentials, attention is paid first and foremost to the less developed non-European parts of the Mediterranean where there are a number of key interconnected processes of economic transformation associated with: (a) agrarian change, water supply, food production and food dependency; (b) the impact of oil and gas exploitation and industrialisation; (c) the debt crisis; and (d) structural adjustment, unemployment and the informal economy. In the final section the factors that distinguish the modernisation of the north Mediterranean market economies are discussed.

THE HISTORICAL ROOTS OF THE CONTEMPORARY ECONOMIC SCENE

The Mediterranean is a place where three great civilisations meet one another: the Latin world centred in Rome; the Islamic world, which originated in Mecca, and whose influence extended westwards as far as Morocco and northwards to Turkey and beyond; and the Greek or Orthodox world centred in Greece and in Constantinople, which, in 1453, became Istanbul and Moslem. What resulted from the encounters of these civilisations and their differential modernisation and development was a complex and varied economic geography made up of a series of dense layers that were successively laid down and disrupted in alternating phases of peaceful development and of conflict and conquest (Boustani and Fargues 1990; Gizard 1993). At its zenith, the Roman Empire encircled and unified the Mediterranean, as we saw in Chapter 5. The fall of the Western Empire gave way to an era of Arab domination that lasted until the Crusades, while the collapse of the Eastern/Byzantine Empire saw the expansion of Ottoman influence, not just along the North African coast, but also north-westwards as far as the gates of Vienna. Not until Napoleon Bonaparte came to power and sought to extend French influence throughout Europe did the Ottoman Empire start to crumble (Chapter 7). A century later Africa was in the hands of European colonial powers. With the end, in this century, of the Balkan wars and of the First World War, Turkey was eliminated as a significant actor in the Mediterranean, while in the Arab world the actions of rival Western imperialist powers and the creation in 1948 of the state of Israel have left a legacy of extreme political fragmentation that is a crucial feature of the contemporary political economy of the Mediterranean world, as Chapter 8 showed.

One important consequence of these economic and geopolitical events was the migration and settlement of distinctive ethnic and religious communities, thereby creating an extraordinarily complex mosaic of peoples, cultures, ethnicities, languages and religions. Examples include the existence of an Arab civilisation in Spain; the settlement of European Christians in Turkey and Egypt; and the existence of North African, Palestinian and Black Sea Jews, of Arab Christians in Lebanon, Syria, Egypt and Palestine, and of Islamic settlers in the former Yugoslavia.

From the outset, the Mediterranean's rich land and sea resources supported remarkable early developments in agriculture and trade and allowed it to emerge as the centre of a succession of hegemonic world economies. Some five hundred years ago, however, when Ottoman power was at its height in the east, Habsburg Spain initiated, with Portugal and with the aid of Italian seafarers, the European conquest of the globe. As the Atlantic was opened up, the coasts of the Mediterranean slowly ceased to be at the centre of global economic power and world decision-making. The centre of economic gravity in the world shifted northwards and westwards, first to Amsterdam and later to Great Britain, while the subsequent industrial revolution, which created such immense disparities between the economies that industrialised and those that did not, at first largely bypassed the Mediterranean or locked its inhabitants into a subordinate role in wider divisions of labour. Among the nations of southern Europe and the Mediterranean, Italy was almost the sole example of significant late nineteenth-century industrialisation. In the twentieth century industrialisation and modernisation did occur, but with different degrees of delay and different trajectories that are reflected in the map of contemporary inequality around the inner sea.

After 1945 the exploitation of rich oil and gas reserves made the Mediterranean and the Middle East a zone of great strategic significance. At the same time, however, decolonisation, the creation of the state of Israel, and the Cold War conflict between the Soviet Union and the United States led to sharp divisions in the Mediterranean, some of which were cemented as a result of moves towards closer military and economic integration. First, most north Mediterranean countries were integrated into the NATO defence alliance created in 1949, though France withdrew in 1966. Greece and Turkey, while members of NATO, have been in frequent conflict, and Spain did not join until 1986. Second, Italy and France and later, as capitalist dictatorships gave way to democracies, Greece, Spain and Portugal were integrated into the European Union (EU) developing intra-Community trade links at the expense of traditional non-Community trade relations, including relations with other Mediterranean countries. Conversely, most of the attempts made by the decolonised southern Mediterranean or the Middle Eastern countries to achieve greater political integration have failed, due to political and economic rivalries and defeat in three conflicts with Israel (in 1967, 1973 and 1982). Dreams of Arab unity persist, but so far little progress has been made in overcoming the fragmentation and Balkanisation that were a conscious legacy of Western imperialism.

The post-World War Two era none the less saw a radical transformation of the economies of the Mediterranean. Agrarian reform, emigration, industrialisation and international tourism have transformed Mediterranean Europe. Oil rents have enriched the producer states and, via mechanisms, of redistribution, had a differential but significant impact on non-oil states. In the Arab world military dictatorships reoriented development and trade with an emphasis on arms and military equipment, large-scale public works and the valorisation of their reserves of gas and oil, while communist Yugoslavia and Albania searched for 'socialist' paths to modernisation and development. Modernisation of the countries to the south and east of the Mediterranean led to the construction of new but always asymmetrical relations with Europe: more than 2 million people from the Maghreb live in Europe, while their economies exchange oil and minerals for food, industrial goods, know-how and technology. Today, however, it is clear that these economies have so far failed to find a durable and autonomous model of development. The contemporary Mediterranean is an expression of the consequent crisis and a region in which a variety of new economic and political strategies of crisis resolution are being played out.

Geographies of development and underdevelopment

In 1992 the four member states of the EU that border the Mediterranean (Spain, France, Italy and Greece) accounted for over 86 per cent of circum-Mediterranean's collective GNP yet contained just over 41 per cent of its population (see Table 9.1). If current demographic trends continue and the relative rates of economic growth do not change, in 2010 one-third of the population will account for well over four-fifths of its wealth. Of the countries with large populations Egypt is the poorest: the average Egyptian receives just $1.75 per day. The French, conversely, are the richest, receiving on average $70 per day or 35 times as much as the average Egyptian, though the Mediterranean regions of France (Languedoc–Roussillon, Provence–Alpes–Côte d'Azur and Corse) fall below the French average.

As Figure 9.1 shows, the countries of the Mediterranean fall into a number of orbits of relative underdevelopment compared with the richer core European countries. Figure 9.1 records Gross National Product (GNP) in 1992 at Purchasing Power Standards for most Mediterranean and European countries. As Purchasing Power Standard measures make allowance for differences in the cost of living in different countries, the disparities are somewhat smaller than those recorded in Table 9.1. Because of differences in measurement and calibration of the data, both Table 9.1 and Figure 9.1 express slightly different data than other tables in the book, e.g. Table 8.3; the broad rankings and magnitudes are, however, generally very similar. As in the case of Table 9.1, Figure 9.1 none the less shows that France ($19 200) and Italy ($17 730) are members of a second inner circle of rich countries surrounded by a series of zones of progressively lower levels of development.

The third zone is made up of countries with a per capita GNP between $10 120 and $14 600. Included here are three of the southern European EU member states and Israel. Israel owes its wealth to its rapid economic growth. Today 74 per cent of the Israeli workforce is in services though it has achieved rapid agricultural, infrastructural and industrial growth. Fast economic growth was a result of two factors: first, the constant arrival of motivated, abundant and often highly qualified immigrants; second, external military and financial support from the United States and Jewish communities (aid and gifts, reparations, special credit arrangements and private investment). In 1992 official development aid was equal to 3.3 per cent of Israeli GDP.

The next zone that lies around the inner circles of rich countries is made up of Cyprus, Malta and Greece. The GNP per head of these countries lies in the range $8010 to $9901. In the next tier lie Algeria, Turkey, Tunisia and the Syrian Arab Republic with a GNP per head of $4907 to $5740, while in the outermost lie a number of Arab countries with a GNP per head of $2779 to $3670.

As these data on economic inequalities suggest, the Mediterranean is at present the

TABLE 9.1 Land, population and wealth around the Mediterranean

	Land area in 1992 in sq. km	Total population in 1993 ('000)	GNP at market prices in 1993 (Atlas method) in millions of current US$	GNP per capita in 1993 (Atlas method) in current US$
Albania	27400	3 389	1 152	340
Algeria	2 381 740	26 722	47 366	1 770
Cyprus	9 240	726	7 539	10 380
Egypt	995 450	56 434	37 160	660
France	550 100	57 472	1 292 600	22 490
Gibraltar	10	28	–	–
Greece	128 900	10 365	77 547	7 480
Israel	20 620	5 219	72 638	13 920
Italy	294 060	57 121	1 133 440	19 840
Jordan	88 930	4 081	5 325	1 300
Lebanon	10 230	3 855	–	–
Libya	1 759 540	5 044	24 353[1]	5 560[1]
Malta	320	361	2 877	7 970
Morocco	446 300	25 945	26 559	1 020
Portugal	91 950	9 841	88 120	8 950
Spain	499 440	39 481	536 214	13 580
Syria	183 780	13 696	14 726[3]	1 150[3]
Tunisia	155 360	8 656	15 034	1 740
Turkey	769 630	59 597	177 120	2 970
Yugoslavia Former	255 400	23 928[2]	70 348[2]	2940[2]
Bosnia and Herzegovina	51 000	4 110	–	–
Croatia	55 920	4 780	–	–
Macedonia	25 430	2 075	1 701	820
Slovenia	20 120	1 991	–	–
Yugoslavia, Serbia	102 000	10 566	–	–

Notes:
[1] Libya 1989
[2] Yugoslavia 1990
[3] Syria 1991

Source: Elaborated from World Bank (1995)

interface between the developed North and the underdeveloped South – an intermediate zone between Europe, on the one hand, and Africa and southern and central Asia, on the other. This North–South cleavage has assumed increasing importance in the last few years, first, because of the end of the Cold War and the end of East–West conflicts in the Mediterranean, and second, because of the profound economic and political crisis in the southern countries where jobs are in short supply, the land cannot feed its inhabitants, water is scarce, debt and external dependence are widespread, frontiers are contested and governments face challenges from Islamic movements.

DIFFERENTIALS IN HUMAN DEVELOPMENT

GNP is a measure of the value of the goods and services a country's inhabitants produce in any

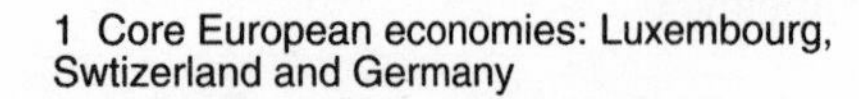

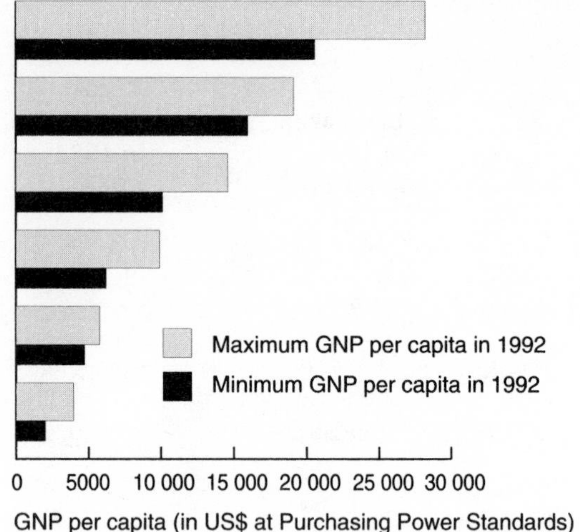

FIGURE 9.1 Orbits of development and underdevelopment in Europe and its neighbours
Source: Elaborated from UNDP (1994)

one year or, in other words, of the wealth they create. As the United Nations Development Programme (UNDP) indicates, the accumulation of wealth is not necessarily a good indicator of human development (UNDP 1995). The reason why is that a number of important determinants of the quality of human life depend more on the way in which a society is organised and the manner in which wealth is created, distributed and used rather than simply on the amount of wealth created. The quality of life depends, for example, on the existence of meaningful personal relationships, friendship, kindness and affection as well as the respect for certain human rights. The extent to which all individuals can live in freedom from fear and want, realise their capabilities and enjoy a meaningful life depends on the degree of economic security and the extent to which access to resources, food, shelter, income, employment, health care and education is shared. The UNDP accordingly argues that the goal of enriching human life implies increasing life expectancy, reducing infant and child mortality and increasing adult literacy. These determinants of the quality of life depend, however, not simply on whether a country is rich or poor but on whether its wealth is used to invest in universal health and education services and to reduce poverty. A related goal is that of not jeopardising the quality of the lives of future generations and their capacity to meet their needs. To meet this objective, current generations must choose sustainable paths to development that do not destroy the capacity of the earth to support human life.

What this UNDP argument shows is two things. First, wealth is a means to an end rather than an end in itself. Second, the needs of the whole population – the universalism of life claims – should be put at the centre of a measure of development. As GNP per head does not indicate whether wealth is used to improve human welfare, the UNDP proceeded to construct an alternative Human Development

Index (HDI). The HDI is a composite index derived by combining indicators of (a) life expectancy; (b) educational attainment (adult literacy rates and the mean years of schooling); and (c) income (real GDP per head measured in Purchasing Power Standards with successive marginal increases over the world average income per head discounted). The HDI scores for Mediterranean countries are set out in Table 9.2. The HDI is computed for nation-states but the UNDP has also used it to explore inequalities within countries and has developed Human Development Indicators adjusted for disparities between men and women and for differences in the distribution of income.

The figures for the Mediterranean countries show that the divide in human development is less than the divide in GNP per head and that the rank order of economies on the two indices differs. In the 1994 UNDP *Human Development Report* (UNDP 1995), 173 countries were examined. Mediterranean countries were ranked from 6th (France) to 111th (Morocco). A small number of countries were ranked significantly higher on the HDI than on GNP per head (Syria, Albania, Greece and Cyprus). A number of others did far worse than their wealth would imply (Libya and Algeria). Algeria's GNP per head was, for example, the 72nd largest in the world, while Albania was ranked 86th, yet life expectancy was 65.6 years in Algeria compared with 73 in Albania, adult literacy 60.6 per cent compared with 85 per cent and mean years of schooling 2.8 compared with 6.2, reflecting the general tendency for communist regimes to invest more in health and education. Algeria's underperformance on the HDI index (–37) is common to oil states in Islamic countries: Oman (–54), United Arab Emirates (–52), Iraq (–41), Saudi Arabia (–36), Qatar (–36), Bahrain (–25) and Kuwait (–23) all do less well than their wealth would suggest due to their underperformance on indicators of life expectancy and educational attainment. In the case of Algeria, the poor results stem from several factors. First, investment in people was sacrificed to industrial growth, as we shall see. Second, the indicator was adversely affected by the low status of women. On the one hand, the confinement of women to domestic tasks leads to a high fertility rate (5.0 children per woman, though this figure is much lower than the 7.4 which prevailed at the start of the 1960s) and high child mortality rates (62 per 1000). On the other hand, as with Morocco, Tunisia and other Islamic countries, Algeria is characterised by very high rates of adult illiteracy and of female illiteracy in particular. In 1992 51 per cent of women of 15 and over could neither read nor write. The comparable male figure of 26 per cent was also poor but much better than the figure for women. It is important, however, to make a distinction between two sets of Arab countries. In a first group, made up of the Lebanon and the Maghreb, fertility rates have declined, especially in urban areas where female employment is increasing, with rates of 3.4 children per woman in Lebanon, 3.9 in Tunisia, 4.6 in Morocco and 4.8 in Algeria in 1985–90. In the second, made up of Gulf oil states, fertility remains high in part as oil revenues are redistributed to support families with children and female labour is imported, with fertility rates of 7.1 in Saudi Arabia, 7.4 in Bahrain, and 8 in the United Arab Emirates.

Weak as the performance of a number of Mediterranean economies is, the situation is far better than it was 30, 20 or even 10 years ago. As Table 9.3 shows, the score on the Human Development Index of all Mediterranean economies has increased significantly. Absolute standards of living and the quality of life have improved, though at different speeds in different countries at different times. Morocco, for example, which had a score of 0.198 in 1960, reached 0.383 in 1980 and 0.549 in 1992, while Egypt rose from 0.210 in 1960 to 0.551 in 1992, Algeria from 0.264 to 0.553, and Tunisia from 0.258 to 0.690. These countries have yet to reach the Italian score of 0.755 for 1960, though as the Italian case indicates, numerical increases grow in difficulty the higher the starting score on the HDI.

TABLE 9.2 Human development in the Mediterranean

	Life expectancy at birth (years) in 1992	Educational attainment in 1992[1]	Real GDP per capita ($ at PPS) in 1991	Adjusted real GDP per capita ($)	HDI in 1992	HDI world rank	GNP per capita rank minus HDI rank
Albania	73	2.11	3 500	3 500	0.714	76	10
Algeria	66	1.40	2 870	2 870	0.553	109	−37
Cyprus	77	2.35	9 844	5 257	0.873	26	4
Egypt	61	1.20	3 600	3 600	0.551	110	12
France	77	2.78	18 430	5 345	0.927	6	7
Greece	77	2.34	7 680	5 221	0.874	25	10
Israel	76	2.58	13 460	5 307	0.900	19	6
Italy	77	2.45	17 040	5 340	0.891	22	−5
Jordan	67	1.98	2 895	2 895	0.628	98	1
Lebanon	68	1.92	2 500	2 500	0.600	103	−20
Libya	62	1.57	7 000	5 207	0.703	79	−38
Malta	76	2.15	7 575	5 219	0.843	41	−9
Morocco	62	1.25	3 340	3 340	0.549	111	−10
Portugal	74	2.15	9 450	5 252	0.838	42	−5
Spain	77	2.42	12 670	5 303	0.888	23	0
Syria	66	1.61	5 220	5 140	0.727	73	21
Tunisia	67	1.50	4 690	4 690	0.690	81	4
Turkey	67	1.88	4 840	4 840	0.739	68	10

Note: [1] An index based on adult literacy rates and the mean years of schooling per inhabitant
Source: After UNDP (1995)

TABLE 9.3 Trends in human development, 1960–92

	Human Development Index (HDI)			
	1960	1970	1980	1992
Albania	–	–	–	0.714
Algeria	0.264	0.323	0.476	0.553
Cyprus	0.579	0.733	0.844	0.873
Egypt	0.210	0.269	0.360	0.551
France	0.853	0.871	0.895	0.927
Greece	0.573	0.723	0.839	0.874
Israel	0.719	0.827	0.862	0.900
Italy	0.755	0.831	0.857	0.891
Jordan	0.296	0.405	0.553	0.628
Lebanon	–	–	–	0.600
Libya	–	–	–	0.703
Malta	0.517	0.615	0.802	0.843
Morocco	0.198	0.282	0.383	0.549
Portugal	0.460	0.588	0.736	0.838
Spain	0.636	0.820	0.851	0.888
Syria	0.318	0.419	0.658	0.727
Tunisia	0.258	0.340	0.499	0.690
Turkey	0.333	0.441	0.549	0.739

Source: UNDP (1995)

TRENDS IN THE GEOGRAPHICAL DISTRIBUTION OF WEALTH

The current map of economic inequality in the Mediterranean is the result of a range of processes with different temporalities. Figure 9.2 shows how the position of different national economies changed in the period from 1962 to 1992. A number of different models of change can be identified.

First France, which contains a number of Mediterranean regions, was throughout this period the leading economic power. Over the period as a whole, its position relative to other Mediterranean economies strengthened.

Second, there was a striking improvement in the relative position of most of the Mediterranean/southern European EU member states. Italy, which started with a per capita GNP equal to just under 140 per cent of the Mediterranean average in 1962 and which stood at the same level after the mid-1970s' oil crisis, soared to almost 220 per cent in 1992. Spain and Portugal grew relatively fast in the 1960s and early 1970s and then experienced particularly fast relative growth in the second half of the 1980s. Of the smaller Mediterranean member states, Greece stands out in that fast relative growth was confined to the 1960s and early 1970s. Malta and Cyprus, on the other hand, achieved relatively fast rates of growth in the late 1970s and 1980s.

Third, North African and East Mediterranean countries that lacked significant oil reserves saw their relative position get somewhat weaker. Egypt, Morocco and Tunisia fall into this group, as does Turkey, though its relative growth exceeded the Mediterranean average in 1989–92.

Fourth, a number of less developed Mediterranean economies achieved above-average rates of economic growth, especially in the 1970s, but then suffered major reversals in the 1980s. In the cases of Libya and Algeria, changing national economic fortunes were closely related to movements in the price of oil and to the successes and failures of oil- and gas-related industrialisation. In the case of Yugoslavia, relatively fast economic growth occurred in the 1970s. In 1980, however, there was a sharp reversal with rapid relative economic decline leading finally to violent national disintegration.

Having outlined the main geographical and historical patterns of uneven development in the Mediterranean region, attention will now be paid to a number of critical and inter-related development issues: land reform and the transformation of the countryside; food dependence; the development and impact of oil and gas exploitation; debt and the debt crisis; models of industrialisation; the growth of international tourism; and the role of informal economies.

AGRARIAN CHANGE AND THE COUNTRYSIDE IN THE NON-EUROPEAN MEDITERRANEAN

In the period since 1945 there has been a profound transformation of the countryside (see Chapter 13 for more on this). Just three decades

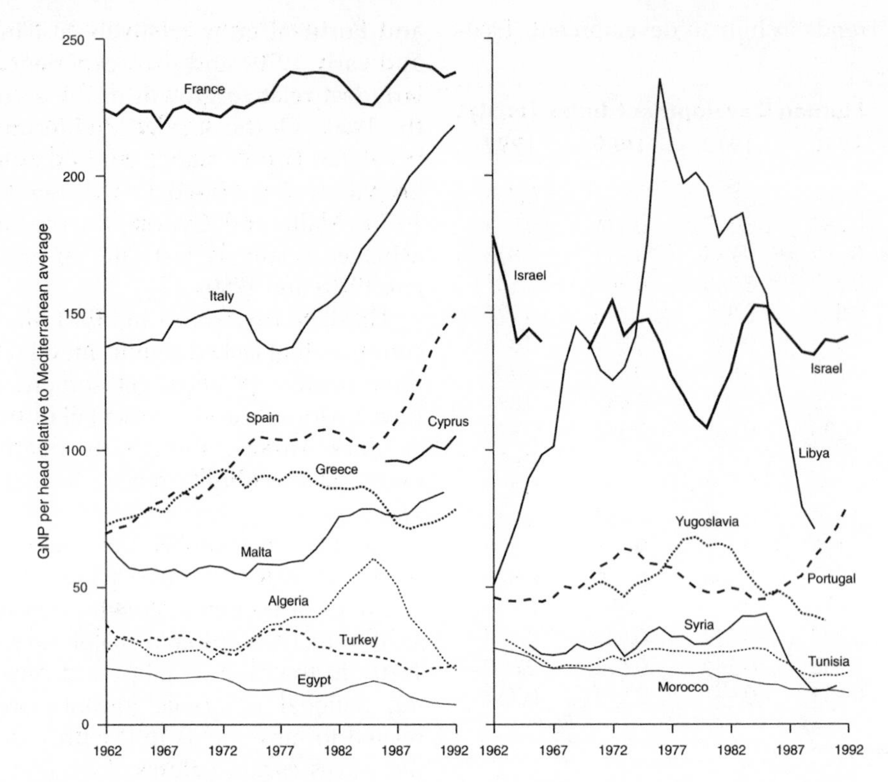

FIGURE 9.2 Uneven development: trends in GNP per capita relative to the Mediterranean average, 1962–92

Source: Elaborated from World Bank (1994)

ago the majority of the population was rural in many countries. Table 9.4 shows that in 1965 Albania, Morocco, Algeria, Egypt, Syria and Turkey all had more than half of the workforce employed in agriculture (once again the lack of consistency across official data sources must be pointed out, hence the slight differences in agricultural employment data between this table, Table 8.1, and Tables 13.1 and 13.2). In North Africa and the Middle and Near East most members of the agricultural workforce in 1965 were sedentary peasants. Most produced cereals, though there was some tree crop cultivation in the mountains and market gardening in irrigated areas near urban settlements. Pastoralism, which had formerly dominated the steppes and the desert areas, was in decline. Over time many nomads adopted a more sedentary existence due to the attraction of oil industry wages, employment on large infrastructural projects, the creation of states whose frontiers disrupted traditional pastoral routes, the use of lorries rather than camels to transport goods, and the attempts made by public authorities to create permanent settlements by constructing schools and health centres, along with recent droughts which led to the slaughter of livestock. Just how numerous pastoralists are is difficult to ascertain: in Anatolia there were 110 000 camels in 1950 but only 8000 in 1982. Today significant numbers are found only in Iran and Afghanistan which lie beyond the Mediterranean.

At present agriculture is in proportionate terms much less important. In 1992 it accounted for 47 per cent of the workforce in Turkey, 46 per cent in Morocco and Egypt, 23 per cent in Syria and 18 per cent in Algeria. Most of the population is urban rather than rural, often with large percentages in big cities of 1 million

Table 9.4 Distribution of employment: agriculture, industry and services, 1965–92

	Percentage distribution of labour force								
	Agriculture			Industry			Services		
	1965	1980	1992	1965	1980	1992	1965	1980	1992
Albania	69	56	56	19	26	19	12	18	25
Algeria	57	31	18	17	27	33	26	42	49
Cyprus	40	26	15	27	33	21	33	41	64
Egypt	55	46	42	15	20	21	30	34	37
France	18	9	6	39	35	29	43	56	65
Greece	47	31	23	24	29	27	29	40	50
Israel	12	6	4	35	32	22	53	62	74
Italy	25	12	9	41	40	32	34	48	59
Jordan	37	10	–	26	26	–	37	64	–
Lebanon	29	14	14	24	27	27	47	59	59
Libya	41	18	20	21	29	30	38	53	50
Malta	8	5	3	41	42	28	51	53	69
Morocco	61	46	46	15	25	25	24	29	29
Portugal	38	26	17	30	37	34	32	37	49
Spain	34	17	11	34	37	33	32	46	56
Syria	52	32	23	20	32	29	28	36	48
Tunisia	49	35	26	21	36	34	30	29	40
Turkey	75	58	47	11	17	20	14	25	33

Source: Elaborated from World Bank (1994)

or more people. In 1990 61 per cent was urban in Turkey, 56 per cent in Tunisia, 82 per cent in Libya, 52 per cent in Algeria, 50 per cent in Syria, 46 per cent in Morocco and 44 per cent in Egypt (World Bank 1995). Despite the movement of large numbers of people from the countryside to the cities, the countryside is more densely populated than ever due to the magnitude of rates of natural increase.

Agricultural modernisation has, however, failed to produce sufficient food with the result that most Mediterranean countries are dependent on food imports. With current rates of population growth, this dependence may well prove long term and is indicative of a range of deep-rooted structural and economic weaknesses and inequalities which agrarian reform has failed to overcome.

The new elites that came to power from 1952 onwards as independence was secured in southern and eastern Mediterranean countries sought to transform land ownership. Traditionally most land was state-owned. The development of private property was a nineteenth- and twentieth-century phenomenon. In Turkey and the Middle East the Ottoman and Iranian Empires often granted land to noble families and tribal chiefs. Similar policies were pursued by Western imperial powers. The distribution of land that resulted was very unequal: in Syria in 1950 1 per cent of owners owned three-quarters of the cultivated land, while one-half of peasants were landless. In Egypt from Pharoah to Muhammad Ali (who governed Egypt 1805–49) the central state controlled land and water. The state could confer land in usufruct to local communities or dignitaries. Private property was established as a result of a series of laws passed from 1837 onwards establishing a capitalist colonial agrarian system devoted in part to the production of cotton for European factories. In 1952, when a *coup d'état* overthrew the puppet monarchy that Great Britain had established, 0.4 per cent of landowners owned 35 per cent of land. In the Maghreb European colonisation led to an increase in large landed property. At the end of the colonial period foreigners owned 2.7 million hectares (one-third of the cultivated area) in Algeria, 1 million hectares or 12 per cent in Morocco and 700 000 hectares in Tunisia (Mutin 1995a).

Throughout North Africa and the Near East the same agrarian system prevailed. The state, noble families, large landowners and foreign companies owned approximately one-third of the cultivated land. At the other end of the spectrum, Koranic inheritance laws and demographic growth led to the division of small holdings into minuscule plots. Wage earning was present on modern, usually foreign-owned, farms especially in the Maghreb, but what predominated was a 'neo-latifundia' system in which landless or near-landless sharecroppers worked on large estates for absentee owners receiving one-fifth of the product of their labour.

In the 1950s and 1960s land reforms were implemented in most Arab countries with the aim of reducing rural injustice, increasing production, recovering (where relevant) foreign-owned land, eroding the power of former ruling elites and creating a new class of middle peasants who would support the new regimes (King 1977, pp. 371–437). All of the reforms, which were inspired by the 1952 Egyptian model, placed limits on the size of holdings and reallocated land in the form of individual lots with the requirement that the beneficiaries join cooperatives. In Egypt the reform ended in the early 1970s without achieving its objectives. Over 1 million feddans (16 per cent of cultivated land) was reallocated in lots of 2 to 5 feddans to some 342 000 fellahs (10 per cent of rural families) who owned less than 5 feddans. Very large holdings of more than 200 feddans were broken up, and as Table 9.5 shows, there was a large increase in the number of small owners of less than 5 feddans. A class of rich peasants (4.6 per cent of owners with over 46 per cent of the cultivated land) none the less replaced the old landowning classes. Additional legislation attempted to improve sharecropping contracts and raise minimum daily wages, but these provisions were often evaded. In 1985 44 per cent of rural families lived beneath the

Table 9.5 Changes in land ownership in Egypt, 1952–85 (per cent)

Size classes (1 feddan = 0.42 ha)	1952		1985	
	Owners	Area owned	Owners	Area owned
Less than 5 feddans	93.4	35.4	95.4	53.9
From 5 to 10 feddans	2.8	8.8	2.4	10.5
From 10 to 20 feddans	1.7	10.7	1.2	10.2
From 20 to 50 feddans	0.8	10.9	0.7	11.5
From 50 to 100 feddans	0.2	7.2	0.2	7.4
More than 100 feddans	0.2	27.0	0.1	6.5
Total (thousands)	2 801	5 984		
Landless peasants (thousands)	1 463			

Source: Mutin (1995a, p. 139)

poverty line and 60 per cent of the agricultural population owned nothing, due in part to a simple shortage of land.

In the non-Arab Near East the situation is more varied. In Turkey no radical change took place, though the distribution of land was less unequal: in 1980 82 per cent of farms were smaller than 10 hectares and occupied 41 per cent of the cultivated land area. Turkey is a country of small peasants and direct owner cultivation where village chiefs, large landowners and rich peasants remain politically powerful. Israel is unique in that it put collective rural structures in place: in 1991 20 per cent of the rural population worked in *mochavot* (which were originally villages with individual land ownership but collective open-field modes of exploitation), 42 per cent in *mochav* (small family holdings with compulsory adhesion to a cooperative that provides collective services) concentrated in the central plain, and one-third in *kibbutz* which are concentrated in irrigated and frontier areas.

The situation in the Maghreb differed due to the need to reappropriate foreign-owned land. The mechanisms there differed. In Morocco colonial land was purchased slowly on mutually agreed terms from 1956–74 by existing Moroccan landowners, thereby reinforcing the position of large landowners. In Algeria recuperation was more radical: in 1962–63 22 000 colonial farms were nationalised and turned into 2200 self-managed units. A 1972–76 agricultural reform transformed 1.3 million hectares of private land, that its owners did not cultivate directly, into small cooperatives. At the start of the 1980s the state held 45 per cent of agricultural land (3.3 million hectares). The economic results of land reform fell short of expectations. As a result, greater encouragement was given to the private sector, and in 1988 a number of public demesnes were granted to private farmers (see Mutin 1995a).

In the Maghreb and the Mashreq the fundamental contemporary problem is that agriculture can no longer feed its population. Clearly, climatic conditions are a constraint, as is the fragility of natural environments. In particular, erosion and the desertification of the Mediterranean steppes result in losses of agricultural land. As elsewhere, agricultural land is also lost to urban and industrial uses: in Egypt the loss of irrigated land in the Nile valley to urban uses since 1950 is equal to nearly one-half of the new land irrigated as a result of the construction of the Aswan Dam (Mutin 1995a). Such losses are the more severe as the cultivated land area accounts for a small part of the overall land surface. As much of the immense land area of the Arab countries is desert where permanent or temporary aridity is a major fact of life, large areas are uncultivated and unsettled. As Table 9.6 shows, 97 per cent of the land area of Egypt is uncultivated, 91 per cent of Libya, 82 per cent of Algeria and 66 per cent of Israel. The prospects of increasing the cultivated surface substantially are, however, small. Of the non-European Mediterranean countries, only Turkey has increased the cultivated land area in the recent past. Yet with population growth, increased incomes and increased per capita food consumption, the demand for food grows.

The main response to the food production problem has been the intensification of agriculture and, in particular, irrigation. In the Mashreq most emphasis was placed on large river management schemes, though a more economical use of water is in many places a second possibility. In Egypt the completion of the Aswan Dam in 1975 led to intensification of agriculture on existing land (as it made water available throughout the year) and reclamation of new land. In Syria there are large irrigation projects centred on the management of the two international rivers (Orontes and Euphrates) that traverse the country. On the Euphrates the immense Tabqa Dam was completed in 1973 as a prelude to irrigating 640 000 hectares. In the case of the Euphrates, however, there are tensions with Turkey upstream over the release of sufficient water and with Iraq downstream over Syrian retention of water. In Israel the completion of the National Water Carrier in 1967 created an integrated water supply network drawing on all its surface water resources, its aquifers and water from the occupied West Bank (see Chapter 15). Three-quarters of the water is used for agriculture, and 41 per cent of the agricultural land area is irrigated (see Table 9.6) allowing Israel to export

TABLE 9.6 Land use in Mediterranean countries, 1991

	Arable land	Permanent crops	Permanent pasture as % of total land area	Forest and woodland	Other land	Total land area ('000 hectares)	Irrigated land as % of arable and permanent crop area
Albania	21	5	15	38	22	2 740	60
Algeria	3	0	13	2	82	238 174	5
Cyprus	12	5	1	13	69	924	23
Egypt	2	0	0	0	97	99 545	100
France	33	2	20	27	18	55 010	6
Gaza Strip	24	39	0	11	26	38	50
Greece	22	8	41	20	9	12 890	31
Israel	17	4	7	6	66	2 033	41
Italy	31	10	17	23	20	29 406	26
Jordan (1986–88)	4	0	9	1	86	8 893	16
Lebanon	21	9	1	8	61	1 023	28
Libya	1	0	8	0	91	175 954	11
Malta	37	3	0	0	59	32	8
Morocco	20	1	47	20	12	44 630	14
Portugal (1986–88)	26	9	9	32	24	9 195	20
Spain	31	10	21	32	7	49 944	17
Syria	26	4	42	4	23	18 392	12
Tunisia	19	13	28	4	36	15 536	5
Turkey	32	4	11	26	27	76 963	9
Yugoslavia	27	3	25	36	9	25 540	2

Source: FAO (1992)

fruit and vegetables to pay for imports of cereals, meat and sugar. At present, however, Israel is depleting its renewable water resources, while two-thirds of the latter come from outside of its 1948 frontiers. In Morocco ambitious irrigation projects were implemented in lieu of land reform: 65 dams have been constructed and 850 000 hectares irrigated.

Considerable efforts have been made to increase output, and the irrigated land area and the range of crops cultivated have increased. Food security is, however, elusive. Cereal production occupies more than half of farmland in North Africa and the Near East, yet yields are low, averaging 18–20 quintals per hectare. In 1991, for example, wheat yields stood at 11 quintals per hectare in Algeria, 13 in Libya, 17 in Syria and Tunisia, 19 in Morocco, 20 in Lebanon and Israel and 22 in Turkey. In Egypt the average yield was 46. The figures for southern Europe were: 13 in Portugal, 24 in Spain, 26 in Greece and 35 in Italy. In France a hectare yielded 67 quintals (Conseil Économique et Social 1993, p. 73). The reasons for low yields are numerous but include the cultivation of steeply sloping land, erosion, limited rainfall and high percentages of fallow along with an under-emphasis on seed selection and modern agronomic practices.

Increases in yields have, however, occurred. In 1961/5 to 1986/90 agricultural output in North Africa and the Near East almost doubled (see Table 9.7). At the same time, however, the population of these countries tripled. Wine production (in Cyprus and in the former French colonies of Algeria, Morocco and Tunisia) declined from 17 million hectolitres in 1960 to 2 million today. Citrus fruit production increased: oranges, for example, increased from 2.1 to 4.8 million tonnes with Egypt and Morocco in the lead. Vegetable output increased more than threefold with significant exports from Turkey and Egypt. Increases in the production of sugar beet and cane sugar were substantial but other than in Turkey were insufficient to meet domestic requirements. Cotton is a traditional product that increased little with a stagnation of output in Syria, a decline in Egypt and an increase in Turkey. Sheep reared on cereal farms or by steppe pastoralists increased from over 76 million to over 130 million with most (40 million) in Turkey, but still fell short of consumption.

Agriculture has failed, for the most part, to meet the food needs of the Arab world due to

TABLE 9.7 Agricultural production in North Africa and the Near East

	1961/65 average	1986/90 average	1992	Average annual percentage rate of growth from 1961/65 to 1986/90
Cereals ('000 tonnes)	32 617	58 897	64 277	2.4
Wheat ('000 tonnes)	15 855	33 480	37 046	3.0
Oranges ('000 tonnes)	2 115	4 605	4 772	3.2
Vegetables ('000 tonnes)	16 761	52 845	45 068	4.7
Cotton fibres ('000 tonnes)	910	1 157	1 184	1.0
Sugar beet ('000 tonnes)	3 965	16 243	19 875	5.8
Raw sugar ('000 tonnes)	976	3 227	3 448	4.9
Oil-seeds ('000 tonnes)	748	1 679		3.3
Sheep and goats ('000 head)	76 884	124 768	130 781	2.0
Beef cattle ('000 head)	22 148	24 664	24 784	0.4
Milk ('000 hectolitres)	7 217	13 386	11 696	2.5
Poultry meat ('000 tonnes)	278	1 789	1 853	7.7

Source: Médagri (1994); Mutin (1995a, p. 170)

the difficulties of the environment, the failure of reforms to come up to expectations and the undervaluation of this critical sector (Abdel–Fadil *et al.* 1993). Agriculture was not central to the development strategies of these countries: on the one hand, in order to to hold down wages and support industrialisation and the exploitation of oil and gas, the prices of cereals were kept down at levels that just covered costs and discouraged investment; on the other hand, too little was done to intensify dry farming of cereals and improve the status and role of the fellah.

With the exception of Turkey and Israel where exports match imports, output growth simply did not match the rapid growth in the demand for food, which was itself a result of population growth, increasing food consumption per head, urbanisation and changes in consumer behaviour as households abandoned traditional diets for Western models. The consequence is a dependence of most Arab countries on massive food imports. National output meets just 30 per cent of needs in Algeria and 50 per cent in Egypt. The Maghreb and Mashreq are massive importers of cereals which account for more than 60 per cent of the calorie intake of their populations. In 1992 their wheat purchases amounted to some 20 per cent of the world market which was of the order of 107 million tonnes per year. Growth of demand for cereals is remarkable: 10mt in 1974, 21 in 1980, 32 in 1985, 40 in 1990. To these add coffee and tea, oil, sugar and, recently, meat.

These imports are an economic burden and a major cause of trade deficits. Just three Arab countries have significant exports of agricultural products: Morocco which exports citrus fruits and salad crops, and Syria and Egypt which export cotton. All, however, import far more than they export. In Egypt and Algeria food imports account for 30 per cent of total imports. Increased dependence on food imports has a number of further consequences. As most governments have long-standing low food price policies, the switch to food imported at world market prices has produced large public subsidies. In the mid-1980s these subsidies amounted to 15–33 per cent of current government expenditure. This increase in dependence is, moreover, concentrated on a small number of food-exporting countries (United States, Canada and the European Union). The tendency is for governments to attempt to limit imports and to reduce the subsidisation of food (as is required under structural adjustment programmes agreed with the International Monetary Fund) at the risk, however, of a loss of legitimacy and a threat to social and political peace.

In the next few decades continued population growth will, other things being equal, cause deficits to increase further. To reduce these deficits and increase employment there is a likelihood that there will be greater specialisation (if there is enough water) in the intensive production of Mediterranean fruit, vegetables and flowers. On small areas of land in the Mediterranean with its natural climatic advantages, intensive fruit and vegetable production produces 10 to 20 times as much revenue as cereal production, and provides more employment. A greater specialisation in export agriculture, however, will make these countries more vulnerable to the vagaries of international markets. At the same time, it will add a southern challenge to the northern one already faced by their north Mediterranean neighbours: at a point at which northern capital is being moved out of less remunerative temperate farming and into fruit and vegetable production, south and eastern Mediterranean countries may contribute to an increase in output and competition in product markets in which southern European farmers have a strong specialisation (Conseil Économique et Social 1993, pp. 109–12).

MEDITERRANEAN AND MIDDLE EASTERN OIL

Discovered in the Arab world in the inter-war years, hydrocarbons were exploited on a large scale after the Second World War. In 1994 the Middle East and North Africa accounted for 60 per cent of known world oil reserves and 20 per cent of world gas reserves. In that year, with an embargo on Iraq after the Gulf War, the Arab world still produced 950 million tonnes of oil out of a world total of 3165 million. At 30 per cent of the world total, this figure was somewhat lower than the 32 per cent reached in the late 1970s. Of Arab output, however, 82 per cent was exported. Arab exports accounted for 44 per cent of world exports. At present 22 per cent of the US market, 63 per cent of the West European market, 72 per cent of the Japanese market and a large share of other Asian markets depend on Arab oil. As a result, Arab hydrocarbons and hydrocarbon export routes are highly significant not just for economic, but also for strategic and geopolitical reasons.

Within the Arab world, as Mutin (1995b) points out, the distribution and production of oil are very unequal (see Figure 9.3). In 1994 Algeria produced 56 million tonnes from Hassi-Messaoud and Edjeleh, first exploited in 1956, but at current rates of extraction just 20 years' supply remain. Algeria also produces 56 billion cubic metres of gas and derivatives (liquid petroleum gas). Of this total, 30 million tonnes was exported in liquefied form or via a gas pipeline to Sicily and via the Messina Strait to mainland Italy. Egypt is a small producer whose main reserves are in the Sinai Desert which was recovered from Israel after the Camp David agreements, and Tunisia is self-sufficient, while Libyan production has varied significantly. In 1970 Libyan output reached 161 million tonnes, in part because the pre-Gaddafi regime gave the oil companies a free hand to exploit Libyan oil which is of high quality, abundant, accessible and located west of Suez. In 1994 output was lower, standing at 67 million tonnes, due to the Libyan government's desire to limit production.

The main producers are in the Arab Middle East. In 1994 the Middle East produced 776 million tonnes. If Iran is added, this figure rises to 954 million tonnes. Iraq produced just 24 million tonnes compared with a potential annual output of 250–300 million tonnes. Of the Middle East total, 722 million tonnes were produced in the Arabian peninsula. The peninsula, made up of Qatar, Oman, Yemen, the United Arab Emirates, Kuwait and Saudi Arabia, accounted for 22 per cent of world production in 1994 and 46 per cent of known world reserves. Of these reserves one-quarter were in Saudi Arabia. Alongside oil the Middle East accounted for 9 per cent of world gas reserves.

Middle Eastern oil is important not just because of its quantity. The oil is easy to extract

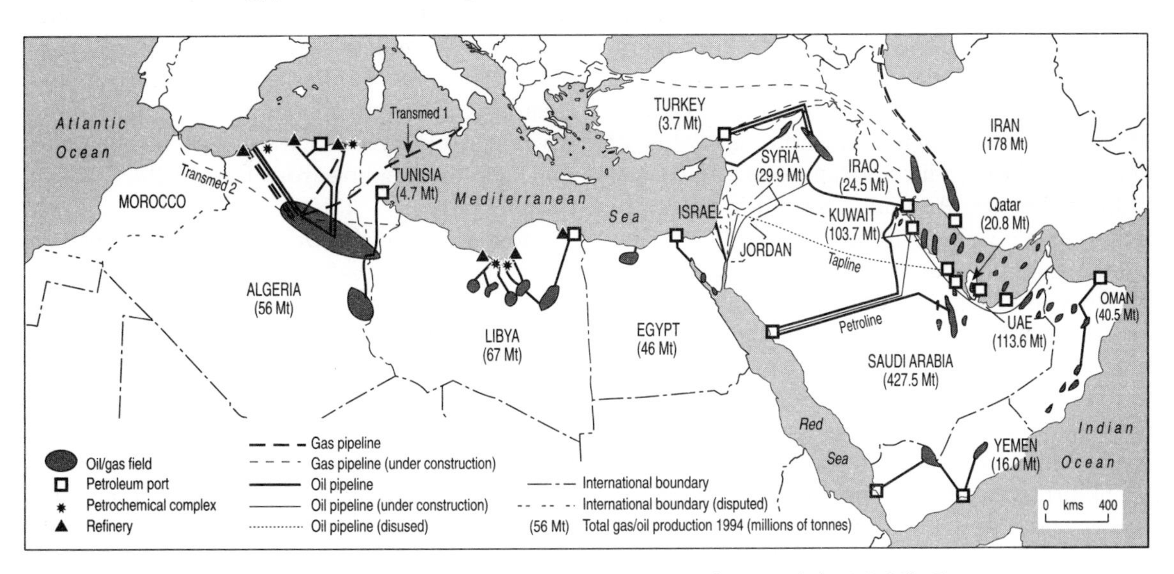

FIGURE 9.3 Geography of hydrocarbons production in North Africa and the Middle East

and the costs of a barrel are the lowest in the world. Arab oil costs one-fifteenth of North Sea oil, one-tenth of North American oil and one-fifth of Latin American oil. What is more, it is easy to vary production levels in line with market conditions. The most striking example of flexible production is Saudi Arabia which produced over 500 million tonnes in 1980, 173 in 1985 and almost 330 in 1990, rapidly increasing its production after the invasion of Kuwait in August 1990 and the implementation of sanctions against Iraq.

Arab oil is, in short, a critical component of supply in a number of advanced countries. In the last three decades, however, the security of these supplies has been threatened by a sequence of conflicts: the Six-Day War (1967), the Iran–Iraq conflict (1980–8) and the Gulf War (1990–1). These geopolitical conflicts have had a particularly severe impact on the organisation of the oil sector in the Middle East and on oil prices.

Middle Eastern oil exports depend on land and sea routes that involve the use of strategic passages through the Hormuz Strait and the Suez Canal (see Fig. 9.3). North African oil was less affected as the routes are maritime and Mediterranean. In the case of North Africa, oil and gas pipelines transport energy products to Mediterranean ports for export mainly to Europe but also to the United States. In Algeria a 9300-km network of five oil and four gas pipelines links the Saharan oilfields at Hassi Messaoud and Edjeleh and the gasfields at Hassi R'Mel to Mediterranean ports and coastal industrial zones. Annually, these pipelines can transport 92 million tonnes of crude oil and condensates compared with a 1994 production of 56 million tonnes, and 61 billion cubic metres of gas compared with a production of 56 billion. Oil, condensates and gas are transported to Arzew-Bethioua near Oran, Bejaia in Kabylie or Skikda in Canstantinois on the Algerian coast; oil is transported to the Tunisian port of La Skhirra, chosen for security reasons at the end of the colonial era; and gas is transported along the Transmed pipeline to Sicily which was opened in 1984. A further Transmed pipeline to Spain is under construction crossing Morocco and the Straits of Gibraltar. Algerian gas is therefore exported either by pipeline or by sea after liquefaction at Bethioua or Skikda. A similar situation prevails in Libya where gas and oil are transported from some 30 fields south of the Grande Syrte Gulf where the oil is partially refined at Marsa-el-Brega and Tobrouk and exported from ports that can receive 150 000 tonne oil tankers.

In the 1960s the Mediterranean was also the main oil route to Europe and North America for Middle Eastern oil. Most oil was taken through the Suez Canal though some was transported along the Tapline pipeline from Saudi Arabia to Sidon in the Lebanon and along pipelines from Iraq to Tripoli and Banias in Syria (see Fig. 9.3). After the Six-Day War the Suez Canal was closed and the Tapline was cut. In 1975 conflict between Iraq and Syria led the latter to stop the flow of Iraqi oil across its territory. As a result, new routes were developed. Most important was the sea route around the Cape of Good Hope whose use led to the development of 500 000 tonne supertankers and a dramatic increase in the importance of the Hormuz Strait. At the end of the 1970s half of the oil traded in the world went through this strait. With the Iran–Iraq conflict, however, this traffic was jeopardised. Due to the insecurity of this route and the 1975 reopening of the Suez Canal, new pipelines were constructed (a pipeline being cheaper if its length is one-quarter or less of the sea distance). Included were the Petroline and the 1100-km pipeline from Iraq to the Turkish port of Dörtyol.

Movements in the price of oil in the last 25 years have also had dramatic effects on the development of not just the oil producers but also oil-importing countries (see Fig. 9.4). Until the 1950s the oil industry was completely dominated by seven large oil companies. Called the 'seven sisters', this group was made up of the American firms, Standard Oil of New Jersey–Exxon, Standard Oil of California, Mobil, Gulf Oil and Texaco, the British firm British Petroleum and the Anglo-Dutch firm Royal Dutch Shell. These companies and the smaller

French Compagnie Française des Pétroles were granted concessions for which they paid a royalty equal to 12.5 per cent of the price they posted. Due to a fall in the posted price the producing countries established the Organisation of Petroleum Exporting Countries (OPEC) in Baghdad in 1960 on the initiative of Venezuela, while in 1967 the Arab oil exporters created the Organisation of Arab Petroleum Exporting Countries (OAPEC). In the 1960s OPEC managed to increase its members' share of the oil revenues, though prices and production were determined by the oil majors. At the same time, new companies sought to break the monopoly of the seven sisters. Included were independent American firms (Getty or Occidental), national firms such as Ente Nazionale Idrocarburi (ENI) established in Italy to develop the hydrocarbons sector, and national companies in the oil-producing countries such as Sonatrach in Algeria.

A major change occurred in the 1970s when prices started to rise and the producers secured control over price setting (see Fig. 9.4). Then, in a four-month period starting in October 1973 after the Yom Kippur war, prices almost quadrupled. At the root of the price increase were a number of factors: the desire to put pressure on Israel to withdraw from the territories it had occupied; a feeling that the oil producers were insufficiently well remunerated under the agreements with the companies that extracted the oil; and a reduction in the purchasing power of the dollars earned by the oil-exporting countries because of the American decision to end the convertibility of the dollar into gold (on which the Bretton Woods system depended) and its subsequent devaluation. In 1979 there was a second oil shock coinciding with the Shiite revolution in Iran which saw a tripling of oil prices in two years. Increased prices coincided with a further shift in the distribution of oil revenues between producing and consuming countries: at the end of 1974 the oil producers were getting 85 per cent of the oil profits and a royalty of 18 per cent in part as a result of nationalisations (in Algeria in 1971, Libya in 1972, Iraq in 1972 and Iran in 1973) and share

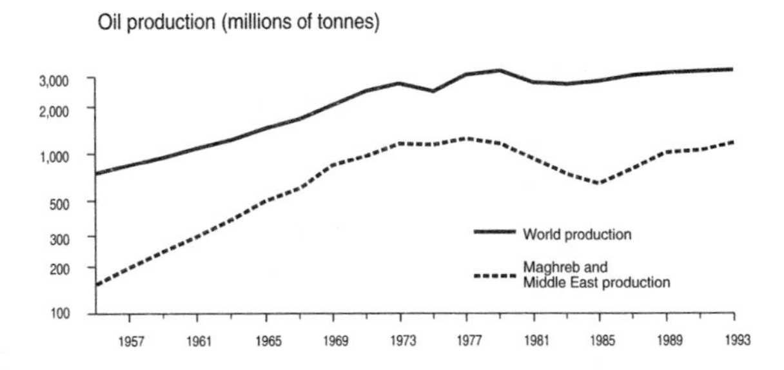

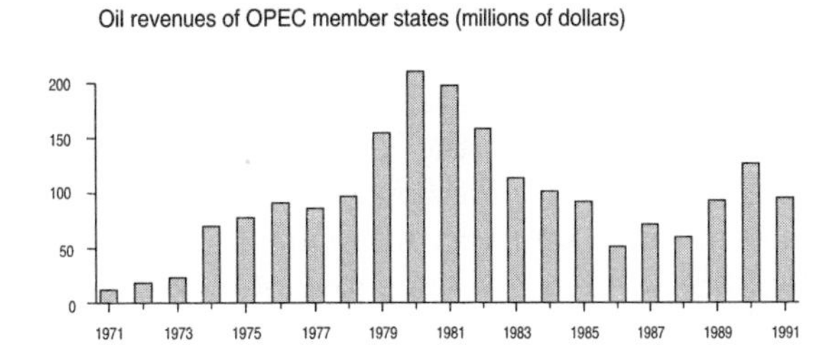

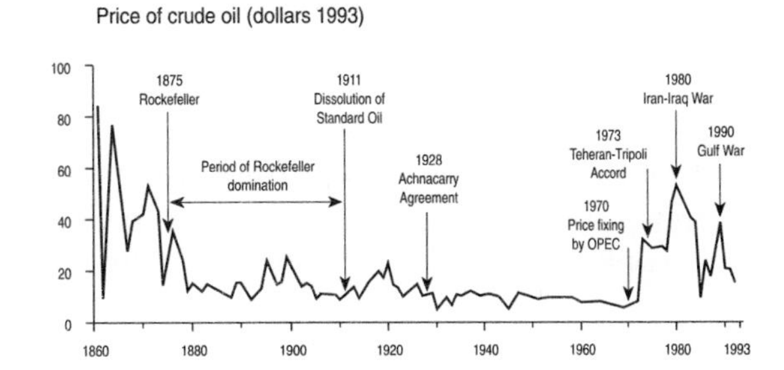

FIGURE 9.4 Trends in oil output, prices and revenues

Source: After Brunel (1995); Mutin and Durand-Dastes (1995)

acquisitions which continued until 100 per cent control was established.

The combination of price increases, the reapportionment of oil revenues and nationalisation/share acquisition led to a spectacular increase in the oil revenues of the oil-producing states (see Fig. 9.4). Having received $5 billion in 1970, the Arab states received $51 billion in 1974 and $205 billion in 1980. After the second oil shock, energy-importing countries responded with more determined energy conservation measures. At the same time the output of non-

OPEC countries (Great Britain, Norway, Mexico, etc.) increased. In 1985 there was a counter-shock: oil prices and sales fell sharply. In 1986 Arab oil revenues stood at just $45 million. Despite subsequent increases and the receipt of $98 billion in 1992 and $82 billion in 1994, many of the oil-producing states were left with large trade deficits and profound economic difficulties.

OIL AND DEVELOPMENT

The exploitation of oil had a profound but very uneven impact on the development of the countries of the Mediterranean Arab world. With the exception of Algeria and Iraq, oil is found in countries with very small populations. After the increase in oil revenues, GDP per head in these countries was amongst the highest in the world. In 1980 when French per capita GDP stood at $12680, the figures for Qatar were $33420, United Arab Emirates $32210, Kuwait $19200, Saudi Arabia $14250, and Libya $10640, while that for Egypt was just $500. The frontiers created by the advanced Western countries concentrated wealth in countries with few people and people in countries with little oil wealth.

Oil and industrialisation

The oil-exporting countries were, however, very dependent on a single product whose price was very sensitive to international market and political conditions. In general, oil accounted for 90 per cent of their export earnings. A strong exchange rate damaged the competitiveness of industrial activities, while domestic agriculture was eroded in the face of a flood of imported food. A large share of national wealth was invested in defence. Today defence expenditure stands at 10 per cent of GDP in Iraq, Syria, Oman, Yemen, Saudi Arabia, Qatar and Jordan and 8 per cent officially in Israel. A number of countries (Algeria with its '*industries industrialisantes*' centred on growth-pole concepts; Iran and Iraq) did make great efforts to diversify their economies and to increase their economic autonomy, though this path proved of questionable value.

The model was Egypt: in the 1950s public enterprises spearheaded industrial growth (steel at Helwan, and fertilisers at Aswan) and the Aswan Dam project was launched. Most of the countries that chose to industrialise sought to use oil and gas as the raw materials, oil revenues as the money capital and the technology of Western companies as the know-how on which to base processes of industrialisation and urbanisation. A number of developments took place:

1. The export of transformed products with higher value-added rather than raw materials. In the case of oil, refineries were constructed near centres of extraction rather than markets. Algeria can refine 26 million tonnes per year, half its output. Gas export led to the development in Algeria from 1964 onwards of liquefaction at Arzew-Bethioua and Skikda.
2. The development of petrochemical industries: in the 1970s Algeria sought to develop the ethylene cycle producing ethylene, methanol and polyethylene which serve to make plastics and polyester fibres at Arzew-Bethioua (resins, methanol and ammoniac) and Skikda (ethylene and thermoplastics), with Libya following suit at Marsa el Brega and Ras Lanouf and Iraq at Bassora. Considerable importance was also attached to the production of fertilisers (Arzew in Algeria). These petrochemical industries are located in immense port-related industrial complexes (2000 hectares at Skikda and 1500 hectares at Arzew-Bethioua).
3. The development in the Gulf of capital- and energy-intensive aluminium and special steels industries using imported raw materials and abundant and cheap local energy. Another capital-intensive sector favoured was ship-building and repair.
4. The establishment of import-substituting industries producing consumer goods to serve the domestic market. Most important were construction materials industries which grew

rapidly due to the scale and number of large infrastructure projects (roads, airports, ports, cities, housing and water supply) and rapid and often speculative urban growth. In this first phase attempts were also made to produce food and drinks and a range of consumer goods. In a second stage, Algeria in the Maghreb and Egypt, Iraq and Syria in the Mashreq set up plants designed to substitute for imported parts and sub-assemblies for vehicle and mechanical assembly industries.

5. The orientation of industries established to meet domestic needs towards export markets taking advantage of their cost advantages. A classic case is the Tunisian clothing industry which employs part-time female homeworkers at wages equal to 5–10 per cent of those of European workers. Clothing accounts for almost two-thirds of Tunisian industrial exports.

Migration, remittances and transfers

With the exception of Algeria and Iraq, oil is found in sparsely populated countries. Greater oil production, increased oil revenues and industrial diversification led to a demand for labour far in excess of national resources. Excess demand for labour along with immense disparities in wealth and income within the Arab world led to substantial rotational migration after 1973, as young men from populous but poor countries went to work for periods of 3 to 6 years in the oil states. Prior to 1973 there were fewer than 1 million migrants in the oil states; all were from Arab countries. In 1985 there were more than 5.5 million foreign workers in the Arabian peninsula (one-third of the resident population) of whom 4 million were Arab (from Jordan, Syria, Lebanon, Egypt and Yemen) and most of the others were Asian. About 1 million Egyptians worked in Iraq though these immigrant workers have now left. Libya counted 700 000 Egyptian, Tunisian, Turkish and Asian workers in a population of 4.1 million. Of the exporting countries Egypt, with no earlier tradition of emigration, had 3 million of its citizens abroad (6 per cent of the population and 10–15 per cent of the labour force). Syria and Lebanon in the Levant exported 1.5 million, of whom 600 000 were Palestinian though many Palestinians were expelled from the oil-producing states as a result of the Gulf War.

Temporary emigration generated large remittances of earnings. In 1986 declared transfers from the oil states (which fell short of actual transfers) amounted to $8.6 billion. The major recipient was Egypt ($3.2 billion). These sums played a major role in improving the balance of payments of labour-exporting countries. In the case of Egypt, in 1994 remittances were more important than the receipts from its three most important foreign exchange-earning activities put together: tourism, the sale of oil and the revenues from Suez Canal tolls, though the advantage is less than appears as much remittance income is spent on imported consumer goods. Additional remittances come from North African and Turkish emigrants to France, Germany, Benelux and, more recently, Italy and Spain.

Alongside the remittances of emigrants there were also significant transfers of development aid between rich oil-producing states and the less developed countries in the Maghreb and Mashreq.

As Table 9.8 shows, net private transfers, which are composed largely of remittances in all countries other than Israel, and official transfers play an important role in closing the gap between imports and exports. Egypt's 1992 deficit, for example, which stood at 7.8 per cent of GDP, was converted into a current surplus as a result of remittances equal to 15.4 per cent of GDP and official transfers of 2.9 per cent. In the case of Israel, private transfers equal to 3.9 per cent of GDP and official transfers worth 6 per cent of GDP more than offset its trade deficit.

Petrodollars, the debt crisis and structural adjustment

In the oil-producing states the growth of oil revenues created large trade surpluses which were translated into: increased imports of weapons, consumer and capital goods and technological

TABLE 9.8 Balance of payments, transfers and capital flows in Mediterranean countries, 1970 and 1992

	GDP at market prices (US$ million)		Exports of goods and services		Imports of goods and services		Current account balance		Net private transfers		Net official transfers		Net long-term capital		Other net capital inflows		Net change in reserves	
			in US$ as a percentage of GDP															
	1970	1992	1970	1992	1970	1992	1970	1992	1970	1992	1970	1992	1970	1992	1970	1992	1970	1992
Albania		675		13.8		98.9		−85.1		22.2		55.4		6.1		−1.6		−2.0
Algeria	4 876	47 866	23.2	25.4	30.5	26.3	−7.3	−0.9	4.0	3.0	0.8	0.0	1.2	−0.2	−0.3	−1.8	1.7	−0.1
Cyprus		6 854		54.8		58.8		−4.0		0.3		0.1		2.6		−2.2		3.2
Egypt	7 682	35 556	12.5	34.5	18.8	42.3	−6.3	−7.8	0.4	15.4	4.0	2.9	0.0	1.5	1.7	2.6	0.2	−14.7
France	142 869	1 322 090	17.7	30.8	17.2	29.8	0.5	1.0	−0.5	−0.2	−0.2	−0.4	0.1	1.1	1.3	−2.4	−1.3	1.0
Greece	9 964	77 810	11.0	19.7	18.7	30.8	−7.7	−11.1	3.5	3.1	0.0	5.2	2.8	2.8	1.2	−0.3	0.2	0.2
Israel	5 603	69 762	25.0	29.8	46.7	39.3	−21.6	−9.6	8.0	3.9	3.6	6.0	12.1	−1.2	−1.9	−1.2	−0.1	2.1
Italy	107 485	1 220 640	17.6	22.0	17.0	23.8	0.6	−1.8	0.5	0.0	−0.3	−0.4	0.3	−0.7	−0.9	1.8	−0.2	2.0
Jordan		5 138		54.1		92.3		−38.2		17.0		6.8		2.9		12.5		−1.0
Lebanon		5 546		18.1		76.8		−58.7		3.6		0.0		−1.6		60.8		−4.2
Libya	3 994		63.5		41.0		22.5		−3.5		−2.8		3.5		−2.2		−17.5	
Malta	228	2 735	61.0	101.1	83.4	103.4	−22.4	−2.3	9.5	0.3	10.4	3.1	0.9	−3.5	8.8	3.9	−7.3	−1.6
Morocco	3 956	28 253	17.8	23.4	22.8	33.9	−5.0	−10.5	0.9	7.7	0.9	1.3	2.9	3.0	0.7	1.8	−0.4	−3.3
Portugal	7 031	95 029	19.2	27.3	28.4	35.7	−9.2	−8.4	6.9	5.0	0.0	3.2	4.0	−0.9	−1.3	0.9	−0.5	0.2
Spain	37 569	576 311	13.0	19.0	14.6	23.2	−1.5	−4.2	1.8	0.5	0.0	0.6	1.8	3.5	0.2	−3.4	−2.2	3.0
Syria[1]	2 140	17 236	15.2	26.5	18.9	25.8	−3.7	0.7	0.3	2.0	0.1	1.4	0.5	−0.2	2.3	−3.4	0.5	−0.4
Tunisia	1 439	15 575	23.1	39.3	30.8	49.6	−7.7	−10.2	1.6	3.7	2.4	0.6	3.5	3.6	1.5	1.2	−1.4	−0.6
Turkey	12 652	159 095	6.1	15.9	9.1	19.1	−3.0	−3.1	2.5	2.0	0.1	0.6	2.6	1.4	−1.1	0.1	−1.1	−0.9
Yugoslavia, Former Federal Republic of[2]	13 732	88 234	17.7	24.3	24.4	38.1	−6.7	−13.8	4.0	11.1	0.0	−11.3	1.4	−0.7	0.7	4.9	0.6	−1.5

Notes: [1] Syrian figures are for 1970 and 1991
[2] Yugoslav figures are for 1970 and 1990
Source: Elaborated from World Bank (1994, 1995)

know-how; immigration and large migrant remittances; and, most of all, massive private holdings of petrodollars. Attempts were made to create new financial centres, to create an Arab Monetary Fund and to develop Arab banks. Most petrodollars, however, were invested in financial assets managed by international financial institutions creating immense flows of interest payments to the asset owners and their countries.

The international financial institutions in which these dollars were invested recycled them as loans to developing countries. A number of developing countries around the Mediterranean and throughout the world borrowed heavily to finance industrialisation and economic modernisation projects. The foreign exchange was used to purchase the capital goods and know-how needed to develop infrastructures and export industries whose earnings, it was hoped, would enable the loans to be repaid. The countries concerned did not reckon with the impact of monetarism introduced first in the United Kingdom and the United States under the leadership of Thatcher and Reagan. Monetarism had two effects. First, it led to a profound international recession which meant that the markets and therefore the exports and export earnings of the countries that chose debt-financed industrialisation grew slower than expected (due to the recession and also due to the protectionism it prompted in advanced countries). Second, it pushed up interest rates and therefore dramatically increased the cost of servicing the debts they had incurred (see Lipietz 1987).

The consequence of these two factors was the debt crisis which had a profound effect on a number of Mediterranean economies. As Table 9.9 shows, in 1974–92 there was a major increase in international indebtedness, with a dramatic increase in the economic cost of servicing this debt up to 1986 (1992 in the case of Algeria). In a number of Arab countries (other than Egypt, much of whose debt was written off in return for supporting the US and its allies in the Gulf War), there was a depletion of reserves relative to outstanding debt. In 1986, for example, 59 per cent of Algeria's export earnings were required simply to service its debt, 51 per cent of Morocco's and 43 per cent of Egypt's.

In the face of the resulting economic crisis, the International Monetary Fund was able to secure compliance with dramatic programmes of structural adjustment. The consequences were numerous (Destreman and Signoles 1995; Gizard 1993; Valmont 1993). First, there were reductions in public expenditure on health and education. Second, there was reduced recruitment into the public administration. Third, subsidies for food and other essential goods were reduced with in many cases serious political consequences. In April 1989, for example, when the Jordanian government bowed to the wishes of the International Monetary Fund (IMF) and announced increases in the prices of basic commodities, there were five days of rioting, and the government was dismissed. In August 1996, in deference to the IMF, Jordan's government lifted subsidies on bread, raising its retail price by 300 per cent and sending shock waves through a country in which 20 per cent of the population was already below the poverty line. Fourth, there were reductions in the public funds used to finance investment, loans and subsidies to public enterprises. In Egypt, for example, these funds declined from 6 per cent of GDP in the early 1980s to 3 per cent by the end of the decade.

Employment, poverty and the informal economy

Structural adjustment and the Gulf War put the question of employment at the centre of the political debate. As a result of high population growth, on the one hand, and slow economic growth, the weakness of the industrial system and an over-expansion of activities dependent on transfers, on the other, countries with large populations are suffering from serious structural unemployment and poverty. After the start of the 1960s the lack of jobs was offset by the export of the relative surplus population to the oil states and Western Europe. After the Gulf War, however, large numbers of emigrants

TABLE 9.9 Debt and Mediterranean development

	Total external debt (US$ million)			Debt service as % of exports of goods and services			Interest as % of exports of goods and services			Reserves as % of debt		
	1974	**1986**	**1992**	**1974**	**1986**	**1992**	**1974**	**1986**	**1992**	**1974**	**1986**	**1992**
Albania	0	0	624			2		0	2			4 238
Algeria	3 366	22 634	26 349	14	59	76	4	18	18	74	17	13
Egypt	2 815	46 342	40 431	15	43	23	3	22	10	25	4	29
Lebanon	49	850	1 812			7			5	6 080	481	253
Malta	26	222	603	1	2	2	0	1	1	1 721	597	217
Morocco	1 364	17 889	21 418	7	51	31	2	23	15	37	3	18
Portugal	1 542	16 642	32 046	4	41	22	1	15	8	412	56	76
Syria[1]	593	12 919	16 513	8	24	20	1	11	6	103	4	
Tunisia	1 000	5943	8 476	7	32	23	2	12	8	44	6	11
Turkey	4 633	32 832	54 772	12	41	36	5	19	15	48	9	14
Yugoslavia, Former Federal Republic of[2]	5 604	21 501	16 294									

Notes: [1] Syrian data are for 1991 and not 1992
[2] Yugoslav data are for 1990 and not 1992
Source: Elaborated from World Bank (1994)

returned, especially to Jordan and Palestine, while the closing of EU frontiers, with more restrictive approaches to family reunion and closer control of illegal entry, reduced the possibilities of entry into Europe.

As opportunities to emigrate declined (notwithstanding the significant scale of clandestine entry to work in the informal sector in Italy and Spain), and as structural adjustment programmes reduced public sector employment and salaries, the surplus population grew quantitatively and qualitatively due to the gap between available work and the educational qualifications of the potential workforce. As Castells and Portes (1989) have pointed out:

> The moving boundaries of the informal sector will be determined by the dynamics of social struggles and political bargaining, which involve, but are not limited to, the changing conflict between capital and labour. The social challenge posed by these developments lies in the choice between the advantages of a new society based on the relationship between unrestrained capital and primary social networks and those of a society in which public institutions extend control over the logic of capital by incorporating it into a new social contract with workers and entrepreneurs operating outside the legal realm.

Traditionally, declared unemployment was low in North Africa and the eastern Mediterranean as the unemployed were supported and helped by their families and tribal and local communities, while underemployment was high especially in rural areas where plural activities and seasonal work were common (see also Mingione 1991). With the contraction in the 1980s of the oil sector, the economic crisis and the implementation of structural adjustment programmes that had a profound effect on employment and income, declared unemployment increased, along with informal work outside of agriculture. At the same time there was an increase in poverty in part due to a decline in state redistribution. Greater poverty was reflected in turn in worsening nutrition, mortality, health, education and infrastructures.

Declared unemployment increased rapidly. In the late 1980s it reached 12 per cent in Morocco, 18 per cent in Egypt, 16.8 per cent in Tunisia and 20.5 per cent in Algeria. What happened was that increasing numbers of people came to recognise themselves as unemployed. At the centre of this phenomenon were groups whose expectations were dented by structural adjustment. An indication of this situation is the fact that the highest rates of unemployment were in the cities, amongst the young and in particular amongst qualified school leavers and graduates who in the past would have been recruited almost automatically into the public administration and the public sector (especially in Algeria and Egypt). In Morocco in 1991 100 000 qualified young people, of whom one-third were graduates, were out of work; 75 per cent of them had been out of work for more than a year, adding further to the consequent increase in social tension.

At the same time there was a significant increase in the size of the (non-agricultural) informal sector. Table 9.10 shows that between the late 1970s and late 1980s the informal sector increased its share of non-agricultural employment from 21.8 to 25.6 per cent in Algeria, 43.5 to 46.9 per cent in Egypt using a direct estimate (58.7 and 65.3 per cent using a corrected measure) and 38.4 to 39.3 per cent in Tunisia but with a low of 36 per cent in 1980.

The informal sector is quite diverse, being made up of:

1. micro-enterprises with permanent staff (apprentices, wage earners, etc.) but which are not registered or fall below a certain size threshold;
2. self-employed individuals, employees of the public administration who hold second jobs to supplement salaries reduced as a result of structural adjustment, family enterprises that do not provide permanent employment (family helpers and domestic workers) and itinerants in search of an income rather than a job; and

TABLE 9.10 Trends in the share of the informal sector in non-agricultural employment, 1975–89

	GNP per head in US$ in 1990	Rate of informal employment in percentages			
		1975–80		1985–89	
Algeria	2 330	1977	21.8	1985	25.6
Tunisia	1 450	1975	38.4	1989	39.3
Morocco	970			1982	56.9
Egypt	610	1976	43.5[1]	1986	46.9[1]
		1976	58.7[2]	1986	65.3[2]
Mauritania[3]	500	1980	69.4	1988	75.3

Notes: [1] Direct estimates
[2] Estimates corrected for multiple job holding by public functionaries
[3] Data for Mauritania is for comparison
Source: After Charmes *et al.* (1993)

3. individuals involved, in countries with administered prices and exchange rates (such as Algeria), in parallel markets and markets for the sale of stolen goods which often provide good earnings for idle young men turned occasional traders.

Of these categories of employment, the ones that increased most during the crisis of adjustment in the second half of the 1980s were itinerants, domestic workers, and low productivity and low-paid tertiary sector workers. The way in which surplus labour was absorbed was therefore through a relative expansion of the most precarious types of informal employment.

THE MODERNISATION OF SOUTHERN EUROPE

The modernisation of southern Europe was also associated with agricultural modernisation and a transition from an agrarian to an urban society, major infrastructural investments, emigration and reliance on a flow of remittances to cover structural trade deficits, and industrialisation (Burgel *et al.* 1992; Sapelli 1995). In southern Europe, however, change occurred well in advance of that in the North African and east Mediterranean world, and took place in a different and more propitious economic and political context. At each point in time, moreover, there was a large cross-Mediterranean divide in the level of development; while in the 1980s this gap widened significantly as a result of a substantial increase in growth rate differentials between a number of southern EU member states and the rest of the Mediterranean (see Fig. 9.2). Increasingly the Mediterranean is the Rio Grande of Europe (Montanari and Cortese 1993).

The specific nature of the countries that fringe the northern Mediterranean is a product of their diverse yet overlapping histories and geographies. After the Napoleonic Wars the whole of the Mediterranean was peripheralised and left behind. In the period before the First World War, however, there was faster growth and industrialisation in northern Italy and Iberia, while in the Balkans the end of the war brought fundamental territorial reorganisation. In the inter-war period attempts to catch up economically in a period of global economic crisis were associated with the development of authoritarian states, nationalism and attempts to revive ancient glories. Spain, Portugal and Italy had fascist governments. Greece and Turkey experienced right-wing authoritarian rule. All of these countries were characterised by a rhetoric of national solidarity, repressive corporatist state–society relations and neo-mercantilist state action to strengthen their national economies, though with significant differences between countries and between the Iberian peninsula, Italy and the Balkans. After the

Second World War the southern European states were reintegrated into the American-dominated international division of labour, and a number of them formed the southern European and Middle Eastern flank of NATO. At different rates in different countries, the economic significance of political rule diminished as state action was used to supplement rather than replace the market. Immediately after the war a stable parliamentary regime was established in Italy where subsequent rapid economic growth led to its convergence on a north-European model and its joining the ranks of the seven leading industrial countries in the 1970s. In Greece and Turkey unstable parliamentary regimes alternated with military rule, while in Spain and Portugal authoritarian regimes remained in place. In the 1960s a phase of comparatively rapid economic growth led to an increase in the influence of modern industrial and commercial classes. These groups, who had close connections with international capital and who favoured the liberalisation of economic and political life, came into conflict with the traditional landowners and merchants who supported mercantilist policies and authoritarian rule. In the middle of the 1970s this conflict led to a crisis of the dictatorships (Poulantzas 1976), to the establishment of more stable parliamentary regimes and the election of (neo-liberal) socialist or social-democratic governments in Greece, Spain and Portugal. In 1981 Greece joined the EU, and in 1986 Spain and Portugal followed suit (Arrighi 1985).

The rapid growth of Mediterranean Europe (see Fig. 9.2) was closely related to the rapid growth of output, productivity and employment in core West European countries in the post-war golden age which was itself associated with the adoption of new principles of work organisation, mass production and oligopolistic competition, on the one hand, and the expansion of income and demand, on the other. Growth had a number of important effects on the semi-peripheral areas of southern Europe. First, it led to large-scale flows of labour to the core economies of western Europe (in France, Germany and Switzerland) and later to more local areas of industrial growth particularly in northern Italy but also in other parts of southern Europe itself.

Second, as Chapter 14 demonstrates, it led to large flows of tourists in the opposite direction. Due to increases in income, the establishment and lengthening of paid holidays, the development of mass air transport and motorways, and the activities of travel agents, tour operators, developers and local authorities who all sought to commercialise the cultural and climatic assets of the Mediterranean, the number of tourists increased dramatically, as did tourist expenditures and investment in hotels, holiday apartments, transport infrastructures, pleasure ports and other leisure facilities. The revenues from tourism are, however, highly unevenly distributed between the countries of the Mediterranean, introducing another element of north–south contrast. For instance, tourist revenues for 1990 were around $20 billion in France ($21.6 bn), Italy ($ 19.7 bn) and Spain ($18.7 bn), but much lower in all other countries – in descending order Turkey $3.3 bn, Greece $2.6 bn, Egypt $1.5 bn, Israel, Morocco and Cyprus all $1.3bn, Tunisia $1.0 bn, etc.

Third, there was significant industrial development in the Italian north and the industrial provinces of Spain, and a number of waves of externally controlled investment in the Italian Mezzogiorno and the less-industrialised parts of Spain, Portugal and Greece as international companies sought access to local markets, or sought to extract and process local resources and to use them in internationalised processes of production.

The remittances of emigrants, the expenditures of tourists, foreign investment and later Structural and Cohesion Fund transfers helped pay for net imports of consumer and capital goods and the development of infrastructures and factories. Development in southern Europe was associated, however, with a series of profound inequalities and dualisms. Aymard (1985) has argued that the roots of this unevenness lay in a series of hierarchies (in infrastructural provision, agrarian structures, the degree of commodification and the structure of accumulation) which structured space in ways that did not coincide with state boundaries but with larger (the Mediterranean, the Balkans, etc.) or smaller units (as in the antinomies of city/*contado*, plains/mountains, coast/hinterland, etc.). In the case of Greece and the Balkans these hierarchies lay in external domination; in Iberia they lay in the inverted relation between the industrialised peripheries (Catalonia, the Basque Country, and the north coast) open to the exterior and a less developed centre (Madrid) that held political power; in the case of Italy they lay in the differential trajectories of the states that unification brought together. In the face of these conditions, strategies of modernisation were adopted which imposed the costs on 'those sectors that were already inferior and dependent, locally and internationally'. Italian development, for example, involved a use of the divide between the North and the Mezzogiorno as an engine of growth that allowed the core to achieve high rates of growth and join the advanced countries of Europe but at the expense of a profound north–south disequilibrium and southern dependence.

Outside of the core areas of southern Europe, employment growth was not sufficient to absorb rural labour. Agricultural modernisation therefore led first to labour export when there were jobs elsewhere and later to growing urban unemployment. In April 1992 in the EU unemployment stood at 9.4 per cent but differed sharply from one region to another. In Southern Italy and Spain there were many areas with unemployment rates in excess of 20 per cent (Andalusia 27 per cent, Extremadura 26.3 per cent, Sicily and Basilicata 21.8 per cent and Campania 21.3 per cent), though unemployment was lower in Portugal outside of the Lisbon area and the Alentejo. In the EU 8 per cent of men were out of work compared with 11.5 per cent of women and 18.1 per cent of people under 25. Geographical variations were once again substantial. Amongst under–25 year olds unemployment exceeded 50 per cent in many parts of the Italian Mezzogiorno (72.7 per cent in the Sicilian province of Caltanissetta and over 60 per cent in the provinces of Naples, Enna and Agrigento), and exceeded 30 per cent in two-thirds of Spanish regions (EUROSTAT 1993).

At the same time the underground economy has expanded. As it is made up, on the one hand, of criminal activities (drug trafficking, prostitution and dealing in stolen goods) and, on the other, of the undeclared work of illegal immigrants and of irregularly employed members of extended families, social protection and direct taxation are by-passed. As a consequence, wage costs are held down, yet governments lack the resources to invest in health, education and other public services. Estimates suggest that 40 per cent of Spanish production of shoes involves illegal work. A similar situation prevails in the shoe and clothing industries of north-west Portugal. In Italy ISTAT estimates suggested that the output of the underground economy was equal in 1992 to 19 per cent of GDP (64 per cent in hotels and 56 per cent in the retail sector). Unemployment, illegal employment and precarious employment are, in short, key features of contemporary southern Europe and have a profound impact on the nature of Mediterranean societies.

CONCLUSIONS

The aim of this chapter was to develop a set of empirical and theoretical connections between geographies of inequality, on the one hand, and the geographies of modernisation and development in the Mediterranean, on the other. After outlining the differential trajectories of the Mediterranean EU economies, the east Mediterranean, the Mashreq and the Maghreb, attention was paid first to the less developed North African countries and the Middle East, and second to southern Europe. An account of the unresolved problems of food dependence, emigration and industrialisation, on the one hand, and the crisis of debt and structural adjustment, on the other, enabled a link to be established between the trajectories and phases of development of the first group of countries and the failure of their strategies of modernisation. The record of southern

Europe was better, though modernisation was associated with the reproduction of sharp dualisms and strong territorial imbalances. More recently, however, with the slowdown of growth in Europe, the rise of neo-liberalism, and the global competitive challenge from Eastern Europe and newly industrialised countries, mass unemployment, dependence and exclusion have come to the fore on the northern shores of the Mediterranean. In the final analysis it is these trends that lie at the root of the political divisions that have led in extreme cases to the fragmentation of states and conflicts over security. At the centre of these problems of the whole Mediterranean zone is a set of contradictions between the transnational dynamics of global and European integration and the mobilisation, management and control of the resources of water, land, shelter, work and food on which survival and development depend. As in the past a resolution of particular problems will almost certainly depend on the creation of a framework which can regulate these wider contradictions.

References

ABDEL-FADIL, M., AYUBI, N., OUALALOU, F. and HERMASSI, A. 1993: *Stato ed Economia nel Mondo Arabo*. Turin: Edizioni della Fondazione Giovanni Agnelli.

ARRIGHI, G. 1985: Introduction. In Arrighi, G. (ed.), *Semiperipheral Development: The Politics of Southern Europe in the Twentieth Century*. Beverly Hills, CA: Sage, 11–27.

AYMARD, M. 1985: Nation states and interregional disparities of development. In Arrighi, G. (ed.), *Semiperipheral Development: The Politics of Southern Europe in the Twentieth Century*. Beverly Hills, CA: Sage, 40–54.

BOUSTANI, R. and FARGUES, P. 1990: *Atlas du Monde Arabe: Géopolitique et Société*. Paris: Bordas.

BRUNEL, S. 1995: *Le Sud dans la Nouvelle Économie Mondiale*. Paris: Presses Universitaires de France.

BURGEL, G., DACHARRY, M., DAVY, L., DRAIN, M., GACHELIN, C., LHÉNAFF, R., LIEUTAUD, J., RIVIÈRE, D., SIVIGNON, M., THUMERELLE, P.-J., and VIGARIÉ, A. 1992: *La C.E.E. Méditerranéenne*. Paris: SEDES.

CASTELLS, M. and PORTES, A. 1989: World underneath: the origins, dynamics and effects of the informal economy. In Portes, A., Castells, M. and Benton, L. (eds), *The Informal Economy*. Baltimore, MD: Johns Hopkins University Press, 11–37.

CHARMES, J., DABOUSSI, R. and LEBON, A. 1993: *Population and Employment in the Countries of the Mediterranean Basin*. Geneva: ILO/MIES Working Paper 91/3E.

CONSEIL ÉCONOMIQUE ET SOCIAL 1993: *L'Agriculture Française et l'Agriculture des Autres Pays Méditerranéens: Complémentarités et Concurrences*. Paris: Direction des Journaux Officiels.

DESTREMAU, B. and SIGNOLES, P. 1995: Le difficile ajustement d'économies différenciées en rapide mutation. In Troin, J.-F. (ed.), *Maghreb Moyen-Orient Mutations*. Paris: SEDES, 5–84.

EUROSTAT 1993: *Unemployment in the Regions of the Community in 1992. Rapid Reports: Regions*, No. 2. Brussels: Commission of the European Communities.

FAO 1992: *FAO Yearbook: Production, Volume 46*. Rome: Food and Agriculture Organisation of the United Nations.

GIZARD, X. 1993: *La Méditerranée Inquiète*. La Tour d'Aigues: Éditions de l'Aube, and Paris: DATAR.

KING, R. 1977: *Land Reform: A World Survey*. London: Bell.

LIPIETZ, A. 1987: *Mirages and Miracles. The Crises of Global Fordism*. London: Verso.

MÉDAGRI 1994: *Annuaire des Économies Agricoles et Alimentaires des Pays Méditerranéens et Arabes*. Montpellier: CIHEAM, Institut Agronomique Méditerranéen de Montpellier.

MINGIONE, E. 1991: *Fragmented Societies: A Sociology of Economic Life beyond the Market Paradigm*. Oxford: Blackwell.

MONTANARI, A. and CORTESE, A. 1993: South to North migration in a Mediterranean perspective. In King, R. (ed.), *Mass Migrations in Europe: The Legacy and the Future*. London: Belhaven Press, 212–33.

MUTIN, G. 1995a: Campagnes en crise. In Troin, J.-F. (ed.), *Maghreb Moyen-Orient Mutations*. Paris: SEDES, 133–78.

MUTIN, G. 1995b: Les hydrocarbures du monde arabe: une richesse inégalement répartie. In Troin, J.-F. (ed.), *Maghreb Moyen-Orient Mutations*. Paris: SEDES, 291–325.

MUTIN, G. and DURAND-DASTES, F. 1995: *Afrique du Nord, Moyen-Orient, Monde Indien, Géographie Universelle*. Paris, Berlin and Montpellier: RECLUS.

POULANTZAS, N. 1976: *The Crisis of the Dictatorships: Portugal, Greece, Spain*. London: New Left Books.

SAPELLI, G. 1995: *Southern Europe since 1945: Tradition and Modernity in Portugal, Spain, Italy, Greece and Turkey*. London: Longman.

UNDP (United Nations Development Programme) 1994: *Human Development Report: 1994*. New York: Oxford University Press.

VALMONT, A. (ed.) 1993: *Économie et Stratégie dans le Monde Arabe et Musulman*. Paris: Editions EMAM.

WORLD BANK 1994: *World Tables*. Washington: International Bank for Reconstruction and Development.

WORLD BANK 1995: *World Tables*. Washington: International Bank for Reconstruction and Development.

10

THE EUROPEAN UNION'S MEDITERRANEAN POLICY: FROM PRAGMATISM TO PARTNERSHIP

ALUN JONES

INTRODUCTION

Although the European Community had concluded trade agreements with several Mediterranean states during the 1960s and early 1970s, it was only in 1972 that some degree of coordination was introduced in the Community's external trade relations with these countries. The series of agreements which were concluded after 1972 under what the Community described as a 'Global' Mediterranean Policy (GMP) were designed to secure some political stability in the region on the basis of free trade and EU financial aid for economic development. However, the successful applications for EU membership by Greece, Spain and Portugal created significant difficulties for the Community after 1986 in sustaining this 'global' approach. Moreover, renewed applications for membership of the Community by other Mediterranean states such as Cyprus, Malta, Turkey and Morocco compounded the problems. By the early 1990s, therefore, the Community was forced to reassess its policy towards the region, leading to the launch of a 'New Mediterranean Policy' (NMP) in June 1990 and more recently culminating in an agreement to create in the long term a Euro-Mediterranean Economic Area. It is the scope of this relatively brief chapter to trace this intensifying relationship between the EU and the Mediterranean region, recognising that this relationship is an important factor in many of the themes and topics addressed by other chapters in this book. For a preliminary and summary indication of individual countries' relationship to the EU, reference can be made back to Chapter 8, specifically Tables 8.2 (final column) and 8.4.

THE EUROPEAN UNION'S MEDITERRANEAN POLICY, 1957–72

Although the Treaty of Rome which created the European Economic Community (EEC) did not delegate any foreign policy powers, the obligation on the part of the EEC to evolve a common commercial policy ensured that it would inevitably come into contact with non-member states. In addition, it can be argued that the Treaty itself was based to some degree upon overseas interests and commitments (Van der Lee 1967). At the time of its signing, four of its signatories had substantial colonial interests especially in Africa (Twitchett 1976). France, the member state with the largest overseas empire, demanded that its colonial dependencies be associated in some form with the Community (Cosgrove 1969), a demand which may well have been due to the financial burden of maintaining these overseas colonies and the prospect that the EEC would share in France's overseas economic development programmes. Hence the formal links between the newly established Community and the Mediterranean region were in essence based upon the colonial ties between France and the Maghreb countries and between Italy and Libya. France, in fact, insisted that the Treaty of Rome should commit the Community to establish some form of economic association with the Maghreb.

At the beginning of the 1960s, the only possible provisions for special relations between the EEC and 'third' countries, apart from full membership, was the formula of association (Tovias 1977). Legally, the EEC's external trade policy could only be based upon Articles 237 and 238 of the Treaty, which related to full membership and association of the Community respectively. These Articles nested comfortably within the guidelines for international trade agreed through GATT – the General Agreement on Tariffs and Trade. European states could apply to become full members of the Community provided they met certain economic and political criteria, comparable to those which pertained to its members.

Throughout the early 1960s the EEC responded to a variety of requests from Mediterranean states for association agreements; this demonstrated an increasing 'European policy' on the part of the Mediterranean states themselves rather than any coherent external relations policy towards the Mediterranean on the part of the EEC.

The first Mediterranean country to sign an agreement with the EEC was Greece, which concluded an association accord in July 1961 (Tsoukalis 1981). This was followed soon after by a similar association agreement between the EEC and Turkey; the latter not wishing to lose any political advantage to its neighbour (Haseki and Andritsakis 1976). Both agreements were concluded under Article 238 of the Treaty of Rome, with the eventual prospect for both countries of full membership of the Community under Article 237. The Community's view of these agreements was that economic prosperity through freer trade would foster political stability in both countries. However, a number of key events dramatically forced the Community to reappraise its association policy towards the Mediterranean. The first of these was the *coup d'état* in Greece in 1967 which not only resulted in the Community freezing the association agreement with that country but also highlighted the dangers for the EEC of maintaining too close relationships with third countries. The Arab–Israeli war of 1967 sowed further seeds of doubt in the Community about its relationship with the Mediterranean. Consequently, the EEC sought an alternative, and more politically cautious, means by which it could maintain relations with countries in the region. Articles 111–114 of the Treaty provided an appropriate solution to this issue, allowing the Community to sign preferential trade agreements with third countries. This removed the necessity of close political ties between the Community and third countries as embodied in Article 238.

Developments within the EEC during the late 1960s, in particular the creation of the Common Agricultural Policy (CAP), combined with the growth in levels of trade between the Community and the Mediterranean (a threefold increase 1960–70), forced the Community into a 'stock-taking' of its Mediterranean agreements (Lambert 1971). Although the major limiting factor to any comprehensive external relations policy was the need to comply with GATT rules, this did not mean that the Community kept closely within these international regulations. The agreements which the EEC signed with Tunisia and Morocco in 1969, for example, fell well short of GATT criteria, and as a result acted as a 'detonator' in the establishment of a system of preferential agreements with Mediterranean states. The applications for EEC association by both Spain and Israel in 1968, with the obvious political implications that these presented to the Community, rendered preferential agreements even more necessary, especially as the EEC was the destination for over one-third of Spain's and one-quarter of Israel's total exports. Despite a great deal of American protest, on the grounds of trade diversion away from the US and the EEC's role in the irresistible trend towards regionalism in the global trading economy, the Community after 1969 signed agreements with several Mediterranean states, including Spain, Israel and Egypt, on the basis of preferential treatment. These concessions not only marked a significant stage in the strengthening of EEC trade links with the Mediterranean countries but also demonstrated a decreasing degree of conformity with GATT regulations and an increased Community desire to create an economic sphere of influence in the Mediterranean region. This was further demonstrated not only by the renegotiation of agreements with countries such as Lebanon (1972), but also by the conclusion of association agreements with Malta (1971) and Cyprus (1972).

By 1972, then, the Community had successfully concluded, or was about to conclude, agreements with all countries in the Mediterranean region except Libya, Albania and Syria. It soon became obvious, however, as Shlaim (1976, p. 4) remarked, that 'this disjointed incrementalism would have to give way to a more systematic and coherent approach ... which would take into account the problems and needs of the region as a whole'.

The emergence of a 'global' Mediterranean policy after 1972

Although the EEC's Global Mediterranean Policy (GMP) was developed throughout the 1970s, the underlying concept can be traced back to 1964 when the Italian government issued a statement setting out the need for an overall policy towards the region. The issue was also raised in the European Parliament in that same year and received a reply from the European Commission in March 1964. From the Mediterranean countries' point of view, the similarities of their economies and the necessity for the Community to adopt a common approach to the Mediterranean were the significant conclusions drawn from a conference held at Nîmes (France) in May 1964 under the aegis of the FAO. By 1972 the idea of an overall approach to the Mediterranean (known as 'global' in Community parlance) had become a realistic prospect for several reasons:

1. The proposed enlargement of the EEC to include the United Kingdom, Ireland and Denmark meant that additional protocols would need to be signed to existing agreements with Mediterranean states.
2. Many of the agreements themselves were due for renegotiation (for example, the Moroccan and Tunisian agreements were due to expire in 1974).
3. The mosaic of agreements which had been concluded were signed at various stages of European economic development and union and in many respects reflected the emerging nature of the Community's institutions and policies.
4. There were far-reaching economic, political and strategic reasons for the EEC to implement an overall policy. Economically the Mediterranean region had become an important trading bloc for the EEC, even more important in terms of trade than the US and Canada. In addition, the Mediterranean region was an important zone for European investment and the Community was a growing market for Mediterranean exports: on average the EEC took over 35 per cent of the region's exports (Ehrhardt 1971). Politically the region was unstable, not only as a result of inter-state tensions, as with Greece and Turkey and the wider Arab–Israeli dispute, but also as a result of the presence of the superpowers in the region. As Blake (1978,

p. 256) noted, both the US and the Soviet Union had realised 'the strategic and political value of the Mediterranean sphere in the global confrontation between east and west'. The Community therefore believed that by developing a common approach to the region the EEC could become a viable alternative to the bi-polar power structure which hitherto had had the unfortunate effect of 'splitting the Mediterranean into conflicting spheres of interest' (Gasteyger 1972, p. 7).

5. The Mediterranean region was also strategically important to the EEC since almost 20 per cent of Community oil imports originated there (Tsoukalis 1977), and was also vital in the geography of oil transit from the Middle East petroleum fields (see Chapter 9).

In October 1972 the Heads of State and government of the EEC met in Paris to discuss the precise details of a GMP. It was agreed that the GMP would have three principal elements: free trade in manufactured goods, the removal of restrictions on a substantial part of agricultural trade, and financial and technical cooperation. Although the GMP was implemented in late 1972 it was not until 1975 that the first agreement under the GMP was signed – ironically as it turned out, with Israel. There were several reasons behind the slow progress:

1. The GMP was causing some dissension among EEC members, especially between the UK, on the one hand, and Italy and France, on the other. The UK government was particularly concerned about the strain that the GMP would place upon UK–US relations as a result of GATT infringements. In addition, the UK was eager to move the EEC towards establishing a broad developing–world policy which would include British overseas interests.
2. The Middle East war of October 1973 and the subsequent quadrupling of oil prices by OPEC meant that EEC Mediterranean policy could not be isolated from policy towards the Arab countries (known as the Euro-Arab dialogue). Consequently, it became more difficult for the EEC to adopt an even-handed policy towards the region in terms of north–south Mediterranean interests, and towards Israel and the Arab states of the region. An added complication was that the legal framework on which the GMP was based, i.e. the prospect of full EEC membership for European states in the Mediterranean and preferential trade accords for non-European states, turned the EEC's Mediterranean policy into one addressed mainly to the Arab countries of the area (Tsoukalis 1977, p. 422).
3. Another major factor delaying the progress of the GMP was the somewhat short-sighted belief by the Community that a uniform approach could be adopted to the region. Marked socio-economic differences existed (and of course still exist) between Mediterranean countries. If we take levels of wealth as measured by GNP per capita, the variations can be contrasted by examining various groupings of countries (all figures are for 1978 and are in US $): the 'European' Mediterranean (Spain 3520, Portugal 2020, Greece 3270, Yugoslavia 2390, Malta 2160, Cyprus 2110); North Africa (Morocco 670, Algeria 1260, Tunisia 950, Egypt 400); and a more heterogenous group (Israel 4120, Turkey 1210, Syria 930 and Jordan 1050). Such diversity led Robertson (1976, p. 334) to suggest that any global policy was sure to be a 'sketchy framework'. Furthermore, the demand for a comprehensive concept raised the question of the geographic–geopolitical delimitation of the Mediterranean region itself.
4. Difficulties were also encountered in the degree to which the Community was prepared to make concessions for agricultural and manufactured imports from the Mediterranean. The principal problem arose from the fact that almost all the Mediterranean countries competed for access to the Community market for similar products. For example, textiles, a sector in which the Community was already facing structural

difficulties, accounted for a large share of total manufactured outputs from the Mediterranean states in the late 1970s (11 per cent of Spain's manufactured exports; 14 per cent in the case of Morocco; 18 per cent for Egypt; 23 per cent Cyprus; 34 per cent Portugal; 53 per cent Malta). On the agricultural front, citrus fruits, olive oil, wine, fruit and vegetables were products over which the Community had been encountering problems of surplus production and where marketing arrangements to protect French and Italian producers under the CAP were coming under increasing strain.

The agreements which the Community signed with the Mediterranean states under the GMP (Israel 1975, Maghreb states 1976, Mashreq states [Egypt, Syria, Lebanon, Jordan] 1977), the reaffirmation to create a customs union with Cyprus, Malta and Turkey, discrepancies between EEC treatment of the Maghreb compared to the Mashreq states, and the applications by Greece, Spain and Portugal in 1977 (after the successful overthrowing of dictatorships) to become full members of the Community, all led observers to remark that the GMP appeared to be one of policy juxtaposition rather than coordination (De la Serre 1981).

Greek and Iberian membership of the EEC was regarded by the Community as essential for the political and strategic stability of the Mediterranean region as well as to protect these countries' fledgling democracies; an objective that was felt unequivocally by the Community to outweigh all the economic, social and financial problems of membership (Seers and Vaitsos 1982). Moreover, the impact upon the Community's relations with other Mediterranean states was deemed not to be of sufficient political concern to derail Community enlargement plans. Indeed, Greece's accession in 1981 went ahead without any consultation with the Maghreb or Mashreq countries. Spanish and Portuguese accession to the Community in 1986 further weakened the understanding on which the Mediterranean policy had been based and tended to confirm a view held by many southern Mediterranean states that the Community's Mediterranean policy was largely biased in favour of European Mediterranean states. As the Community's largest Mediterranean trading partner, Spain's membership severely threatened the export possibilities for agricultural and manufactured goods from countries in the Maghreb and Mashreq since not only would Spanish producers have free access to Community markets but also support under the Community's CAP (Hine 1985). The dynamic effects of Community membership for the Iberian economies were regarded as particularly threatening for the rest of the Mediterranean since the prospect of Community self-sufficiency in key sectors would be raised, thereby increasing the possibility of the EEC closing its markets to imports from the rest of the Mediterranean. In the agricultural sector the GMP had encouraged Mediterranean states to come to depend upon the Community as an export market. For example, 96 per cent of Morocco's olive oil, 93 per cent of its tomatoes and 92 per cent of its potato exports were destined for the Community market in 1986. Similarly 100 per cent of Tunisia's potato exports, 90 per cent of its vegetables, and 87 per cent of its olive oil exports were for EEC consumers.

European Union policy towards the Mediterranean region since the mid-1980s

Since the mid-1980s the Community's policy towards the Mediterranean region has thus been faced with a number of challenges. Important among these have not only been the southern enlargement of the Community to include Spain and Portugal, but also the democratisation of Eastern Europe and this region's establishment of closer relations with the Community, which has led to suggestions that the Community might loosen ties with the Mediterranean. Of equal importance was the Community's Single Market programme, which was perceived by Mediterranean countries as the completion of a 'fortress Europe'. Nor should we forget the economic, political and military conflicts in the Mediterranean countries themselves, notably those outlined in Chapter 8. Collectively, these factors prompted the Community to reassess its policy towards the Mediterranean region at the end of the 1980s.

Consequently, in 1990 the Community agreed a new Mediterranean policy based upon six principal features:

1. support for structural adjustment in the economies of the Mediterranean region;
2. encouragement of private investment by European companies in the region;
3. increasing member state and Community financial assistance to the area;
4. maintaining (and improving where possible) access for Mediterranean products to the Community markets;
5. consultation with the Mediterranean states over progress towards a Single Market in the Community and its wider implications;
6. strengthening in a formal way the Community's political and economic dialogue with the region.

The Community allocated some Ecu 4405 millions (in both grants and loans from the European Investment Bank) to these measures for the period 1991–96. Just under half of this total aid package was devoted to country-specific financial assistance (Egypt, Morocco, Algeria, Tunisia, Syria, Jordan, Israel, Lebanon) and the rest for more broadly based assistance including four Mediterranean networks. These networks are known specifically as Med-Urbs, Med-Campus, Med-Invest and Med-Media and are designed to increase cooperation between countries in the Community and those in the Mediterranean. Med-Urbs aims to contribute to the improvement of the quality of life of urban populations and to the strengthening of local democracy by stimulating decentralised cooperation between local authorities of the EU and the Mediterranean countries. Several priority areas have been established within this network including environmental protection, energy resource management and urban transport schemes. In this last context the cooperation between local authorities in Brussels, Istanbul, Lisbon, Tunis and Valencia over the integration of light railways into urban transport planning is held up by the Community as a model of this type of network activity. Med-Campus involves cooperation between research institutions in the EU and the Mediterranean states on key scientific questions such as the causes, mechanisms and consequences of desertification, the development and application of renewable energy sources, and research into and management of water resources (see Chapter 15). Med-Invest is designed to support the creation of small and medium-sized enterprises through the promotion of meetings between potential business partners and administrative assistance in the marketing of products. The Med-Media initiative meanwhile was designed to facilitate training of TV, radio and print media journalists.

Despite these new developments in the Community's relations with the Mediterranean, the Community has felt that there is a need to go further in order to promote economic growth and thereby ensure political stability in the region. The European Commission, for example, announced in 1994 that results from the initiatives taken in 1990 had been mixed, adding that 'the instruments used and the policies pursued have been too narrow in scope and insufficiently effective in comparison with the needs of the region' (European Union 1994). Faced with renewed instability in the region (Algeria, Yugoslavia), the Commission at the end of 1994 proposed the creation of a 'Euro-Mediterranean Economic Area'. The Commission in its report to the Council and European Parliament believed that a free trade area between the Community and the Mediterranean could be achieved by the year 2010. If achieved, this would represent the largest free trade area in the world. The trade area would involve reciprocal free trade in all manufactured goods, preferential and reciprocal access for agricultural products of interest to both parties, and free trade among Mediterranean states themselves (European Commission 1994). The Commission document spelt out clearly the reasons for a renewed emphasis on this region, in particular the continued interdependence between the EU and the states of the Mediterranean and the recognition that the Community cannot be isolated or sheltered from political instabilities in the region.

THE BARCELONA CONFERENCE ON EU POLICY TOWARDS THE MEDITERRANEAN

Following on from the Commission's blueprint, ministers from all the member states of the European Union and the Mediterranean (except Albania, Libya and former Yugoslavia) met in Barcelona in November 1995 to set out the aims of a new Mediterranean policy with the general objective of turning the region into 'an area of dialogue, exchange and cooperation guaranteeing peace, stability and prosperity' (European Union 1995). The conference agreed upon a Euro-Mediterranean partnership, the principal features of which are strengthened political dialogue on a regular basis, the development of economic and financial cooperation, and measures to combat poverty and promote greater understanding between cultures. The terms for the meeting mirrored those already taken by the EU in relation to central and eastern Europe – though in this case the prospect of membership of the EU was not on the agenda. The conference, from the EU's perspective, marked the beginning of an ambitious policy of cooperation with the south which would form a 'counterpart to the policy of openness to the east' and would give the 'EU's external action its geopolitical coherence' (European Union 1995). The EU's view is that freer trade and financial assistance will create stability and increase prosperity in the southern and eastern Mediterranean, which in turn will underpin the ongoing (though turbulent) Middle East peace process and promote political pluralism, as well as help to damp down some of the 'root causes' of emigration.

The agreement over political dialogue underpins much of the partnership, with Mediterranean states committing themselves to adhere to the principles of international law in respect of human rights and fundamental freedoms, the territorial integrity of states, refraining from developing military capabilities beyond their legitimate defence requirements, and cooperating with the EU over the prevention of terrorism, international crime and drugs trafficking.

The economic and financial aspects of the Euro-Mediterranean partnership are founded upon the progressive establishment of a free trade area, the implementation of appropriate economic cooperation, and a substantial increase in the level of financial assistance offered by the EU to the Mediterranean states. The year 2010 has been set as the target date for the gradual establishment of a free trade area. To ease the free movement of manufactured goods and the liberalisation of agricultural produce, the Mediterranean states have agreed to introduce measures for rules of origin and certification. Economic cooperation is envisaged in the fields of energy supply, fisheries policy, protection of the Mediterranean environment, and agricultural modernisation.

The Barcelona conference agreed that the creation of a free trade area would require massive injections of European finance. Some Ecu 4685 millions have been allocated for the Euro-Mediterranean partnership for the period 1995–99 with additional assistance set aside in the form of loans from the European Investment Bank.

Partnership in the field of social, cultural and human resources specifically addresses the problem of migratory pressures, illegal immigration into the EU and the need for agreements between the EU and the Mediterranean states over procedures to readmit their nationals who are in an illegal situation. These migratory issues are the subject of Chapter 11. The Barcelona Convention outlines priorities for cooperation between the EU and the

Mediterranean states in the fields of industrial and agricultural development, fisheries, energy, transport, environment, science and technology, tourism and health policy. In farming, for example, the aims are to encourage the diversification of agricultural production away from those products where the EU's marketing arrangements are under pressure (olive oil, certain fruit and vegetables), to support agricultural practices that are more respectful of the environment, and to promote integrated rural development schemes which not only recognise the need to improve basic services and associated economic activities but are mindful of the difficulties facing those regions which are affected by the eradication of illicit crops.

Environmental issues are regarded as a key aspect of the Convention and are to be the subject of further long-term action between the EU and the Mediterranean states which will be coordinated by the European Commission. In the next few years attention will be devoted to assessing the nature and extent of environmental problems in the Mediterranean and deciding upon the necessary course of policy action. In a clear effort to avoid political controversy at an early stage over pollution sources and country culprits the Convention conveniently agreed upon a shopping list of issues to tackle, and these included river and coastal management, forest protection, natural heritage and site conservation, soil degradation, desertification and erosion. These issues are discussed in Chapters 16 and 17.

While the EU states were at pains to emphasise cooperation, it is clear that the relationship remains unequal. For the EU, the conference was primarily to buy security. Indeed, even the immigration component of the talks was regarded as a security rather than a political (let alone a social) issue. EU support for democratic reforms in the southern Mediterranean states is grounded in the belief that this will reduce the potential for large-scale migrations. However, whether trade concessions and EU cash transfers will be sufficient in this regard remains to be seen.

CONCLUSIONS

The future direction of EU policy towards the Mediterranean region has thus been laid out for the medium term. A Customs Union agreement with Turkey, which came into force in 1996, and association agreements with Malta and Cyprus (dating back to 1971 and 1973 respectively) pave the way for possible full membership while a free trade area embracing the whole region is the ultimate goal (Redmond 1993; Wessels and Engel 1993). To this end, in 1995 six Mediterranean states signed or were seeking to sign new agreements with the EU. Tunisia, Israel and Morocco have done so while Egypt, Jordan, and Lebanon have begun talks with the European Commission. The agreement with Turkey extends most of the EU's trade and competition rules to its economy, and brings the country into the single market programme. The customs union established the closest trade and political relationship the EU has with any non-member country. In principle, the agreement should prepare Turkey for full membership of the Union at a later (though unspecified) date. For the Turkish government accession to the Union remains the committed goal. As the Mediterranean region's most populous Moslem state and its largest and most sophisticated economy, an Islamist threat to the country would have far-reaching implications for the EU.

The Community's Mediterranean policy has thus altered considerably since the 1960s. EU policy towards the Mediterranean in the 1990s and beyond recognises not only the ways in which the Union itself is changing but also the increased importance of the Mediterranean region for trade, labour, and as a location for international terrorism based upon arms and drugs trafficking; though the region is not unique in this regard. The EU will have to perform a careful balancing act between encouragement for economic reform, political association and eventual membership of the EU for countries in central and eastern Europe whilst at the same time sustaining commitment to the economic reform process in the Mediterranean on which the Barcelona

Convention is based. While stability in the region remains the committed goal of the European Union, it may in the long run have to pay a bigger financial and political cost for it.

References

BLAKE, G. 1978: Settlement and conflict in the Mediterranean world. *Transactions of the Institute of British Geographers* **3**, 255–8.

COSGROVE, C.A. 1969: The Common Market and its colonial heritage. *Journal of Contemporary History* **4**, 73–88.

DE LA SERRE, F. 1981: The Community's Mediterranean policy after the second enlargement. *Journal of Common Market Studies* **19**, 377–87.

EHRHARDT, C.A. 1971: The EEC and the Mediterranean area. *Außenpolitik* **22**, 20–30.

EUROPEAN COMMISSION, 1994: *Strengthening the Mediterranean Policy of the EU*. Brussels: European Commission Communication Com 94 (427) Final.

EUROPEAN UNION 1994: *Europe and the Mediterranean*. Brussels: European Commission Background Report.

EUROPEAN UNION 1995: *Conférence Euro-Méditerranéenne de Barcelone*. Brussels: European Commission Info Note 52/95.

GASTEYGER, C. 1972: The Mediterranean, Europe and the Maghreb. *Atlantic Papers* **1**, 15–23.

HASEKI, A. and ANDRITSAKIS, A. 1976: The association of Greece and Turkey with the EEC. University of Amsterdam: *Occasional Papers of the Europa Institute* **2**, 5–19.

HINE, R.C. 1985: *The Political Economy of European Trade*. London: Wheatsheaf.

LAMBERT, J. 1971: The Cheshire cat and the pond: EEC and the Mediterranean area. *Journal of Common Market Studies* **10**, 37–46.

REDMOND, J. 1993: *The Next Mediterranean Enlargement of the European Community: Turkey, Cyprus and Malta?* Aldershot: Dartmouth.

ROBERTSON, D. 1976: The EEC Mediterranean policy in a world context. In Shlaim, A. (ed.), *The EEC and the Mediterranean Countries*. Cambridge: Cambridge University Press, 56–67.

SEERS, D. and VAITSOS, C. 1982: *The Second Enlargement of the EEC*. London: Macmillan.

SHLAIM, A. (ed.) 1976: *The EEC and the Mediterranean Countries*. Cambridge: Cambridge University Press.

TOVIAS, A. 1977: *Trade Preferences in Mediterranean Diplomacy*. London: Macmillan.

TSOUKALIS, L. 1977: The EEC and the Mediterranean: is 'global' policy a misnomer? *International Affairs* **53**, 422–38.

TSOUKALIS, L. 1981: Greece in Europe: the tenth member. *The World Today* **37**, 120–6.

TWITCHETT, K.J. 1976: *Europe and the World: the External Relations of the Common Market*. London: Europa for the David Davies Memorial Institute of International Studies.

VAN DER LEE, J.J. 1967: Association relations between the EEC and African states. *African Affairs* **66**, 197–212.

WESSELS, W. and ENGEL, C. 1993: *The European Community in the 1990s: Ever Closer and Larger*. Bonn: Europa Verlag.

11

POPULATION GROWTH: AN AVOIDABLE CRISIS?

RUSSELL KING

INTRODUCTION

In the past it could be said that population functioned as a rather quiet background factor to the human geography of the Mediterranean. Whilst this statement can certainly be challenged – in fact, it will be elaborated upon and modified in the next few pages of this chapter – it does contain an element of truth when compared to the present situation. Since the 1980s it has been increasingly realised that demography lies at the heart of Mediterranean destiny. Population *growth*, in particular, is connected to a whole series of processes which are shaping the future of the region: economic development (or the lack of it), unemployment, urbanisation, international migration and geopolitics. The Mediterranean is now a demographic frontier between two entirely different population systems. This demographic fault-line is also a divide between two very different economic and politico-religious regimes. The relationships between population, economics and geopolitics are of course very complicated and no clear lines of simple causation can be established: all are related to each other – and to the wider dimensions of ethics and human rights – in a complex and multivariate setting.

Easily the greatest concern as viewed from the European side of the Mediterranean is that of a 'population overflow' from the southern to the northern shore. High rates of population increase in North Africa and the Middle East are perceived as the main structural component of this 'threat' but more worrying to European governments in the short term are geopolitical 'triggers' like the Gulf War or the present unstable position in Algeria. For instance, escalation of the Algerian conflict would have two immediate effects: first, a northward exodus of people escaping the general economic and political chaos, including refugees fleeing from possible persecution; and second, the transfer, via both established and new Algerian immigrant communities in France and elsewhere, of the campaign of violence into Europe's heartland.

This chapter has four main sections. In the next section we will briefly examine the role of population in Mediterranean history. Given the length of this history as well as the complexity and uncertainty of some of the relationships, this section will be a set of pointers rather than a complete survey. The second part of the chapter will examine the role of the Mediterranean as a demographic reservoir in the post-war decades, providing migrant labour for the prosperous economies of northern Europe in the 1950s, 1960s and early 1970s, and, to a lesser extent, for the oil-rich countries of the Middle

East during the 1970s and 1980s. The third part will examine current growth patterns amongst Mediterranean populations, focusing in particular on the extremely sharp divide between the European Mediterranean, on the one hand, and the southern and eastern shore countries, on the other. This section will also set out and comment upon United Nations and Blue Plan scenarios of population growth for the future. The final substantive part of the chapter will examine international migration as the 'solution' to the population 'crisis'. The conclusion to the chapter will briefly return to the issue of the centrality of demography in Mediterranean human geography and regional planning and stress once again the interconnectedness of population and many other themes which are the subjects of separate chapters in this book.

In order to keep the discussion manageable, and because many population data (e.g. on migration or growth scenarios) are only available at national levels, the chapter will deal mainly with national entities, recognising the problems of such an analysis pointed out in Chapter 1.

POPULATION IN HISTORY

The Mediterranean has always been a demographically dynamic region. This dynamism is expressed both through time – growth being the key process – and, via colonisation and migration, across space. Another established characteristic of Mediterranean populations is their *urbanism* – a long and continuous tradition of urban life (Houston 1964, pp. 5–6). For thousands of years Mediterranean populations have established themselves largely in and around urban sites; these were the fixed points around which human transformation of the Mediterranean landscape took place. The current urban network, with very few exceptions such as recent tourist settlements, is the direct inheritance of 2000–3000 years ago, in particular from the Roman Empire, as we saw in Chapter 5. Tightly packed settlements derive from the cultural traditions of the area and from specific factors such as shortage of water and the repeated need for defence. For many centuries towns grew by an accretion of dwellings on the spot rather than by horizontal suburban expansion. Densities became extremely high, especially in Islamic towns and in cities inherited from the Middle Ages; many towns did not expand beyond their medieval fortifications until the late eighteenth century (Grenon and Batisse 1989, p. 12).

The distribution of the major cities of the Mediterranean reflects above all the map of both land and sea transport (Braudel 1972, p. 316). This is true both now (Fig. 11.1) and in the past. In the Western Mediterranean and on the north shore of the eastern basin the great cities were all near the sea – Barcelona, Marseilles, Genoa, Naples, Venice, Ragusa (Dubrovnik), Athens, Constantinople, Algiers and many more scarcely less important. In the south and east of the Mediterranean, the great inland cities of Islam are a clear response to the importance of roads through the desert. Naples was the largest city in Christendom in the sixteenth century (it had 280 000 people in 1595) but now its role is but a regional one within the south of Italy: Rome, Algiers and Barcelona are the three big metropolitan centres of the western basin (Fig. 11.1). Athens and Alexandria have achieved similar dimensions to these in the eastern basin, but Istanbul (already 700 000 in 1600) remains the largest metropolitan area on the shores of the Mediterranean. It is exceeded in size today by Cairo (the largest city in Africa) but it is debatable whether Cairo can be regarded as a Mediterranean city. The same could be said of the huge, sprawling Milan conurbation in northern Italy.

Rural population densities have always been closely related to the physical environment and its agricultural potential. Remarkably high densities have characterised areas of irrigation and intensive garden agriculture such as the Spanish *huertas* or the rich volcanic soils of Vesuvius and Etna: both Houston (1964) and Ribeiro (1983) indicate rural densities of up to 700 per square kilometre for these areas. Much lower densities were found in areas of traditional dry farming

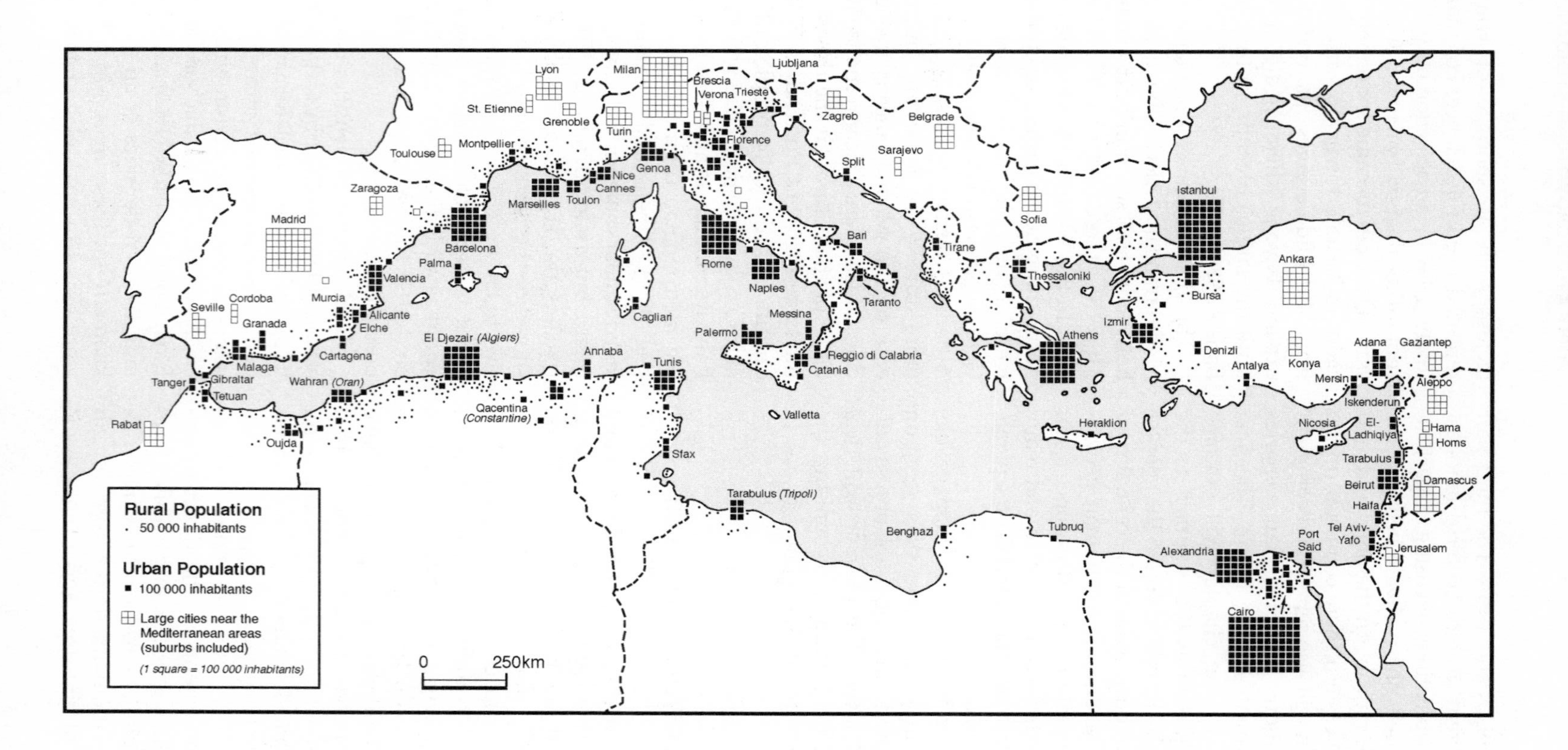

FIGURE 11.1 Population distribution in the Mediterranean Basin, *ca.* 1985

Source: Grenon and Batisse (1989, pp. 10–11)

(wheat, olives, etc.) and still lower densities in hill and mountain districts, fit only for pastoralism. However, many other factors caused variations from this 'human response' model: the differential availability of technology, such as the cultivation techniques spread by the Romans or Arabs; the availability of profitable markets for certain specialised products, including markets external to the Mediterranean; vulnerability to attack, which emptied many fertile coastal plains of their inhabitants who retreated inland and uphill; and land tenure–peasant ownership leading to a more complete transformation of the land, and hence higher population densities, than where farm labourers were marginalised by latifundia or insecure tenancies. Not to be forgotten are cases where the model is reversed, population growth forcing the transformation of the environment by the creation of intensive agricultural ecosystems based on terracing and careful soil husbandry. Mediterranean islands provide some good examples of this.

Calculations of numbers of people in the Mediterranean Basin in the past are severely hampered, if not made virtually impossible, by the paucity of data, by the political fragmentation of the area, and by the familiar problem of how to define the region. Braudel's detailed discussion of the sixteenth-century sources resulted in an estimate of a doubling of the region's population between 1500 (30–35 million) and 1600 (60–70 million) (Braudel 1972, pp. 394–418). After 1600 it took another three centuries or more for the population to double again, reaching 130 million in 1937. Then the rate accelerated: 175 million in 1956 and 330 million by 1980. By this time the region was adding 8 million per year, an annual increment equivalent, as the Portuguese geographer Orlando Ribeiro ruefully pointed out, to the population of his own country (Ribeiro 1983, pp. 19–20). The story of modern, and future, population growth will be picked up later in the chapter.

For many periods in Mediterranean history, and in many parts of the region, population growth has outstripped economic growth. Agricultural production, in particular, has often not matched demographic increase, whilst industrial development and trade have failed to compensate. Instead, the population itself has become an item of trade, exported the world over. The internationalisation of Mediterranean people has been long established. According to Braudel (1972, p. 312), 'there was proverbially and probably literally a Florentine in every corner of the world'. But here one should distinguish between elite migrations of merchants, bankers, colonial administrators, etc., and the humble masses driven by hunger and oppression to search for a better life elsewhere.

In fact, the history of Mediterranean migration contains many types. Rural to urban migration not only relieved pressure from the overcrowded countryside but also fed towns with their 'indispensable immigrants' – necessary for the towns to grow and replenish their populations culled by plagues (Braudel 1972, pp. 334–8). Often these fairly localised migration currents led from nearby mountain areas where agriculture was marginal and where population growth had 'overshot' the ability of the environment to sustain it (McNeill 1992). Other cityward migration flows were longer-range. At Marseilles the typical immigrant was a Corsican; in its heyday Venice attracted an extraordinary polyethnic population; local migrants from the *terra firma*, more distant recruits from Friuli, Romagna and Le Marche, plus significant numbers of Albanians, Greeks, Turks, Armenians and Jews. Not all were poor, uneducated labourers; some were merchants, artisans and industrialists. The Jews, here expelled, there accepted, played a particularly prominent role in the transfer of technology and enterprise around the Mediterranean.

Other migrations took the form of colonisations, often to adjacent territories where people could settle in a relatively familiar climatic and economic environment (Ribeiro 1983, pp. 156–7). The Ancient Greeks expanded westward to Sicily and Calabria (Magna Grecia), and eastward to Asia Minor. In the Middle Ages Venice colonised the Dalmatian coast and island territories further afield. The nineteenth century saw the massive French colonial settlement of Algeria. Fascist Italy colonised Libya with its farmers and agricultural technology. Few of these colonisations were harmonious and not all were permanent; often they met a final backlash. Thus the Greeks were expelled from Turkey in 1922; the French colonial *pieds noirs* fled Algeria in 1962, by which time the Italians too had withdrawn their settlers from various parts of North Africa.

Finally in this brief survey of pre-Second World War migrations, there were emigration flows out of the Mediterranean Basin. They led in two main directions: northwards to Europe and westwards to the Americas. Overwhelmingly these affected migrants from the European Mediterranean: Italians across the Alps to France and beyond, Spaniards to Central and South America, Portuguese to Brazil and the colonies, Italians and Greeks to North America. Lebanese and Syrians also migrated widely before 1950, mainly as merchants and artisans.

In the late nineteenth and early twentieth centuries Italians were by far the most numerous migrants leaving the Mediterranean. Between the late 1880s and 1900 about 250 000 departed each year, roughly equally divided between European and transatlantic destinations (mainly Brazil and Argentina). Most came from northern Italy. After the turn of the century the flow changed character: southern Italians came to predominate and the United States took over as the main destination. Between 1900 and 1915 between 600 000 and 700 000 emigrated per year: an annual departure of about 1 in 50 of the Italian population. Not without reason did Italians call this the Great Emigration (King 1992).

A DEMOGRAPHIC RESERVOIR FOR EUROPE

Emigration, now mainly to Europe, dominated the population geography of the European Mediterranean in the post-war decades, affecting also the Maghreb states and Turkey. High rates of population growth and a stagnant economy were the main 'push factors' for emigration, but equally important was the rise in demand for immigrant labour in the booming industrial economies of north-west Europe. Data on rural employment are very revealing here. Whilst in northern Europe the agricultural population fell by 20 per cent during 1920–50, in the Mediterranean it grew by 15 per cent. Hence by the 1950s the rural economies of the Mediterranean had reached crisis point: technologically still backward, saturated with labour, and rent by conflict between the agrarian classes of landlord, tenant and labourer (see Chapter 13 for more details). Population growth showed few signs of abating, remaining moderately high in southern Europe (Yugoslavia 1.1 per cent, Spain 1.0 per cent, Portugal 0.9 per cent, Italy 0.8 per cent, Greece 0.7 per cent) and very high elsewhere (Algeria 3.0 per cent, Morocco 2.9 per cent, Syria 2.8 per cent, Egypt, Lebanon and Turkey 2.5 per cent, Tunisia 2.2 per cent) – these figures are annual rates of increase for the 1960s. However, to view Mediterranean emigration during the 1950s and 1960s solely as an economic process underlain by differential rates of economic and population growth would be to oversimplify the matter. There were other more subtle and ultimately more powerful processes at work. These influences had to do with the changing nature of Mediterranean societies and the opening up of geographical horizons as perceived by individuals. The spread of mass education and of a 'European' culture transmitted from the 'core' via television, magazines and other media undermined the stable cultural fabric of Mediterranean peasant societies. They would never be the same again.

The social geography of this post-war Mediterranean migration to north-west Europe has been comprehensively researched (see, for example, King 1976; Salt and Clout 1976; White 1993). After the establishment of the Common Market Italians were free to emigrate to France, Germany and the Benelux countries. They also continued to migrate overseas, especially to the United States, Canada, Australia and Venezuela (King 1992). Since Algeria was part of France until 1962, Algerians had the right to move to the 'mother country'. This migration flow continued and even grew after Algerian independence, for France at this time had a rather relaxed approach to immigration from the Maghreb and southern Europe. The total foreign population of France doubled between 1954 (1 765 300) and 1975 (3 442 415). Thereafter it rose more slowly, as recessions and rising unemployment created both an economic and a political climate necessitating strict controls on immigration from outside the European Community. At the 1982 census, foreigners numbered 3 680 100, roughly equally divided between Maghrebis (41.8 per cent of total foreigners) and southern Europeans from Portugal, Spain and Italy (38.6 per cent). Ministry of Interior records for the same year gave figures which were about 15 per cent higher (Ogden 1991).

Mediterranean labour migration to West Germany, the other major 'recruiting' country, was rather differently organised, and the origin countries were also different (Greece, Yugoslavia and Turkey predominated). Strict annual quotas calibrated the flows and, at least until the late 1960s, it was mainly males and not family members who migrated in order to perform the undesirable jobs the German workers increasingly opted not to do. Later, following European Community social legislation, family reunion migration was allowed and Mediterranean ethnic communities quickly took root in German cities as they had done in Paris, Lyons and Marseilles. Similar communities of Mediterranean migrants also became established in Belgian and Dutch cities: here Turks and Moroccans were the main groups.

The period between the 1950s and the 1970s thus saw the 'Mediterranisation' of very many European cities, especially large industrial towns, where immigrants were needed for factory work, and capital cities, where they were required for low-grade service sector jobs. The various immigrant nationalities formed distinct ethnic concentrations which, although rarely attaining the density of true ghettoes, nevertheless lent a marked multicultural tone to certain districts such as inner-city areas, industrial suburbs and some peripheral overspill estates. Moreover, these ethnic clusters were highly dynamic, reflecting upward mobility of migrants to areas of better housing and their progressive shift away from heavy factory or construction work into lighter and more remunerative types of employment.

What were the effects of this massive exodus of Mediterranean labour migrants and their families on the mainly rural sending areas? Again these have been thoroughly documented. What follows is a summary of a few key points which are recurrent in the literature (see, for example, Böhning 1975; King 1979, 1984; Rhoades 1978). First, there is an obvious demographic effect. This is conditioned by whether migrants move abroad permanently or temporarily, and if the latter, for what length of time. Permanent family emigration brings about a lasting reduction in the number of inhabitants in the sending area. Given the age profile of migrant streams – mainly young, reproductively active adults – the residual population is deprived of its child-bearing cohorts and left dominated by elderly people (Fig. 11.2). Even temporary emigration can have some long-term

FIGURE 11.2 Sustained outmigration from rural areas of the Mediterranean over many decades has produced a predominantly elderly residual population, as this scene from the south Italian village of Aliano shows

demographic effects. When young men (or women) go abroad to work for a few years, their marriage age is usually raised (if they are single), or they may not produce so many children due to separation from their wives (it being nearly always the wives who stayed at home in this kind of 'split family' migration). Migrants' exposure to the consumeristic, small-family societies of northern Europe also had a 'demonstration effect' of lowered fertility when they returned, as Gregory (1976) found in Andalusia.

The socio-economic effects of emigration are harder to evaluate. On the positive side, relief of social tension and the 'export' of unemployment have frequently been commented upon. In all the countries where post-war emigration reached 'mass' proportions, the inflow of remittances from emigrants working abroad made a significant, even vital, contribution to the balance of payments. On the other hand, many studies have shown that emigrants were often skilled, innovative and ambitious individuals; their departure deprived the sending society of initiative and leadership, leaving it in the hands of elderly traditionalists (Baučić 1972).

Equal disagreement surrounds the impact of emigrants when they returned to the Mediterranean, as many did in the 1970s and 1980s, driven back by unemployment, nostalgia, the need to look after aged parents or the desire to bring up their children in their 'native' culture. The 'return of innovation' is shown to be largely a myth; the innovative migrants are precisely those who are least likely to return (Böhning 1975; Cerase 1974). For the majority of those who return at working age, finding employment is a great problem. Kinship and nostalgia pull them back to their villages of origin where work opportunities outside of a declining, residual peasant agriculture are often slim. Many 'take refuge' in the petty service sector, sinking some of their migrant savings in a shop, bar or taxi. Reluctantly, others return to farming. Physically, the main impact of return migration has been on rural housing and settlement patterns. Returnees have shown themselves almost pathologically keen to splurge their savings on new luxurious houses. In areas of the Mediterranean with a long and diverse emigration tradition, villages have grown by the accretion of distinct returnee quarters. Some south Italian villages have 'American' quarters built by returnees from the United States in the 1920s and 1930s, to which have been added a 'French district' in the 1950s and another quarter built with German and Swiss remittances in the 1960s and 1970s (King 1988). In Malta and Gozo emigrant exhibitionism reaches amazing proportions: returned migrants decorate their houses with stone kangaroos, maple-leaf motifs and carved eagles according to their particular destination country – Australia, Canada or the USA (King 1980). Or they may name their new houses after the country or city they emigrated to (Fig. 11.3).

CURRENT GROWTH PATTERNS: THE MEDITERRANEAN AS A DEMOGRAPHIC FRONTIER

In the previous section we saw that, until the 1960s, the entire Mediterranean Basin was a region of consistent population growth, with either moderate (European Mediterranean) or high rates of demographic increase. No Mediterranean country was, at that time, close to the state of zero population growth about to be experienced in some northern European countries like West Germany or Denmark.

By the 1980s this situation had dramatically changed with the emergence of what could be termed a new 'South European demographic model' pioneered by Italy and Spain, with Portugal and Greece also included. The essential features of this model are historically unprecedented low levels of fertility and an increasing ageing of the population. The ageing is provoked partly by low fertility and partly by increased life expectancy which in the case of Spain, Italy and Greece (but not Portugal) is set to overtake north European levels.

Let us explore southern European population structure in a little more detail starting with

FIGURE 11.3 Returned emigrants tend to channel much of their carefully saved wealth into new houses, the names and decorations of which sometimes reveal the migration experience of the occupants, as this example from Malta testifies

mortality and life expectancy. Traditionally, life expectancy was highest in the Netherlands and Scandinavia; currently the highest level is held by Japan. South European rates are rapidly encroaching, however. Noin (1993) notes that already in 1980 female life expectancy was over 80 in parts of northern Spain; now such levels apply to virtually all of Spain, Greece and central and southern Italy. A similar spatial pattern is revealed by high male life expectancy – around 75–77 – except that gender differentials in life expectancy are greater (up to 10 years) in regions such as the Basque provinces and north-east Italy where male consumption of cigarettes and alcohol is unusually high. Overall, the high figures for life expectancy recorded in Mediterranean Europe are thought to reflect, above all, the healthy Mediterranean diet based on olive oil and a high intake of fresh fruit and vegetables (Noin 1993), coupled with vastly improved medical and welfare services over the past few decades.

More important than longevity in characterising the new demographic regime of southern Europe is the region's extraordinary low fertility; as a recent newspaper headline put it, 'the incredible shrinking family of southern Europe' (Hooper and Luce 1994). The shrinking was particularly rapid after the mid-1970s; already by 1986 Italy attained the world's lowest fertility at 1.3 children per woman (this index, known as the total fertility rate or TFR, refers to the mean number of children borne by women over their reproductive lives). By 1990 southern European TFRs (Italy 1.3, Spain 1.3, Portugal 1.4, Greece 1.5) were significantly below those for all northern European countries except Germany (1.5). By 1992 Spain had overtaken Italy in the 'race' towards the one-child-per-family norm – 1.23 as against Italy's 1.25 (Hall 1995, p. 44). Given that southern parts of Italy and Spain (Naples, Sardinia, Andalusia) are known to have significantly larger families than the national averages, TFRs of around 1.0 are already being recorded in parts of northern Spain and Italy. Particularly dramatic was the 'fertility collapse' of Spain, the TFR halving from 2.5 to 1.25 within the decade 1981–91.

Although causal links have yet to be proven, several possibly relevant factors can be suggested for these ultra-low birth-rates (Faus-Pujol 1995; King 1993). First, and above all, there has been a rejection by women of their traditional sole maternal role. This process has been linked to the secularisation of society and the rejection of the Catholic Church's teachings on contraception and abortion. Male domination of women, to the extent that family size was virtually dictated by the husband with number of children a source of male pride, is breaking down. Second, rising standards of education mean that southern European women are wanting to develop careers and hence postpone both marriage and childbearing; cohabitation and single mothers are still stigmatised. OECD data on the numbers of working women show steep rises between 1970 and 1990 for all four southern European countries – Italy from 6.0 million to 9.0m, Spain from 3.1m to 5.3m, Portugal from 0.9m to 2.0m and Greece from 0.9m to 1.5m (King and Konjhodzic 1995, p. 4). But lack of proper support for working mothers (lack of state nursery facilities and the failure of husbands to help with household and childcare duties) makes it difficult for women to fulfil the dual role of having a career and a family. Third, the increasing materialism of life in southern Europe means that many couples place the acquisition of consumer goods before the starting of a family. Finally, there are labour market factors. Economic restructuring and the casualisation of labour have put many female jobs at risk, thereby causing women to hold on to their jobs rather than risk not getting them back if they leave to have a child. These factors are all hypothetical, but all seem plausible.

Whilst southern Europe's population threatens to implode, that of the eastern and southern shores (Turkey clockwise round to Morocco) continues to grow at a rapid rate. In the 'middle ground', the former Yugoslavia, Albania, Malta, Cyprus and Israel hold generally intermediate demographic positions. Malta, Cyprus and the northern republics of ex-Yugoslavia (especially Slovenia) are following the 'European model' most closely (although there are problems of data availability in Cyprus, split into Greek and Turkish sectors, and in 'post-Yugoslavia'). Albania's population evolution is more transitional between the northern and southern Mediterranean regimes. The case of Israel is unique because of consistently high rates of immigration and a rather different social formation. The starkest contrasts, however, are between the zero-growth, low-fertility countries of the southern EU and the high-fertility regimes of the southern and eastern Mediterranean. The geopolitical implications of this spatial demographic disequilibrium within the Mediterranean Basin were introduced in Chapter 8, and the first column of Table 8.3 exhibited the contrast in annual population growth rates between the northern and southern Mediterranean countries for the early 1990s. Table 11.1 enables this north–south contrast to be examined in greater detail. Various demographic parameters are presented for individual Mediterranean countries. Current fertility is measured in two ways – crude birth-rate and total fertility rate. Medium-term population growth is assessed by the proportion of under-15s – the parents of the forthcoming generation, and United Nations mean-variant projections are given for 2000 and 2025. Close examination of the table reveals the extraordinary contrast between the stagnant or even shrinking (see Italy) population of the European sector and the fertile populations of the eastern and southern shores which are set to virtually double over the period 1990–2025, i.e. within little more than one generation. Taking the countries from Turkey clockwise round to Morocco, the increase will be from 194 million to 374 million.

Other revealing measurements and projections of Mediterranean population growth have been made by the Blue Plan team (see Grenon and Batisse 1989, pp. 46–57). These authors point out that, as a whole, the population of the Mediterranean increased by 68 per cent during 1950–85, an annual rate of increase of 1.5 per cent. The growth rate peaked in the 1960s, since when it has decreased steadily, mainly because of the changes in the European countries noted already. Projected annual rates of increase for the whole Mediterranean Basin are 1.3 per cent during 1985–2000 and 0.9 per cent during 2000–2025.

Grenon and Batisse are at pains to stress regional variations in demographic profiles and projected growth rates within the Mediterranean basin. They delineate three sub-regions as follows: A = Spain, France, Italy, Greece, Yugoslavia; B = Turkey, Syria, Egypt, Libya, Tunisia, Algeria, Morocco; and C = Malta, Cyprus, Albania, Israel and Lebanon. Table 11.2 sets out the population totals for these three sub-regions for various years between 1950 and 2025, based again on United Nations median-variant data. The projection figures on Table 11.2 are slightly different from those in Table 11.1, for two reasons. First, the Blue Plan defini-

Table 11.1 Demographic data for Mediterranean countries, *ca.* 1990

Country	CBR	TFR	% popn <15	Popn (m) 1990	Estimated popn (m) 2010	2025
Portugal	12	1.5	21	10.4	10.8	10.6
Spain	11	1.3	20	39.0	41.2	40.5
France	14	1.8	20	56.7	58.8	58.6
Italy	10	1.3	17	57.7	55.9	52.3
Greece	10	1.5	20	10.1	10.2	9.7
Malta	16	2.2	24	0.4	0.4	0.4
Yugoslavia (ex-)	14	1.9	23	23.9	25.8	26.3
Albania	25	3.0	32	3.3	4.3	4.8
Turkey	30	3.7	38	58.5	83.4	102.7
Syria	43	6.7	49	12.8	25.9	41.4
Lebanon	28	3.6	39	3.4	4.9	6.2
Egypt	38	4.5	40	54.5	81.8	105.4
Libya	37	5.2	44	4.4	7.1	9.3
Tunisia	29	4.1	39	8.4	11.5	13.6
Algeria	35	5.4	46	26.0	39.1	49.3
Morocco	34	4.5	42	26.2	37.3	46.2

Notes: CBR = crude birth-rate (number of births per thousand population per year)
TFR = total fertility rate (mean number of children per woman over reproductive life)
Source: 1991 World Population Data Sheet. New York, Population Reference Bureau

tion of the Mediterranean does not include all parts of all countries bordering the sea (e.g. France). Second, the Blue Plan scenarios are based on projections from the mid-1980s rather than 1990 (Table 11.1); the former give less weight to the fertility collapse of southern Europe, the scale of which only began to be apparent in the late 1980s.

Figure 11.4 graphs the 'north–south' data for population growth in the Mediterranean for the same period as Table 11.2. The diagram shows that the southern shore overtook the northern in 1990. The north's total population is predicted to remain more or less constant across the period 1995–2015 but then to commence a potentially rapid decline; of course an increase in birth-rate (or immigration) could change this prediction. Much more important in the overall Mediterranean population scenario is what happens to the countries in sub-region B, especially the big ones like Turkey, Egypt and Algeria. Although in many respects the twenty-first-century scenarios are alarming, there are positive signs for long-term population control (Fig.

TABLE 11.2 Population trends in sub-regions of the Mediterranean, 1950–2025

	Population (m)				% growth	
	1950	1980	2000	2025	1980–2000	1980–2025
Sub-region A	140	180	194	199	7.8	10.6
Sub-region B	67	142	226	329	59.2	130.3
Sub-region C	5	10	14	19	40.0	90.0
Total	212	333	433	547	30.0	64.3

Note: For countries in each sub-region see text
Source: Grenon and Batisse (1989, p. 48)

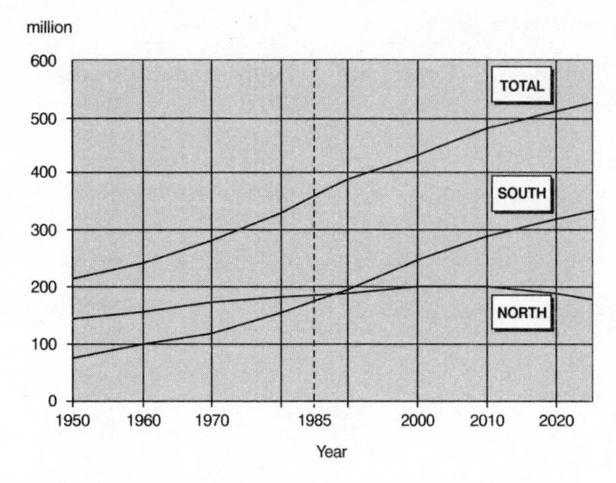

FIGURE 11.4 Population trends and projections in northern and southern Mediterranean countries, 1950–2025

Source: Grenon and Batisse (1989, p. 49)

11.5). Grenon and Batisse (1989, p. 49) stress that these countries have three key features in common:

1. They have all now entered a phase of declining fertility, this decline having accelerated since the 1970s (Morocco's TFR, for instance, fell from 7.2 to 4.5 in 25 years); of the main countries in this group, Morocco, Tunisia and Turkey are leading the way in this transition.
2. Nevertheless, by European standards fertility remains high at 4–6 children per woman.
3. There is a sharp difference in fertility between urban and rural areas (that in Cairo and Alexandria, for instance, being one-third below the Egyptian national average).

IS MIGRATION THE SOLUTION TO THE CRISIS?

The population trends and projections described in the previous section have perhaps less significance as absolute values than they do in terms of the variable abilities of the southern and eastern Mediterranean countries to actually support their rapidly growing populations, particularly in terms of providing employment opportunities. All the indications are that, at least in the short and medium terms, the growth in the labour force of these countries will outstrip the ability of their economies to provide employment. This is not just a matter of population dynamics but also reflects changing social circumstances, notably the increasing proportion of women wanting to enter the labour force as a result of improved educational standards (Charmes *et al.* 1993).

The evolving demographic pressure on the labour market can be examined by focusing on the number of 15–19 year olds in 1985 and 2025, this age-group constituting the supply side of labour coming on to the market at the end of full-time education. As Table 11.3 shows, this cohort is projected to decrease in Spain, France, Italy, ex-Yugoslavia and Greece – in Italy by more than 30 per cent. All the other countries show projected increases, up to threefold in Libya and Syria. Over the whole region there will be an increase of nearly 7.5 million 15–19 year olds during the period in question, an increase of 22.5 per cent.

TABLE 11.3 Population aged 15–19 in Mediterranean countries, 1985 and 2025 ('000)

			change	
Country	**1985**	**2025**	**no.**	**%**
Spain	3 288	2 950	−338	−10.3
France	4 165	3 419	−746	−17.9
Italy	4 560	3 174	−1 386	−30.4
Greece	777	672	−105	−13.5
Yugoslavia (ex-)	1 802	1 667	−135	−7.5
Albania	316	447	+131	+41.5
Cyprus	49	64	+15	+30.6
Turkey	5 401	7 364	+1 963	+36.3
Syria	1 094	3 115	+2 021	+184.7
Lebanon	326	419	+93	+28.5
Israel	394	509	+115	+29.2
Egypt	4 755	7 444	+2 689	+56.6
Libya	367	1 132	+765	+208.4
Tunisia	824	973	+149	+18.1
Algeria	2 430	4 210	+1 780	+73.3
Morocco	2 532	2 971	+439	+17.3
Total	33 080	40 530	+7 450	+22.5

Source: Blue Plan 'average scenario' data in Grenon and Batisse (1989, p. 56)

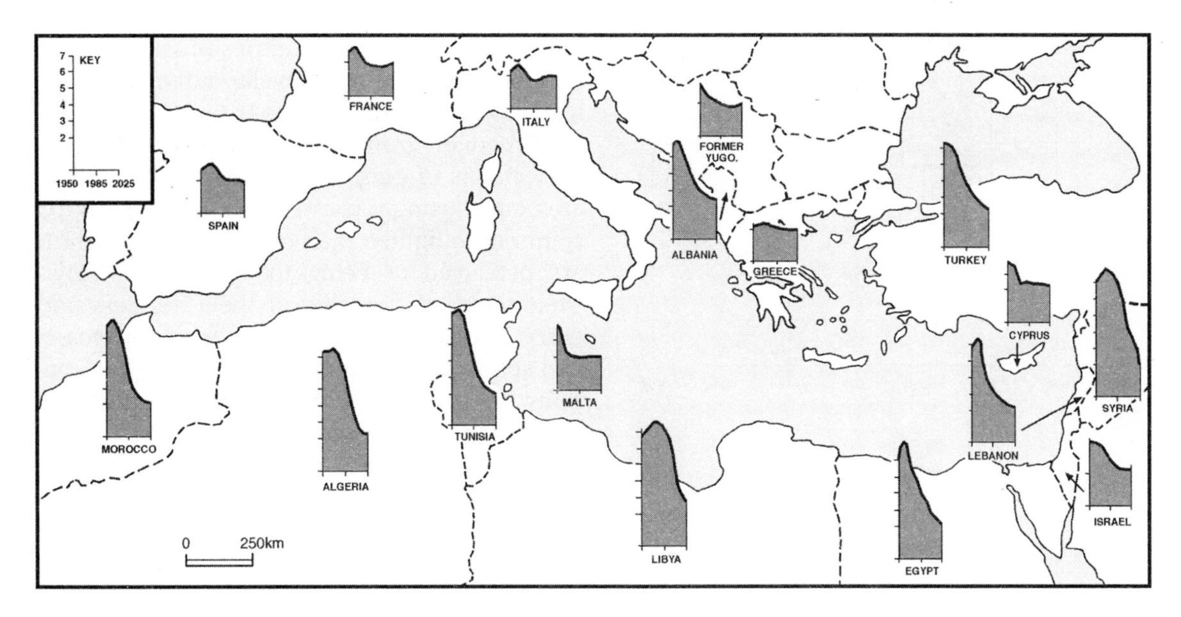

FIGURE 11.5 Total fertility rate (number of children per woman): trends and projections by country, 1950–2025

Source: Grenon and Batisse (1989, p. 55)

Figure 11.6 shows the contrast in another way by comparing labour market entrants (numbers aged 15–24 at 10-year intervals) with labour market departures (numbers aged 55–66) for two groups of countries, one clearly of low fertility and one of high. The graphs show withdrawals outnumbering entries in southern Europe in about 20 years from now, whilst in the southern and eastern Mediterranean countries entries into the labour market run at about three times departures throughout the period from 1980 to 2010, falling to about twice after 2020.

These data portend strong, even irresistible, pressure for south–north migration across the Mediterranean Sea. In fact, the Mediterranean is the contact-zone for migration forces from potential migration fields which are much wider: the whole of Europe to the north and the whole of Africa to the south (Montanari and Cortese 1993). But not all movements are oriented south–north: lateral east–west movements have also occurred, both in the wake of the 1970s oil crises, when some of the migration to northern Europe (e.g. of Portuguese and Turks) was diverted into contract labour for the oil-producing countries of the Gulf (Libya meanwhile drew on surpluses of labour from neighbouring North African states), and in the post-communist era with Albanians seeking to move in large numbers into Italy and Greece, and refugees from the break-up of Yugoslavia driven to seek sanctuary in countries such as Italy and Austria, and, within the former federation, in Slovenia and Croatia.

Nevertheless, the main intra-Mediterranean migrations in the last 20 years have been from south to north. A simple indication of this is the fact that Moroccans now form the largest immigrant groups in both Spain and Italy (Table 11.4), and Algerians continue to be the most numerous group in France, where they are concentrated especially in Marseilles and the south. Greek data on immigration are very imprecise, but Albanians and Egyptians are known to be present in large numbers (Fakiolas and King 1996). Portugal's main immigrant groups – from Brazil, Cape Verde, Angola, Mozambique – reflect much more its colonial heritage. In all southern European states, an accurate estimate of the size of the immigrant nationalities or of the annual scale of their flows is impossible

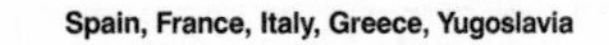

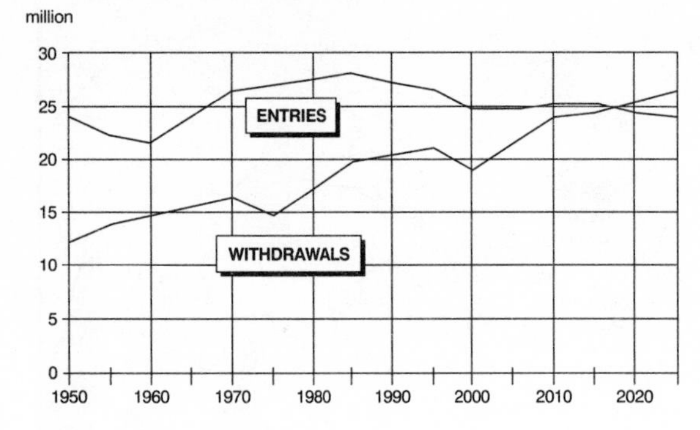

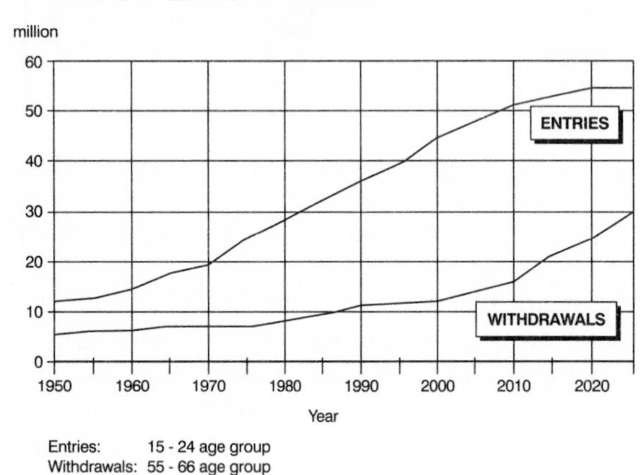

FIGURE 11.6 Entries to and withdrawals from the labour market of two groups of Mediterranean countries, 1950–2025

Source: Grenon and Batisse (1989, p. 57)

because so much of the movement is clandestine. Periodic 'regularisation' schemes for illegal immigrants help to 'fix' the populations and where this is backed up by a system of residence permits, as in Spain and Italy, this helps to provide some reasonable estimates (Table 11.4). However, even in these two countries, there are large numbers of illegal immigrants to add to the figures in the table, perhaps increasing the totals by 30–50 per cent. In Greece the majority of immigrants are classified as illegal, making a nonsense of any official figures.

Spain, Portugal, Italy and Greece are only now coming to terms with their new status as immigration countries – traditionally, as we have seen, they were countries of emigration – and have been slow to develop rational policies towards immigrants. All four are members of the Schengen group for facilitating cross-border movements of people within the EU, but they are being strongly pressed by the northern EU countries to tighten their external borders which are perceived as being too permeable. However, thorough policing of their frontiers and entry points is impracticable: long sea coasts and sparsely populated mountain ranges (especially in Greece and Italy), plus the importance of 'easy entry' at ports and airports for the tourist trade, make a proper sealing of the borders to non-EU migrants impossible. Around the coast of southern Italy, for example, boatloads of Albanians are reported to arrive almost every night; similar flotillas of 'boat people' arrive from Tunisia in Sicily. Some Moroccans even swim to Spain – many do not make it. Between January 1991 and August 1994 Greece deported over 700 000 Albanians; many subsequently re-entered, only to be repatriated again in a continuous cycle of entry and expulsion (SOPEMI 1995, p. 93). Another common entry strategy is to arrive on tourist visas and overstay to find work in the flourishing submerged economy. Unlike the earlier generation of southern European migrants who found formal-sector jobs in the industrial economies of northern Europe, the 'new immigrants' in southern Europe work mainly in the informal sector. Here, they are willing to work for wages which would be unacceptable to local workers: they accept such low rates because they are still several times the level of earnings in their own countries (King 1996). Amongst the wide variety of activities performed by immigrants in southern Europe, the following may be mentioned: seasonal crop harvesting, fishing, casual construction work, hotel and restaurant work, work in small factories and workshops, street-hawking. These are overwhelmingly male jobs. Females work as domestic servants and nannies. Several of these labour market niches are monopolised by single nationalities. In Italy, for example, Moroccans have for some time been established as street-sellers of carpets and rugs, whilst in

Table 11.4 Main immigrant nationalities with residence permits in Spain and Italy ('000)

Groups of countries	Spain (>5,000 in 1993)			Groups of countries	Italy (>15,000 in 1993)		
	1983	1988	1993		1985	1990	1993
Mediterranean:				Mediterranean:			
Morocco	4.1	11.9	61.3	Morocco	2.6	78.0	97.6
Portugal	22.7	31.6	32.3	Yugoslavia (ex-)	13.9	29.8	72.4
France	16.2	25.2	25.5	Tunisia	4.4	41.2	44.5
Italy	9.7	14.1	15.9	Albania			30.8
Other European:				France	23.7	24.4	27.0
UK	31.9	64.1	58.2	Egypt	7.0	19.8	24.6
Germany	25.3	39.4	34.1	Spain	12.6	14.4	17.0
Netherlands	9.1	15.0	11.1	Greece	28.8	21.0	16.5
Belgium	6.6	10.6	7.6	Other European:			
Switzerland	4.9	7.5	5.8	Germany	37.2	41.6	39.9
Sweden	4.2	7.2	5.0	UK	27.9	26.6	29.1
Other:				Poland		17.0	21.1
USA	10.7	16.7	14.3	Romania		7.5	19.4
Peru	1.4	2.6	10.0	Switzerland	18.2	20.0	18.2
Dominican Rep.	1.0	2.0	9.2	Other:			
Philippines	5.0	8.2	8.4	USA	51.1	58.1	64.0
China	1.1	3.5	7.8	Philippines	7.6	34.3	46.3
Venezuela	5.7	8.4	7.0	Senegal	0.3	25.1	26.4
Colombia	2.4	4.1	6.2	China	1.6	18.7	22.9
Chile	3.6	6.2	5.9	Brazil	4.7	14.3	21.1
India	3.2	5.2	5.7	Sri Lanka	2.5	11.5	19.7
				Somalia	1.8	9.4	19.6
TOTAL	210.4	360.0	430.4	TOTAL	423.0	781.1	987.4

Note: Data refer to 31 December for each year
Source: SOPEMI (1995, pp. 203, 208)

western Sicily Tunisians man the fishing boats of Trapani, Marsala and Mazara (for these and other examples see King and Konjhodzic 1995; King and Rybaczuk 1993; Simon 1987).

In recent years immigration control has become a major geopolitical issue within the Mediterranean Basin. Within individual countries such as France and Italy immigration has evolved – or rather has been manipulated – into a situation in which it is also a flashpoint for domestic politics. Right-wing political parties such as Jean-Marie Le Pen's *Front National* and Gianfranco Fini's *Alleanza Nazionale* make electoral capital out of draconian measures against immigrants from outside the EU. The political strength of these groups has tended to significantly edge the centre-right governments in power towards increasingly repressive measures against immigrants.

Is northward migration a solution to the crisis of Mediterranean population growth along the southern and eastern shores? Both the language and politics of trans-Mediterranean migration are generally cast in the negative light of the necessity of 'control' against 'threats' of mass immigration from countries whose demographic characteristics are 'third-world' and whose cultures are non-European, predominantly Islamic (Courbage 1994). This defensive mind-set of control against non-European, non-Christian 'outsiders' narrowly sees the 'solution' as the 'problem'. A broader view would see some role for a properly managed migration regime whereby more southern migrants are admitted to northern shores. Such a regime would allow North African labour, for instance, to ease certain bottlenecks in the labour markets of southern European states (in agriculture, or the tourist sector perhaps) whilst relieving pressure on the limited employment opportunities in the poorer sending countries. Another approach, much discussed in EU and OECD circles in recent years, is to concentrate attention on targeting aid at development schemes in the migrant sending countries, hence reducing the need to emigrate at source. Yet another approach, pressed by the Maghreb states in particular, is for greater market access into Europe for their export produce, thereby boosting employment at source and reducing pressures for emigration. Questions of scale make all of these policies doubtfully effective as single or even combined solutions to the 'crisis'; nevertheless, the alternatives seem few, except to wait for the pace of population growth to slacken right off, as it should do eventually. The crisis is hence unlikely to be long term: the challenge is to manage it in the short and medium terms.

CONCLUSIONS: POPULATION AND GEOPOLITICS IN THE MEDITERRANEAN

From both a demographic and an economic point of view, the Mediterranean marks the frontier between the North and South of the globe; Montanari and Cortese (1993) speak of the 'Mediterranean Rio Grande'. Much more than the real Rio Grande, the Mediterranean is also a religious frontier, between Christianity and Islam, and a geopolitical divide between democratic Europe and less democratic, if not authoritarian, states to the south.

Given the population growth patterns described in this chapter, it appears that the future of the Mediterranean Basin must be more and more structured by the demographic perspective and by the migration imperative. Of the 170 million extra people living around the shores of the Mediterranean by 2025, 68 per cent will be Arabs, 22 per cent Turks. According to the scenarios used in Table 11.1, both Turkey and Egypt will have in the region of 100 million inhabitants in 2025, approximately twice the size of any of the major southern European states at that time. The Maghreb will form another population bloc of 100 million or more. Employment creation *in situ* cannot possibly keep pace with the rate of population increase in these countries. In Algeria in the early 1990s, for instance, annual job creation, 90 000, fell far short of the number of new labour market entrants, 250 000 (Safir 1995, p. 62). Pressures for migration are proving impossible to suppress. Even a regime of strict control, such as the one the EU is trying to impose, will be circumvented by desperate and determined migrants. They will be helped in their desire to migrate by the geography of the Mediterranean, by pre-existing migrant communities and networks, and by the 'privatisation' of migrant transport and transit by enterprising and ruthless traders in human traffic.

Regarding cross-Mediterranean migration, there appear to be three choices for policy. One is the choice of non-policy – for the receiving countries to pay lip-service to trying to control the inflows but to tacitly accept that a certain amount of clandestine entry will take place and to turn a blind eye to this. Such has been the situation to date, more or less. The second is the choice of 'police rather than policy'. Taken to its extreme, this would involve a massive increase in border guards and of checks at entry points, as well as repressive monitoring of the population through identity card and residence permit checks. These measures would be difficult administratively and expensive both in financial terms and in terms of human and political relations. The third option is to move towards a more imaginative and cooperative policy whereby the viewpoints of both the sending and receiving countries are taken into account and, above all, the safety and rights of the migrants are respected. According to Fuller and Lesser (1995, p. 53) the strong incentives for a north–

south 'bargain' on Mediterranean migration policy could stimulate cooperation along economic, political and military lines with positive implications for stability in relations between Europe (and the wider Western world) and the world of Islam. This is very much a United States foreign policy view. From a more local perspective one can ask whether the European governments north of the Mediterranean really have an alternative to helping their neighbours to the south. In this sense the 'problem' (of migration and population growth) actually becomes the 'solution' to a set of broader geopolitical and developmental issues.

Thus migration provides the *entrée* into the possible resolution of a whole set of north–south policy challenges across the Mediterranean. Much will depend upon how the approach to policy formulation is handled. Certainly the EU's Mediterranean policy must be a model of genuine cooperation rather than an echo of a colonial past. It must be a coordinated and integrated approach in which regional growth, foreign investment, aid, employment promotion, trade and migration all have a role; the slogan 'trade and aid – but not migration' is too narrow and ignores the realities of the situation, namely that migration is taking place and will continue to do so. Awareness of the need for a 'Mediterranean policy' seems to be strongest in those parts of Europe – the south – where there is a history of relations with the Maghreb states and the problems, such as those of clandestine migration, are felt more directly. The successive presidency of the EU with France, Spain and Italy over the mid-1990s has certainly helped to focus policy thinking on to these issues.

References

BAUČIĆ, I. 1972: *The Effects of Emigration from Yugoslavia and the Problems of Returning Emigrant Workers*. The Hague: Nijhoff.

BÖHNING, W.R. 1975: Some thoughts on emigration from the Mediterranean basin. *International Labour Review* **111**, 251–77.

BRAUDEL, F. 1972: *The Mediterranean and the Mediterranean World in the Age of Philip II*. Vol. 1. London: Collins.

CERASE, F.P. 1974: Migration and social change: expectations and reality. A case study of return migration from the United States to Italy. *International Migration Review* **8**, 245–62.

CHARMES, J., DABOUSSI, R. and LEBON, A. 1993: *Population, Employment and Migration in the Countries of the Mediterranean Basin*. Geneva: International Labour Office, Mediterranean Information Exchange System Working Paper 93/1E.

COURBAGE, Y. 1994: Demographic transition among the Maghreb peoples of North Africa and in the emigrant community abroad. In Ludlow, P. (ed.), *Europe and the Mediterranean*. London: Brassey's, 47–88.

FAKIOLAS, R. and KING, R. 1996: Emigration, return, immigration: a review and evaluation of Greece's postwar experience of international migration. *International Journal of Population Geography* **2**, 171–90.

FAUS-PUJOL, M.C. 1995: Changes in fertility rates and age structures of the population in Europe. In Hall, R. and White, P. (eds), *Europe's Population Towards the Next Century*. London: UCL Press, 17–33.

FULLER, G.E. and LESSER I.O. 1995: *A Sense of Siege: The Geopolitics of Islam and the West*. Boulder, CO: Westview.

GREGORY, D.D. 1976: The Andalusian dispersion: migration and socio-demographic change. In Aceves, J.B. and Douglass, W.A. (eds), *The Changing Faces of Rural Spain*. New York: Schenkman, 63–95.

GRENON, M. and BATISSE, M. 1989: *Futures for the Mediterranean Basin: The Blue Plan*. Oxford: Oxford University Press.

HALL, R. 1995: Households, families and fertility. In Hall, R. and White, P. (eds), *Europe's Population Towards the Next Century*. London: UCL Press, 34–50.

HOOPER, J. and LUCE, E. 1994: The incredible shrinking family of southern Europe. *The Guardian*, 25 May, 22.

HOUSTON, J.M. 1964: *The Western Mediterranean World: An Introduction to its Regional Landscapes*. London: Longmans.

KING, R. 1976: The evolution of international migration movements concerning the EEC. *Tijdschrift voor Economische en Sociale Geografie* **67**, 66–82.

KING, R. 1979: Return migration: a review of some case studies from southern Europe. *Mediterranean Studies* **1**(2), 3–30.

KING, R. 1980: *The Maltese Migration Cycle: Perspectives on Return*. Oxford: Oxford Polytechnic Discussion Papers in Geography 13.

KING, R. 1984: Population mobility: emigration, return migration and internal migration. In Williams, A.M. (ed.), *Southern Europe Transformed*. London: Harper and Row, 145–78.

KING, R. 1988: *Il Ritorno in Patria: Return Migration to Italy in Historical Perspective*. Durham: University of Durham, Department of Geography, Research Papers in Geography 23.

KING, R. 1992: *Patterns of Italian Migrant Labour: The Historical and Geographical Background*. Bristol: University of Bristol, Centre for Mediterranean Studies, Occasional Paper 4.

KING, R. 1993: Italy reaches zero population growth. *Geography* **78**, 63–9.

KING, R. 1996: Migration and development in the Mediterranean region. *Geography* **81**, 3–14.

KING, R. and KONJHODZIC, I. 1995: *Labour, Employment and Migration in Southern Europe*. Brighton: University of Sussex, Research Papers in Geography 19.

KING, R. and RYBACZUK, K. 1993: Southern Europe and the international division of labour: from emigration to immigration. In King, R. (ed.), *The New Geography of European Migrations*. London: Belhaven, 175–206.

MCNEILL, J.R. 1992: *The Mountains of the Mediterranean World: An Environmental History*. Cambridge: Cambridge University Press.

MONTANARI, A. and CORTESE, A. 1993: South to North migration in a Mediterranean perspective. In King, R. (ed.), *Mass Migrations in Europe: The Legacy and the Future*. London: Belhaven, 212–33.

NOIN, D. 1993: Spatial inequalities in mortality. In Noin, D. and Woods, R. (eds), *The Changing Population of Europe*. Oxford: Blackwell, 38–48.

OGDEN, P.E. 1991: Immigrants to France since 1945: myth and reality. *Ethnic and Racial Studies* **14**, 294–318.

RHOADES, R.E. 1978: Intra-European return migration and rural development: lessons from the Spanish case. *Human Organization* **37**, 136–47.

RIBEIRO, O. 1983: *Il Mediterraneo: Ambiente e Tradizione*. Milan: Mursia.

SAFIR, N. 1995: The question of migration. In Holmes, J.W. (ed.), *Maelstrom: The United States, Southern Europe, and the Challenges of the Mediterranean*. Cambridge, MA: World Peace Foundation, 59–83.

SALT, J, and CLOUT, H. (eds) 1976: *Migration in Post-War Europe: Geographical Perspectives*. London: Oxford University Press.

SIMON, G. 1987: Migration in southern Europe: an overview. In *The Future of Migration*. Paris: OECD, 258–91.

SOPEMI 1995: *Trends in International Migration: Annual Report 1994*. Paris: OECD.

WHITE, P. 1993: Immigrants and the social geography of European cities: an overview. In King, R. (ed.), *Mass Migrations in Europe: The Legacy and the Future*. London: Belhaven, 65–82.

12

FIVE NARRATIVES FOR THE MEDITERRANEAN CITY

LILA LEONTIDOU

INTRODUCTION

Urban geography is still weighed down too heavily under the yoke of the ecological perspective and 'life-cycle' theory in which the Anglo-American city type, economically motivated city growth, and the notion of urbanisation centred on the industrial revolution, dominate geographers' thinking. There are cities with a different past, as Max Weber (1966, p. 197) recognised: 'The South European city, particularly of Italy and South France, despite all differences, was closer to the ancient *polis* than the North European city.' Weber's urban types are not predictable solely by historical sequence: he finds the 'patrician' and the 'plebeian' city type in both Antiquity and the Middle Ages. In other words, his theoretical model solidly incorporates *history*.

The question of the relevance of the past for the present and the future of European (and other) cities, far from being answered, has scarcely even been posed in contemporary urban geography, which remains a-historical. However, the combination of elements from the Weberian tradition and Braudel's (1972) *longue durée* is most appropriate for understanding Mediterranean cities, where the past is constantly revived. Why should the urban identities of Athens and Rome be defined with reference to the industrial revolution rather than the ancient *polis*?

It is true, of course, that antiquity has usually been appropriated by authoritarian and conservative regimes in the Mediterranean, which might lead to a sense that any reference to the cultures of the ancient city–states could be deemed 'politically incorrect'. This is absurd: if Mussolini decided to resurrect the glory of Ancient Rome, or conservatives celebrated certain aspects of antique cultures, this does not mean that these cultures involved only monumentalism, coercion, war and power. Besides, the Roman period is not all that antiquity has to offer, and this period contributed much more than what Mussolini saw in it. Discussing antiquity should not be the privilege of nostalgic conservatives alone, especially if interpreting the Mediterranean is the objective.

It is definitely undesirable in this chapter to apply to Mediterranean cities any 'grand narrative' of progress or convergence with cities of the North – or of reversal of their trajectories, for that matter. The rejection of grand narratives brings us to local narratives, which are particularly rich and diverse (Leontidou 1996). This chapter presents five narratives of the urban Mediterranean (the number is far from exhaustive – others could be added). Each involves the interplay of past and present and a recognition that the sweeping events which led to the formation of urban identities in the Mediterranean preceded the industrial revolution, and that the latter is not a particularly relevant epoch in Mediterranean urbanisation, except as a global context. From a historical viewpoint, many of the following five narratives revolve around two post-war transitions involving Mediterranean Europe: the first one took place during the early 1970s and was dramatised by simultaneous political consolidation, as well as major events affecting urban development trajectories (Leontidou 1990); the second transition is still in progress during the present time and relates with the post-socialist era.

ETERNAL AND EPHEMERAL: CITY–STATES, COLONIES, CAPITAL CITIES

The history of the Mediterranean is the history of its cities. This was one of the refrains to emerge from Chapters 5–7. The depth of urban tradition and diversity in this area is unparalleled in any region of the world. This is where ancient and medieval city–states came to flourish and decline, and where today four worlds mingle, meet and clash: Europe, the Balkans, Asia and Africa. Back in history, there was only one world around the Mediterranean: Europe. As in myth, where she was kidnapped by Zeus in disguise (or escaped from orientalism, according to the matriarchal version), Europe moved westwards and, later, northwards. In myth she was a Phoenician princess who moved from Sidon to Crete to give birth to King Minos and set up the glamorous civilisation bearing his name. In history, Europe was never a continent with any specific boundaries. 'She' moved from the Mediterranean shores towards the west and north, until her contour engulfed 'our' Europe today, excluding the shores of Africa and the western shores of Asia about 1000 years after the birth of Christ on those very shores. No other continent has had such shifting boundaries, none other is as ephemeral as Europe in the '*longue durée*', or as eternal as Europe in human cultural development.

In this sense, the present chapter returns to the definition of Europe of previous millennia, investigating urban unity and diversity around the Mediterranean shores. This used to be a unifying sea, or lake: literally a 'sea in the midst of land'. Greeks and Romans floated on it from the Black Sea to Gibraltar and set up their colonies around it: Ionian colonies from Alonai (Alicante today) to Olivia on the northern coast of the Black Sea and Salamis in Cyprus; Dorian colonies on a north/south axis from Epidamnos on the Adriatic coast to Kirini in Libya; and Roman colonies much further afield. All in an Empire stretching from the whole of Iberia to Mesopotamia and from the northern coasts of Africa to Britain by the second century AD. Meanwhile, as we saw in Chapter 5, new cities had been established and settlements were relocated, especially from antiquity until the Macedonian era (Demand 1990). Decisions on their location were taken with methods which only few cultures would understand today: by oracles, as in the years of Alexander the Great, and by consulting astrologers besides maps, as in the age of discoveries (Boorstin 1983).

Ancient history is discontinuous. Cities have risen, declined, been forgotten, until they recurred, transformed and yet familiar with their past, or rediscovering a past they did not know they had. The destruction of the Aegean civilisation of the second millennium BC is still a mystery to historians: earthquakes buried its buildings and extinguished even its language, Linear A and B. Out of the dark ages between this and the descent of the Dorians, the light of classical civilisation came to shine in a few city–states surrounded by a sea of tyranny. Athens rose and defeated the Assyrians, despite their military supremacy. In the ancient Athenian democracy, however, 'barbarians' were excluded from citizenship and consequently not allowed into the army. This type of exclusion weakened defence: Athens became vulnerable and lost its supremacy to Sparta, then to Macedonian expansionism, and then to Rome.

Urban competition meant war at that period. The resilience of Rome through the centuries must be attributed to its effective administration allowing for local autonomy and self-rule in its colonies and the multicultural composition of its army.

Athens was ephemeral, flowering for a mere 106 years during the fifth century BC, between the expulsion of its tyrant, Hippias, in 510 BC and its surrender to a Spartan general after the Peloponnesian wars in 404 BC; but this small city of 200 000 'free men' was at the same time definitely an eternal city, engraved in human civilisation. Its culture, created by a constellation of brilliant intellectuals, scientists, artists and architects who built the Parthenon in 447–438 BC, is a living legacy in Europe today. The Athenians established direct democracy, where citizens had a voice in the *agora* and leaders were just representatives elected to office for a short term. It was a weak democracy, with slaves and also Athenian women excluded from citizenship. The latter, however, as the strong figures in myths revived in the Athenian tragedies and comedies, keep captivating audiences today.

The Athenians established a network of cities based on colonies along Mediterranean shores, but also *demi*, local communities around the city–state, in Attica. According to Aristotle, 'the union (*koinonia*) of several villages (*komai*) is the full *polis*', in a hierarchy of family/village/ *koinonia* within his political philosophy, which claimed that the State is a natural entity (Demand 1990, pp. 14–15). The Attic *demos* rested on civic gratuities and was supervised by local authorities (Weber 1966, pp. 197–8, 202).

In city–states, city and state were identical and sometimes actually coterminous, in contrast to the political capitals of nation–states, as known today, but also in the past (Alexandria, Rome, Constantinople, Cordoba; Toynbee 1967, p. 5). Capital cities first challenged the city–state concept during the Macedonian period. Kings Philip and Alexander established an empire with their first administrative capital city in Vergina and with Dion as their religious capital, in Macedonia and today's Thessaly respectively. Alexander the Great also established many cities in the territories he conquered from Egypt up to India. He may have founded 70 cities (Demand 1990, pp. 154–7), at least seven of which bore his name: Alexandrias on the Nile, in the Persian Gulf, at the junction of the Acesines and Indus rivers, and others between them. However, the emergence of capital cities during the Macedonian and later the Roman Empire did not end the period of city–states.

This short-lived Athenian culture, ephemeral and eternal at the very same time, gave its lights to the Macedonian kings; it was cultivated and celebrated by the Romans; it was resurrected during the enlightenment and the Renaissance, based on Italian cities. Another constellation of intellectuals set the foundations of European culture, especially in Medici Florence in the fifteenth century. Florence rediscovered Athenian culture and developed beyond it to actually become 'the first modern State of the world' (Toynbee 1967, p. 68). Now 'our' Europe reclaims its Hellenic roots in the Athenian democracy, while at the same time castigating contemporary Athens for failing to conform to capitalist standards.

All Mediterranean cities have failed in this respect, one way or another, because this is simply not their objective or their dream. Their pride and success were linked to the sea, to maritime dominance and to the humanism reflected in urban cultures (Malkin and Hohlfelder 1988). Mediterranean continuity was sustained as long as sea routes were important: as Christianity expanded, as explorers discovered new lands, and trade was busy in ports and canals. However, unity was terminated after the subjugation of the Levante to the Ottoman Empire, but especially as road and later rail transport developed and the nation–states emerged. In the story of successive metropolitan leaders in Europe (Jones 1990, pp. 51–7), after the competition between Venice, Pisa, Amalfi, Florence and Genoa (Benevolo 1993), swings to the North started. Already in the twelfth century Bruges had developed; but it was Antwerp which marks the definite shift. This city rose out of nowhere in the sixteenth century, then still rivalled by Genoa exploiting its Spanish connections (Jones 1990, p. 54). From the sixteenth century onwards, however, after the supremacy of Amsterdam, development swung to the North and stayed there, to culminate in the remarkable industrial revolution. This round of urban competition was won by the Northern cities.

The sea gradually became a fragmenting space. The Mediterranean slipped from core to peripheral status from the seventeenth to the twentieth centuries. Strabo's prosperous *magna Grecia* had metamorphosed into the underdeveloped South of Italy by the eighteenth century and the glorious ancient Greek city–states had lost their name, besides their population, by the time of their liberation from Ottoman rule in the early nineteenth century. Athens' population had shrunk from 12 000 (in the 1570s and 1810s, with oscillations in between) to 6000 by 1832, and Piraeus was unnamed and referred to as Porto Leone (Leontidou 1989).

URBANISATION WITHOUT INDUSTRIALISATION

So, what do Southern cities remember, what have they built their identities upon, and where does their civic pride rest? Certainly, not on the traumatic events preceding the development of capitalism in Northern Europe; nor on the industrial revolution. The Mediterranean was a periphery by the turn of the twentieth century, repelling capitalism as it was marginalised by it. Its cities, however, were growing with migrants. Whereas the industrial revolution created the British, German, Dutch and French urban agglomerations, in the Mediterranean urbanisation escalated without industrialisation.

The region south of Europe is not a major demographic concentration, but by the 1970s this was one of the most urbanised and rapidly urbanising regions of the world. Urban growth rates here were rarely less than 4 per cent during the 1960s (Clarke 1980), and are still today at least double the fastest ones of European cities. Egyptian cities grew spectacularly in absolute rather than relative terms though, and Israeli ones have been affected by policy to limit the growth of the three largest cities – Jerusalem, Tel Aviv and Haifa – and encourage the growth of new towns in the desert. The overwhelming concentration of population in the largest cities is topped by Egypt. Cairo had over 5 million people by the mid-1950s and Alexandria over 2 million, while Southern Europe's largest cities were Athens, Rome, Barcelona and Madrid, each with around 3 million inhabitants in the mid-1970s. During the same period, Istanbul with 2.5 million people surpassed Ankara, the artificial Turkish capital of 1.7 million which, however, had grown sixfold between 1950 and 1975, while Istanbul had only grown threefold.

There is a tendency for primacy to be intensified where capital cities are also ports, and for binary models to appear where the capital is within the mainland and the second city is a port. City-size hierarchies are primate in Greece and Portugal, Morocco and Tunisia, but near log-normal (by rank-size rule) in Spain, Italy and Egypt, where two or more cities dominate the hierarchy, and Algeria. Turkey has a tendency towards a binary city-size distribution, and Libya's Tripoli is growing to be a primate city (Clarke 1980, pp. 48–9). Most cities which defy the neat rank-size rule and 'distort' Anglo-American models are on the Mediterranean shores: Algiers, Tripoli, Tunis, Alexandria, Beirut, Tel Aviv, Athens, Barcelona, Lisbon. In general, for almost a century until the early 1970s, it was these few larger cities that grew at much faster rates than any smaller cities and towns. The rural/urban gap dominated the pattern of spatial divisions, even where North/South (Italy), East/West (Spain) or coast/interior (Portugal, North Africa) were the principal axes of uneven regional development. Growth in large cities accelerated everywhere. Smaller cities stagnated and rural areas were depopulated.

The question remains: what has been pulling people to Mediterranean cities, if not industry? Material scarcity in rural areas has been always underlined in explanations of the 'rural exodus' to large cities in Southern Europe and abroad, towards Northern Europe (see, for instance, Hudson and Lewis 1985; Williams 1984). However, the forces for urbanisation and emigration were not just economic 'push' factors. There were cultural forces and 'pull' factors as well for the massive rural exodus from the 1920s to the 1970s. These remained obscure, because researchers kept studying traditional communities and there was a paucity of Mediterranean urban studies well into the 1980s (Kenny and Kertzer 1983). Yet people kept arriving in cities and internal migrants increased. In Rome the native-born population dropped from 47 per cent of the total in 1921 to 20 per cent in the 1960s. In Athens 56 per cent of the 1960 population were post-war migrants; others were earlier migrants and Asia Minor refugees (Leontidou 1990, p. 103). In the absence of mass industrialisation it has been recognised that people were pushed to cities by the hope for work in the informal sector (Garofoli 1992). There was something else, however: the attraction exerted by the culture of urbanism.

Even before written history began, civilisations around the Mediterranean have been always distinctively urban, except in periods of authoritarianism in our century (Leontidou 1989, pp. 263–80; 1990, pp. 256–9). The strength of ancient civilisations was the city and the city-born spirit of exploration, whether geographical or scientific. Mediterranean cultures involved imagination revolving around the city and a clear urban ideal with advanced political institutions and citizenship. Urban identity is still constantly reaffirmed, often aggressively, in all corners of the Mediterranean. Cities compete as city–states once did, though with peaceful means (Leontidou 1995). In a sense, these cities never lost city–state or capital-city status in popular imagination.

Mediterranean cities are not undergoing counter-urbanisation, with exceptions in formerly Fordist places of mass industrialisation like the Italian Northern triangle and the Basque country. They probably never will, despite their dramatic fertility drop during the last decade (Sporton 1993). The lack of Fordist industry precludes deindustrialisation, and the culture of urbanism is still too strong to allow for urban decline.

ISLAMIC AND EUROPEAN CITYSCAPES: INFORMALITY, RELIGION, GENDER

The diversity of Mediterranean cityscapes is not simply a tourist attraction. These are intelligent multicultural milieux incorporating ancient, modern and postmodern elements, and several social groups, classes and cultures in close proximity. Historical depth is visible in the mixture of ancient ruins, medieval cores and relics of walls. Arabian ruins are tenderly renewed not only in Northern Africa but also in Andalusian cities. Antiquity may be preserved or recycled, like ancient stones of temples built into Christian churches in Greece and Italy. The profane spaces of capitalist speculation have since surrounded these sacred spaces, but have never annihilated their presence or eradicated their appeal to residents and visitors alike.

The Mediterranean mixed cityscape is recently admired in our postmodern era. Whether Islamic, Catholic or Orthodox in persuasion, populations who have shaped Mediterranean urban morphology have given to cities three important common characteristics, which set them apart from the Anglo-American ones: the inverse-Burgess spatial pattern, with the affluent classes in the centre and the poor on the periphery; a compact cityscape, where streets are narrow, buildings tall, and suburbs rather close to the centre; and mixed land use rather than zoning of residence and economic activity. A fourth characteristic common to the European side of the Mediterranean, but not to Islamic cities, is a mixture of social classes (but not necessarily racial groupings) rather than the segregation found in Northern societies.

Some of these features are captured in Figures 12.1 and 12.2, which are scenes from two Italian cities. Figure 12.1 shows the compact

FIGURE 12.1 Aerial view of city centre of Vicenza in northern Italy: a traditional, high-status residential environment based on tall, elegant, historic buildings built at high density around narrow streets

city centre of a not atypical North Italian town, Vicenza, with narrow streets, tall and often historic buildings, generally occupied by high-status residents. Figure 12.2 shows low-income settlements on the outskirts of Palermo.

These particularities have been explained by the culture of urbanism and the consequent popularity of the city centre; land property systems and planning histories; and the long-standing development of the informal sector (Leontidou 1990; 1994). The compact city, for example, is illustrated by city size and density. Barcelona accommodates 1.6 million people in 10 sq. km; Athens 886 000 in 4.2 sq. km in its centre. Population densities reach levels over twice as high as in London, for example. Mixed land use and the dispersal of economic activity within the urban fabric have been attributed to the lack of planning, the importance of the informal sector and the family, and the unimportance of Fordism and its corollary – zoning. Particularly in the Islamic city, centrally created zoning and subdivision regulations are alien, as cities are built by codes responsive to subtle local variations, following 'divine law' but allowing increased flexibility in the combination of elements (Hakim 1986, pp.137–8).

The informal sector is a strong and uniform feature in urban economy and land use. Though the industrial revolution never took root here, economic organisation was already quite soph-isticated in an early period, before the guild system was established in Europe. In classical antiquity, slaves were employed in the oikos economy.

> The ancient craftsmen would join together with slaves in a mystery community (as in Hellas) or into a *collegium* (as later in Rome). However, they would not belong to an organization claiming political rights like the guilds in the Middle Ages. (Weber 1966, p. 201)

The post-war development of the informal sector has for a long period constituted another major departure from models familiar in Europe (Leontidou 1990; Mingione 1991). As

FIGURE 12.2 City peripheries are low-income residential environments in most Mediterranean cities, where shanty-towns and squatters have proliferated. This picture shows a slum area on the periphery of Palermo, Sicily.

informal economy has now been rediscovered after the demise of Fordism, it has suddenly been legitimised, and is no longer labelled 'pre-capitalist'.

Family and informality create a cohabitation of work and residence in these non-Fordist capitalist societies which facilitates and is reproduced by frequent homeworking and subcontracting arrangements. In European Mediterranean cities, economic as well as social mixture is also reproduced by an important alternative to community segregation: vertical differentiation, with various classes inhabiting different storeys of multi-storey apartment buildings. As for social classes, they must be seen in a different light from those in Northern Europe: class boundaries are blurred by multiple activity and the absence of a massive proletariat (Leontidou 1990); or by groups such as Martinotti's (1993, p. 150) commuters as well as residents, city users and metropolitan businessmen. City populations are all but static. In cases where social dynamics have been researched, a trend for socio-spatial homogenisation can be observed (Leontidou 1990; pp. 233–5; 1994), which is the opposite of the polarisation trends evident in Northern Europe recently (Musterd 1994).

Mixed and homogenising, these are also divided cities, however. In the past, political opponents were excluded from working and from certain communities and clustered together in defence. The population groups defeated in the Spanish and Greek civil wars were stigmatised, jailed or exiled, even during periods of democratic party politics. The Middle Eastern ethnic conflicts and colonial wars have had the same effect on urban segregation. At present, as civil war hostilities are fading, South European cities have found new 'barbarians' to exclude: the migrants from Eastern Europe and the Third World.

The Middle Eastern cities are inhabited predominantly by Muslims, sharing an Islamic identity and attached to the application of 'divine law' in the process of city building (Hakim 1986). The 'Middle East' is defined here as stretching from Morocco to Iran and from Turkey to Sudan (cf. Clarke 1980, p. 36). Our references are limited to the Mediterranean part of this large region, i.e. North African and South-west Asian cities, usually Islamic–Arabic ones. Segregation principles here are different from European ones, as we saw in the historical contexts of Chapters 6 and 7. The *mahallas* of these cities house people of a common ethnic or socio-economic background, under the administration of a Mukhtar. Despite variations in the names of these quarters in Egypt or Tunis, the principle is the same, and the segregation in *mahallas* has sometimes been ensured by gates, locked and guarded at night. This type of cityscape does not occur in Mediterranean Europe. Jewish ghettoes in Rome, 'Third World' migrant quarters in Piraeus or newly formed immigrant enclaves in some inner cities, are only pale and remote parallels of *mahallas*.

But there are other differences between the cities of North and South Mediterranean. In fact, the importance of Islam in city building should not be underestimated. Medina, the Arabic name for an urban settlement, with Rabad, its suburbs or later additions, is built according to Arabic–Islamic divine law. Among schools of law, it is the Maliki school which has been influential in Mediterranean city building, especially in Morocco and Algeria, combined with the Hanafi school in Tunisia and Libya (Hakim 1986, p. 15). These lead to an effective building language made up of relatively few elements combined in various ways permitting flexibility, diversity and complexity. In order to deserve the name of a Medina, a city must have a mosque, a main through street, and a governor's residence. Then there is the Kasbah, attached to the Medina, at a strategic point serving as a potential refuge after the fall of a city, with its main square for ceremonies. There are also many other elements – squares, gates, places of prayer, cemeteries, water storage and sewer lines – all arranged by the application of divine law rather than any economic rationale.

Islamic cities also still exclude women. Important differences in gender stereotypes between Northern and Southern cities are vividly reflected in urban structure. The gendered space of nineteenth-century Athens was defined by a series of dualisms: male/female, public/private, work/idleness, outdoor/indoor, open/closed in urban space (Varika 1985). Visitors to the more affluent communities, where women's idleness constituted a value, would encounter a male-dominated city, where women were invisible behind walls of neoclassical houses, surrounded by servants. In low-status suburbs, domestic work was especially heavy, given the lack of even the most basic infrastructure, and therefore women were invisible for different reasons while men frequented the lively cafés and taverns. All this changed abruptly with the wars, which took men to the army and women to the factories (Leontidou 1989). However, the same gender principles are echoed in today's women-unfriendly fundamentalist environments, where the private sphere is female and the public one male. The gender division of urban space in Mediterranean Europe varies and changes with the development of the informal sector and levels of homeworking. Elsewhere, the interaction of Islamic and European cityscapes in 'modernising' Islamic cities and in Andalusian towns creates interesting variations.

DIFFUSE INDUSTRIALISATION BEYOND THE THIRD ITALY

Literature on Mediterranean non-industrial urbanisation is riddled with exclamations about its 'abnormality' and references to 'parasitic' cities where 'traditional' economies are present and migrants are not integrated in the 'regular' economic life of such 'precapitalist' cities (White 1984). Stereotypes such as these emanate from northern models of urban development, and have been severely criticised within the context of alternative models for Mediterranean Europe (Leontidou 1990; Mingione 1991). However, in the Middle Eastern city migrants are still considered as marginal and tenuously linked to the urban economy, which is measured once again against Eurocentric norms (Drakakis-Smith 1980, p. 92). The concept of the informal sector, though transformed in Europe and the USA with the development of post-Fordism, still holds the old connotations in the Middle East: a traditional sector in a Third World economy outside modernity and modernisation.

Mediterranean Europe has been transformed since the 1970s. Urbanisation slowed down and uneven regional development patterns changed from urban/rural (Greece and Portugal), North/South (Italy) and East/West (Spain) into new forms. The first observable pattern was the slowdown of emigration, following restrictions upon entry in Northern Europe. Then, after stability and family reunion on a limited scale, return migration began. During the 1990s, finally, immigration from the less developed world and Eastern Europe started (see Chapter 11). A major turnaround from emigration to immigration has changed Southern European urban networks in a profound manner. Rather than causing a new wave of metropolitan growth, these movements benefited small towns. Return migrants preferred towns adjoining their former villages. They became agents, not only of conspicuous consumption, but also of diffuse industrialisation.

Productive restructuring was a parallel and more potent force. As earlier Fordist centres (Bilbao, Milan, Turin, Genoa) deindustrialised, post-Fordist industry developed in localities away from the major traditional industrial centres, and tourism grew along the coasts. New creative regions sprang into being. The most notorious and well-documented among these is, no doubt, the Third Italy. It encapsulates the European version of the passage to post-Fordism just as Silicon Valley encapsulates the American version (Amin 1989; Garofoli 1992; Goodman *et al.* 1989).

Between the two Italies torn by the North–South divide, a flexible workforce was (re)discovered by industrialists in the towns of Emilia Romagna. The Fordist northern triangle was hit by crisis after the industrial disputes of the 'hot autumn' of 1969. Gradually, as productive restructuring allowed for smaller industrial units, that were easier to govern and more flexible in the combination of skilled labour and new technology, entrepreneurs abandoned the three traditional North Italian industrial poles of Turin, Milan and Genoa. Many fragmented their production processes and dispersed them to the smaller towns of the Third Italy. Others created new enterprises in those localities, reinforcing diffuse industrialisation processes. Diffuse urbanisation was interlocked with this process. Italy is becoming, again, *'il paese delle cento città'*, the land of a hundred cities.

So are the rest of the Southern European countries. Patterns of uneven development changed and diffuse urbanisation appeared here during the late 1970s, though for somewhat different reasons in each country. Greek diffuse urbanisation owes much to diffuse industrialisation, tourism, the growth in the informal sector, and incentives for regional industrial development in border regions (Chronaki *et al.* 1993; Hadjimichalis and Vaiou 1990). Spain developed technopoles outside the major urban centres (Ybarra *et al.* 1991), mass tourism along the coasts, and an informal sector combining agricultural and industrial activity in rural areas (Naylon 1992); but it also put into operation an advanced system of administrative decentralisation affecting the cities (García 1993). Portugal saw the development of small firms in several peripheral regions, while the attraction of Lisbon and Oporto is still strong, unlike large agglomerations in the rest of Southern Europe (Gaspar and Butler 1992).

The population turnaround from large to smaller cities, then, was an outcome of different processes in the various countries, but was generally correlated with diffuse industrialisation and with a nexus of associated processes: industrial restructuring, return migration, incentives for regional industrial development, administrative decentralisation, the increase of mass tourism and diseconomies of urbanisation in former Fordist centres. Industrial activity in large agglomerations is declining but tends to be substituted almost automatically by a growth in the service sector, transport and communications. Madrid and Piraeus are cases in point.

Convergence between North and South Europe, forecast by 'urban life-cycle' theorists, never happened. Population moves to smaller cities and to the coast but, throughout the Mediterranean, cities contrast with Anglo-American ones in the absence of counter-urbanisation. Earlier Fordist centres in decline are few in number. Even Italy has seen a diffuse urbanisation model, despite some instances of counter-urbanisation in the North (Dematteis and Petsimeris 1989). The population of large cities is growing at slower rates in the European part of the Mediterranean, but high rates persist in the south: the Islamic Mediterranean is in a process of rural/urban migration and emigration, as well as development in coastal regions. Cultural diversity, war and creeping fundamentalism preclude convergence between the north and the south of the Mediterranean.

SOCIAL FORCES IN CITY BUILDING AND SPATIAL REGULATION

If planners and the public sector actually built northern cities, the Mediterranean cities knew very little of these agents. For almost a century, urban expansion was spontaneously brought about by workers and low-income strata colonising new land, and infrastructure either followed settlement, or did not expand at all to the squatter communities. A major transition took place in the European region, however, after the early 1970s: all cities left behind the period of spontaneous urban development with a spectacular simultaneous peaking and then decline of urban social movements. After this, a variety of urban regulation systems and strategies emerged. In other words, a brief period of dramatic convergence has led to divergence since the 1980s (Leontidou 1994; 1995).

Let us first consider how city building and spatial regulation developed during the inter-war period, before planning was almost universally abandoned. The anti-urban values of fascism were reflected in laws to control migration, and Mussolini's illusions of grandeur in appalling rhetoric for the resurrection of the glory of Rome. Fascists appropriated antiquity, repelling radicals from any study of past cultures. In 1929 Mussolini ordered the planners of Rome to achieve monumental planning within five years, removing and demolishing all structures (and even a hill!) around ancient monuments (Agnew 1995, pp. 47–51; Fried 1973, pp. 31–6). Mussolini's clearing of the centre of Rome would be later repeated in Franco's Spain, though in a more hesitant manner. Italian corruption and compromise also obstructed the full implementation of the dictator's plans, but imperial Rome was uncovered. Mussolini's modernist project contrasts with impressive postmodern undertones in another one of his projects, EUR, built to host a future World's Fair in Rome in 1942 for the celebration of 20 years of fascism. The exhibition was never realised, but EUR stands today with its impressive architecture as a large administrative suburb. No post-war projects can match these inter-war initiatives, but Italy did not give up urban planning as other Mediterranean cities did. There have been impressive examples, such as Bologna's municipal socialism and the urban restoration which has given to the world marvellous museum cities like Venice, Florence and Pisa (Benevolo 1993).

In Spain, planning achievements preceded Franco's regime, which was rather inefficient. Arturo Soria y Mata's Ciudad Lineal was conceived in 1892 and built in 1894–1904 as a 48-km linear city to the east of Madrid. It was then abandoned, however, with trams replaced by the metro (with a station named after the architect) and the original villas by apartment blocks (Hall 1990, pp. 112–13). The Ciudad Lineal was engulfed by urban development.

Greece had brief periods of dictatorship, which did not excel in planning. The major event here was the housing policy of the Refugee Settlement Commission which swept the whole country from the 1920s until the 1950s, for a brief period combined with the work of post-war reconstruction agencies (Delladetsima and Leontidou 1995). The inter-war policy, ironically, also indirectly led to the spontaneous colonisation of urban peripheral land by the refugees, who were then followed by the post-war migrants in creating the Athens working-class suburbs (Leontidou 1989; 1990).

Throughout the Mediterranean, for a long period, fast urbanisation waves drew the cities towards the suburbs, with population spontaneously colonising new land. Their semi-squatting practices constitute a type of urban social movement resisting planning throughout the Mediterranean region. Alternative cultures were backed by a multitude of 'building by night' and of contravening planning bye-laws. Such survival strategies were (re)discovered by urban populations everywhere, from the Portuguese *barrios clandestinos* to the Istanbul *geçekondus*, and from the Athens *afthereta* to the mausoleum and tomb dwellers of Cairo (Drakakis-Smith 1980, p. 98). There are 'slums of hope', which have eventually acquired legal or semi-legal status, but also 'slums of despair', such as the makeshift huts and shanties of the Islamic city.

European urban social movements peaked in the 1970s (Pickvance 1995). This was especially important in Southern Europe, since simultaneous popular mobilisation led to (or coincided with) political consolidation as well as changes in urban development. In Italian cities, workers' unrest in the industrial triangle spread southward, where Roman populations were already mobilised on the issue of housing and infrastructure, and the movements merged (Marcelloni 1979). In Spanish cities, *asociaciones de vecinos* were most militant in Madrid (Castells 1983). Portuguese movements linked residence with decolonisation (Gaspar 1984). Simultaneous mobilisation also included the Mediterranean populations of Paris. Algerians mobilised for work issues, occupying the Renault factories, as well as for urban demands in the *bidonvilles*. The French government responded with a policy combining efforts at integration of Islamic cultures, together with repatriation incentives. In all cases, authorities stepped in to provide alternatives to shacks and control spontaneous urbanisation. The Greek urban movements appeared earlier, and were crushed by demolition and legalisation during the period of the dictatorship. This difference is partly responsible for creeping repolarisation in Mediterranean Europe (Leontidou 1995).

Simultaneous political consolidation (Greece 1974, Spain 1975–78, Portugal 1974), as well as waves of decolonisation (France 1962, Portugal 1975) transformed the political milieu. In addition, profound urban restructuring has taken place since the 1980s, which was diverse according to state policy in each country. The establishment of planning systems was by no means homogeneous throughout the Mediterranean. On the contrary, almost a century of urban convergence was terminated. Some cities were abandoned to relentless speculation without the safety valve of working-class or public housing, while others developed a sophisticated planning machinery often disregarding homelessness and staging special events to attract attention and visibility. It is interesting that the two extremes were held by Greece and Spain, the two countries with socialist governments on the two edges of Mediterranean Europe (Leontidou 1995).

The control of urban social movements during the mid-1970s was thus a major event which drew the various Mediterranean cities apart, into a period of divergence, first in the sphere of spatial regulation, then across the board. At one extreme, speculation triumphed in Greece, moderated in the 1980s by the direction of EU funds towards infrastructure in poor western areas of Athens, but also strange pilot projects (Leontidou 1993; 1995). At the other extreme, planning was resurrected in Spanish cities, after the important administrative regional reform of 1978 (García 1993; Leontidou 1995). Outside the European Union, Balkan cities were also developing some initiatives until they blew themselves apart by the bombs of war. Social movements here ended up in hostilities, religious and ethnic conflict.

Conclusions

What is happening in Europe today is the exact opposite of what convergence theories, 'urban life-cycle' models or evolutionist perspectives have insisted in suggesting: hundreds of different realities emerge. This is much more so in different corners of the Mediterranean. Cities experience different realities from each other and from Northern cities. Urban competition, place marketing, repolarisation and spatial divisions, but also socio-spatial homogenisation versus division and social exclusion, tear European cities apart.

Outside the EU along the Mediterranean shores, urban change is also tainted by fundamentalism, war, segregated migrant enclaves and gender divisions. These appear or vanish successively in a postmodern kaleidoscope revolving and redistributing its crystals in each city in a fast-moving ephemeral reality. It is quite impossible to summarise the diverse patterns in the hundreds of cities in such a divided region as the Mediterranean. The five narratives in this chapter can only be a reminder of this diversity and, perhaps, the introduction to a new set of narratives emerging at the end of our millennium.

Deconstruction and the interlocking of past and present in the narratives have taken us from one extreme of post-Fordist development, held by the North Italian urban triangle or the Spanish urban success stories, to the other end held by Maghreb cities, through the in-between spaces of Athens, Rome, or Lisbon. Their different 'speeds' of integration into Europe show that dualisms are deceptive and form only a small part of the story. There is a multiplicity of interlocking narratives composing Mediterranean urban diversity in our time. Besides being demographically dynamic compared with Northern European cities, Mediterranean cities have ceased to move in parallel trajectories. Diversity and transformation are their only common ground. Fragmentation of an earlier Mediterranean unity was also revealed, or repolarisation in the case of Southern Europe. A drama of Braudel's *longue durée* is re-enacted today, within a very shortened time-span: Mediterranean unity was undermined especially by Ottoman expansionism, but also by transport systems and technological change. Convergence began again in the nineteenth century, only to be undermined in our days, since the 1970s, by an important transition. The 1990s are darkened by war and hostilities;

but they are also brightened by controversial urban success stories after the speeding up of European integration.

In many ways, current urban theory has tried, and failed, to fit urban structures and trajectories to a few abstract models, like the urban life-cycles approach and the relative convergence and evolutionist theories. These are crushed by the diverse and exuberant reality of Mediterranean cities. We have reached a stage where, despite so-called 'mass' culture and 'global' restructuring, Mediterranean cities can only be interpreted locally, even individually. How can we ever 'theorise' each individual city comprehensively, as long as each one is so multi-faceted as to belong to each and every one of the five narratives presented here, as well as many more omitted?

References

AGNEW, J.A. 1995: *Rome*. Chichester: Wiley.

AMIN, A. 1989: Flexible specialization and small firms in Italy: myths and realities. *Antipode* **21**, 13–34.

BENEVOLO, L. 1993: *The European City*. Oxford: Blackwell.

BOORSTIN, D.J. 1983: *The Discoverers*. Harmondsworth: Penguin.

BRAUDEL, F. 1972: *The Mediterranean and the Mediterranean World in the Age of Philip II*. 2 vols. London: Collins.

CASTELLS, M. 1983: *The City and the Grassroots*. London: Edward Arnold.

CHRONAKI, Z., HADJIMICHALIS, C., LABRIANIDES, L. and VAIOU, D. 1993: Diffused industrialization in Thessaloniki: from expansion to crisis. *International Journal for Urban and Regional Research* **17**, 178–94.

CLARKE, J.I. 1980: Contemporary urban growth. In Blake, G.H. and Lawless, R.I. (eds), *The Changing Middle Eastern City*. London: Croom Helm, 34–53.

DELLADETSIMA, P. and LEONTIDOU, L. 1995: Athens. In Berry, J. and McGreal, S. (eds), *European Cities: Urban Planning and Property Markets*. London: E. & F.N. Spon, 258–87.

DEMAND, N.H. 1990: *Urban Relocation in Archaic and Classical Greece: Flight and Consolidation*. Norman: University of Oklahoma Press.

DEMATTEIS, G. and PETSIMERIS, P. 1989: Italy: counterurbanization as a transitional phase in settlement reorganization. In Champion, A.G. (ed.), *Counterurbanization*. London: Edward Arnold, 187–206.

DRAKAKIS-SMITH, D.W. 1980: Socio-economic problems: the role of the informal sector. In Blake, G.H. and Lawless, R.I. (eds.), *The Changing Middle Eastern City*. London: Croom Helm, 92–119.

FRIED, R.C. 1973: *Planning the Eternal City: Roman Politics and Planning since World War II*. London: Yale University Press.

GARCÍA, S. 1993: Local economic policies and social citizenship in Spanish cities. *Antipode* **25**, 191–205.

GAROFOLI, G. (ed.) 1992: *Endogenous Development and Southern Europe*. Aldershot: Avebury.

GASPAR, J. 1984: Urbanisation: growth, problems and policies. In Williams, A.M. (ed.), *Southern Europe Transformed: Political and Economic Change in Greece, Italy, Portugal and Spain*. London: Harper and Row, 208–35.

GASPAR, J. and BUTLER, C.J. 1992: Social, economic and cultural transformations in the Portuguese urban system. *International Journal of Urban and Regional Research* **16**, 442–61.

GOODMAN, E., BAMFORD, J. and SAYNOR, P. (eds), 1989: *Small Firms and Industrial Districts in Italy*. London: Routledge.

HADJIMICHALIS, C. and VAIOU, D. 1990: Flexible labour markets and regional development in Northern Greece. *International Journal of Urban and Regional Research* **14**, 1–24.

HAKIM, B.S. 1986: *Arabic-Islamic Cities: Building and Planning Principles*. London: KPI Ltd.

HALL, P. 1990: *Cities of Tomorrow*. Oxford: Basil Blackwell.

HUDSON, R. and LEWIS, J.R. (eds) 1985: *Uneven Development in Southern Europe: Studies of Accumulation, Class, Migration and the State*. London: Methuen.

JONES, E. 1990: *Metropolis: The World's Great Cities*. Oxford: Oxford University Press.

KENNY, M. and KERTZER, D.I. (eds) 1983: *Urban Life in Mediterranean Europe: Anthropological Perspectives*. Urbana: University of Illinois Press.

LEONTIDOU, L. 1989: *Cities of Silence: Working-Class Colonization of Urban Space in Athens and Piraeus, 1909–1940*. Athens: ETVA and Themelio (in Greek with English summary).

LEONTIDOU, L. 1990: *The Mediterranean City in Transition: Social Change and Urban Development*. Cambridge: Cambridge University Press.

LEONTIDOU, L. 1993: Postmodernism and the city: Mediterranean versions. *Urban Studies* **30**, 949–65.

LEONTIDOU, L. 1994: Mediterranean cities: divergent trends in a United Europe. In Blacksell, M. and Williams, A.M. (eds), *The European Challenge: Geography and Development in the European Community*. Oxford: Oxford University Press, 127–48.

LEONTIDOU, L. 1995: Re-polarization of the Mediterranean: Spanish and Greek cities in neoliberal Europe. *European Planning Studies* **3**, 155–72.

LEONTIDOU, L. 1996: Alternatives to modernism in (Southern) urban theory: Exploring in-between spaces. *International Journal of Urban and Regional Research*, **20**, 178–95.

MALKIN, I. and HOHLFELDER, R.L. (eds) 1988: *Mediterranean Cities: Historical Perspectives*. London: Frank Cass.

MARCELLONI, M. 1979: Urban movements and political struggles in Italy. *International Journal of Urban and Regional Research* **3**, 251–68.

MARTINOTTI, G. 1993: *Metropoli: La Nuova Morfologia Sociale della Città*. Bologna: Il Mulino.

MINGIONE, E. 1991: *Fragmented Societies: A Sociology of Economic Life beyond the Market Paradigm*. Oxford: Basil Blackwell.

MUSTERD, S. (ed.) 1994: A rising European underclass? Special issue of *Built Environment* **20**, 184–268.

NAYLON, J. 1992: Ascent and decline in the Spanish regional system, *Geography* **77**, 46–62.

PICKVANCE, C. 1995: Where have urban movements gone? In Hadjimichalis, C. and Sadler, D. (eds), *Europe at the Margins: New Mosaics of Inequality*. Chichester: Wiley, 197–217.

SPORTON, D. (1993) Fertility: the lowest level in the world. In Noin, D. and Woods, R.I. (eds), *The Changing Population of Europe*. Oxford: Blackwell, 49–61.

TOYNBEE, A. (ed.) 1967: *Cities of Destiny*. London: Thames and Hudson.

VARIKA, E. 1985: Invisible work and conspicuous consumption. In *The New Greek City*, vol. A, Athens: Society for the Study of New Hellenism, 155–66 (in Greek).

WEBER, M. 1966: *The City*. New York: The Free Press.

WHITE, P. 1984: *The West European City: A Social Geography*. London: Longman.

WILLIAMS, A.M. (ed.) 1984: *Southern Europe Transformed: Political and Economic Change in Greece, Italy, Portugal and Spain*. London: Harper and Row.

YBARRA, J.A., DOMENECH, R. and GINER, J.M. 1991: Technological parks: their theory and reality in Spain. *International Journal of Urban and Regional Research* **15**, 383–94.

13

THE MODERNISATION OF MEDITERRANEAN AGRICULTURE

JEFF PRATT AND DON FUNNELL

INTRODUCTION

This chapter will present an overview of the modernisation of agriculture in the Mediterranean region, concentrating particularly on the four countries of Southern Europe which are members of the European Union (Portugal, Spain, Italy, Greece), and four countries of North Africa (Morocco, Algeria, Tunisia, Libya). Before analysing the modernisation processes which have transformed agriculture in these regions, we need to outline a few of the key enduring geographical features which shape the region, and some of the ways in which agricultural production was organised in the period up until the end of the Second World War. This introductory historical and geographical narrative will build on that provided by Dunford in Chapter 9, and reference should also be made to Tables 9.4 and 9.6 in that chapter.

The unity of the Mediterranean is partly a feature of climate (see Chapter 3), and this has an obvious impact on agriculture; indeed, one way of sketching the boundaries of the region is to trace the limits of the cultivation of olive trees (refer back to Figure 1.4). Wheat, vines and olives are the classic crop mix characteristic of Mediterranean agriculture. However, the agriculture practised in each of the countries we shall be examining is also shaped by the existence of other climatic zones, and in this context there are important differences between the northern and southern shores. In southern Europe, Mediterranean agriculture exists in countries (or unified markets) which also contain temperate zones – to the north of the Apennines we find the intensive arable farms of the Lombard plain. In the Maghreb the southern boundaries of the Mediterranean region merge into arid zones. South of the Atlas only low-intensity grazing is possible outside the oases, except in areas where irrigation systems operate. Throughout the Maghreb, the reliability of rainfall, even in winter, becomes a major factor in arable production and the region experiences periods of extensive drought. Consequently, agricultural production was based traditionally on a risk-minimising strategy incorporating dry-farming techniques, irrigation and stock rearing.

The diversity of Mediterranean agriculture is also derived from the marked variations in terrain, the origins of which were discussed in Chapters 2 and 4. Throughout the region there are high mountain chains, many deforested and heavily degraded, where no forms of intensive agriculture are possible, and the limited economic activity revolves around grazing, or the odd belt of cultivated chestnuts. Lower down are the rolling hills and table-lands where rain-fed arable farming, olive groves and vineyards are common – this zone is more predominant in southern Europe than in North Africa. Finally, there are the lowlands and river valleys where more intensive agriculture based on irrigation has been practised for more than 2000 years, and where much of the capital-intensive farming is found today. These areas include citrus production in Andalusia, the 'Golden Bowl' around Palermo and the irrigated lowlands of the Maghreb; cotton production along the Guadalquivir; tobacco in the better-watered Greek Plains; and intensive fruit and vegetable production in favoured locations throughout the Mediterranean.

There are also considerable variations in the agrarian social structures found in the region. In the immediate post-war period the countries of the Mediterranean littoral all had half or more of their active populations employed in agriculture, and the majority of these people were peasants, catering mainly for their own needs. We tend to think of peasants as a traditional and unchanging category of cultivators, but although much of the technology employed may have been ancient, many of the crucial economic and social characteristics of peasantries in the Mediterranean only emerged in the late eighteenth and nineteenth centuries, with the abolition of feudalism and earlier forms of property rights. Peasants used household labour to cultivate areas of land which rarely produced more than the foodstuffs required for home consumption and a small surplus for sale to obtain a few manufactured household goods. The land-holding of each family was small compared to Northern Europe, and often consisted of scattered plots rather than consolidated farms: this being both a consequence of the widespread practice of partible inheritance, and an ecological necessity for the practice of mixed agriculture in areas with very varied terrain. Davis (1973) gives a detailed case study of production patterns amongst small farmers in southern Italy. In the mountains and other marginal areas many households could not even produce basic subsistence needs from the land available, and relied on seasonal migrant labour to survive. Agriculture was unmechanised, and subject to a harsh and unpredictable climate: drought, flood, late frosts and desiccating winds. Even in the best of years, labour requirements were very variable, with long periods of inactivity interspersed with a few very busy months. As we shall see later on, since the 1950s the rural exodus and agricultural modernisation have removed the 'peasant' character of much Mediterranean agriculture, but in marginal areas small-scale, semi-subsistence farming still survives in the hands of elderly farmers supported by pensions and welfare payments (see Fig. 13.1).

Mediterranean agriculture was never completely dominated by peasant forms of production or 'minifundia'. There were also very large estates – the latifundia. In Southern Europe these tended to be either the ancient properties of aristocratic families and the Church, or had been bought out by an urban bourgeoisie following the abolition of feudalism and the advent of nineteenth-century liberal policies. We can find such estates in southern Portugal, southern Spain (Martinez-Alie, 1971) and parts of central and southern Italy (for Puglia, see Snowden 1986; for Sicily, Schneider and Schneider 1976); they were much less common in Greece. These latifundia were characterised by extensive agriculture – cereals, olive groves,

Figure 13.1 Small-scale unmechanised peasant farming still persists in a few marginal environments in the Mediterranean Basin, even in the more developed European countries. This rural scene, near Stigliano in Basilicata, southern Italy, was taken in 1989.

ranching – and were worked either by sharecroppers under a variety of contractual arrangements (such as the central Italian *mezzadria*) or by the seasonal employment of very large numbers of low-paid labourers. The fact that many of these estates were capitalist, in the sense that they were based on wage-labour, does not imply that they were islands of 'modernised' agriculture: on the contrary, they were in chronic need of investment and new technology. All the regions of latifundia experienced rural unrest, with demands for land reform and attacks on the entrenched interests of very powerful agrarian elites. Rural class conflicts were intense in Italy in the 1940s and 1950s, and in Portugal and Spain with the collapse of authoritarian regimes in the 1970s. Within the Maghreb large estates existed in pre-colonial times, usually occupied by the local elite. Many of these large holdings were located within oases or depended upon complex systems of irrigation. The ancient *khettaras*, in which carefully graded underground channels brought water long distances to irrigate palmeries and cereal crops, provide an example of the ability of powerful leaders to organise labour to construct such systems. It was, however, the advent of colonial occupation – the French in Morocco, Algeria and Tunisia and the Italians in Libya – that laid the basis for modern large-scale farming based upon the use of wage labour and integration with international markets.

Alongside peasant holdings and latifundia we also find pastoralists, predominantly shepherds, virtually throughout the region. These are often transhumant, moving their flocks from summer grazing in the mountains, down into the lowlands in winter, where they often grazed on the stubble land left after the previous cereals harvest. Thus, for part of the year they interacted with settled cultivators (not always without friction); for the rest they were unfettered by other uses for the land. The fact that pastoralists tend to occupy the most remote parts of the Mediterranean landscape and seem most closed and traditional in their ways, should not blind us to the fact they are much more integrated into the market than the average peasant, since they can only gain a livelihood by selling a large proportion of the meat and cheese that they produce. A good example is the various communities within the High Atlas whose production systems have been described by Berque (1978) and Miller (1984).

Throughout the world, modern states have treated nomads as unreliable citizens, and attempted to curtail this form of livelihood through restrictions and obligations. Nevertheless, although data for the Maghreb are scarce, it has been estimated that 17 per cent of Morocco's population are nomadic while the figures for Libya are 10 per cent, Algeria 3 per cent and Tunisia 2 per cent. In Southern Europe, government regulations and the enclosure of land along the transhumant corridors have severely curtailed formerly extensive nomadic forms of pastoralism in areas like the Puglian *tavoliere* (see Snowden 1986) or the Pindus mountains of northern Greece (Campbell 1964). However, other forms of shepherding continue to be an important part of many rural economies in Southern Europe (see Fig. 13.2). Total livestock numbers fluctuate very rapidly, reflecting competition from other producers (for example, in Eastern Europe), and also the knock-on effect of EU policies designed to reduce production in

Figure 13.2 Shepherding remains a significant aspect of Mediterranean rural economies, both in a form which is confined to marginal and scrubland environments, as here in the hills of southern Cyprus, and integrated with cropping via transhumance and fallow systems

other sectors of agriculture. The total number of sheep in the EU rose very fast in the late 1980s, and of the total of 100 million, 24 million are found in Spain and 10 million in Italy, while half of the EU's 12 million goats are found in Greece (Eurostat 1995).

The categories latifundia/minifundia/pastoralists provide a useful framework for describing many aspects of Mediterranean agriculture, but two qualifying comments are necessary. The first is to reiterate the point that 50 years ago there was no close connection between production for the market and the kind of technology used – many of the large capitalist estates still operated very labour-intensive production systems. Second, in the category of minifundia we could find both peasantries oriented predominantly towards the rationality of subsistence needs, and some small family farms thoroughly integrated into the market, for example, practising intensive horticulture.

THE TRANSFORMATION OF AGRICULTURE

In turning to the agricultural transformations of the last 50 years, we need to note that the term 'modernisation' needs to be used with some care, for at least two reasons. It carries with it a number of value judgements, not least about progress – how do we weigh the relative merits of changes which have increased productivity while creating high dependence on non-renewable energy sources and producing major forms of pollution? Second, we have to avoid analysis which collapses into an a-historical dichotomy between a static 'traditional' society versus a modern one. No rural societies are ever completely isolated or static: what is at stake is the speed of change, and there is no doubt that changes in rural Mediterranean society since the last war have been dramatic compared to other historical periods. These changes show many features which are common to agricultural 'modernisation' in other parts of the world, but have had a differential impact on the various farming systems (and political regimes) of the region. For our purposes, in rural and agricultural settings, the term modernisation covers the following:

1. The introduction of new technologies, including mechanisation and new methods for maintaining soil fertility and controlling pests.
2. The decline of manual labour, hence the expulsion of large parts of the labour force from agriculture.
3. The increased incorporation of agricultural systems into regional, national and international markets, one consequence of which is a tendency towards greater specialisation of production.
4. Measures to reduce major inequalities within the rural population of particular regions (for example, through land reform), or major inequalities between regions (for example, through development funds).

Although it could be argued that these processes are general to the Mediterranean, there are nevertheless structural differences between North Africa and Southern Europe which mean that the timing of the modernisation process, its impact on existing agricultural systems, and its long-term trajectory take significantly different forms. Three major structural differences stand out.

First, North Africa was colonised by Europe and its economy shaped largely around metropolitan needs, especially in the French territories. Land ownership and production strategies were major priorities for the post-independence governments of these countries, and there has been a history of greater state intervention in the agricultural sector than has generally been true of Southern Europe.

Second, the modernisation of agriculture takes place in a very different economic context north and south of the Mediterranean. The expulsion of labour from agriculture will always be traumatic, and generate social and cultural dislocation or rural exodus. It is less traumatic if it takes place in a context where either there is sufficient growth in the industrial and service sector to absorb the agricultural labour surplus, or the society has sufficient resources to fund transfer payments to cushion the economic circumstances of those trapped into increasingly unproductive and uncompetitive rural sectors. Italy, Spain, Portugal and Greece (in that order) have at different times, and to different extents, had that growth and those resources. EU provisions have been an increasingly important buttress to national policy. The Maghreb has not had the same levels of growth or resources, and has faced much tougher economic decisions, polarisation and dislocation.

The third factor compounds the second: the demographic trends in Europe and North Africa are strikingly different, as emerged in Chapter 11. Population fertility rates are dropping throughout Mediterranean Europe (Italy and Spain now have the lowest in the world), while the rates in the Maghreb are amongst the highest in the world, although now falling. A modernisation process which expels labour from agriculture cannot follow the same path or have the same impact in such different demographic regimes. One result has been migration: about 3 million people from the Maghreb – mostly from a rural, farming background – are currently working in the EU. An indication of the very different economic trajectories within the Mediterranean Basin is given by the fact that a (small) proportion of Maghreb migrants are actually employed doing seasonal low-paid labour in European agriculture – in the vineyards of Languedoc or the tomato fields of southern Italy. One of the critical factors behind this flow of migrants is rural poverty. Despite the considerable achievements in these countries since independence, a combination of marginal environments, demographic explosion and inappropriate policies has meant that there are still considerable differences in income between rural and urban areas. In an authoritative analysis of North African rural development, El-Ghonemy (1993) cites data which suggest that income in rural areas may be only half that in urban areas. Moreover, the absolute levels of income remain low except in Libya where the high GNP per capita reflects the different economic structure associated with oil wealth and a relatively small population.

One way of portraying the transformation process within the Mediterranean is to chart the changes in the importance of agriculture as measured by the proportion of the labour force in that sector. Unfortunately, comparable data sets across the Mediterranean are not very satisfactory, but Figure 13.3, based upon FAO sources, gives a clear indication of the trends. Information for the years 1961, 1971, 1981 and 1991 is available and the graph indicates the differential in the pace of change between countries. The data illustrate the trends discussed above and, in particular, the rapid transformation of Spain and Portugal. Most revealing is the similarity between Greece and Tunisia despite their considerable historical and geographical differences.

Given these differences, our account of agricultural modernisation has to deal separately with the processes unfolding on the northern and southern shores, as was pointed out forcibly in Chapter 9. The next section will give a brief overview of some features of modernisation, illustrated with material from Southern Europe. This is followed by an analysis of the agrarian transformation process within the economic and political framework of the Maghreb. The conclusion returns to the theme of whether agricultural development is following the same trajectory on the two shores of the Mediterranean.

RURAL MODERNISATION IN SOUTHERN EUROPE

Modernisation involves both the spread of market forces, and to a varying extent, forms of state intervention to mitigate the extreme forms of social dislocation which these can generate. The four transformations listed above (see p. 197) are connected, and can be examined starting either from an analysis of changes in production techniques, or from an analysis of the impact of markets. The most direct way into the issues is to examine the industrialisation of agriculture. For a more detailed discussion applied

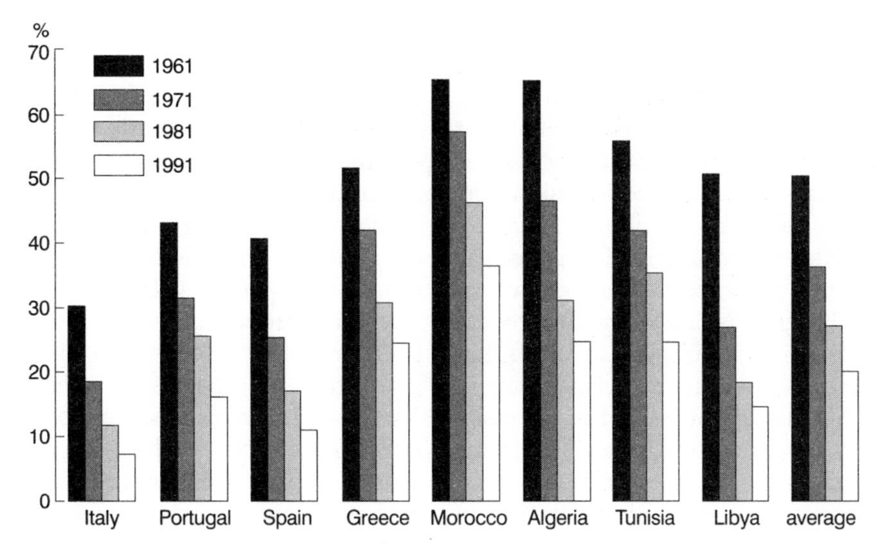

Figure 13.3 Evolution of the labour force in agriculture in selected Mediterranean countries, 1961–91

Source: FAO data

to Tuscany, see Pratt (1994). Older agricultural technologies used very few industrial products in farming. Work implements were manufactured locally, often by the farming household, fields were worked by animal and human labour, while soil fertility was maintained by fallowing, crop rotation and manuring. The control of plant diseases and pests, where possible, was achieved through human labour: for example, in the control of weeds through hoeing and ploughing which were part of the techniques of dry farming developed in Roman times (Stevens 1966). The rapid adoption of industrial products in agriculture has revolutionised all these aspects of production. Mechanisation, especially the tractor and the combine harvester, has transformed farm work; even for the small farmer or market gardener, the rotary cultivator (which can double up as an irrigation pump) has reduced much drudgery (Fig. 13.4). Soil fertility is now maintained principally by the use of chemical fertilisers, while specialist herbicides and pesticides deal with the farmers' old enemies.

This industrial revolution in farming has ramifying consequences. Yields – the productivity of the land – have increased, especially for arable farming (less so for vines or olives). In all sectors of agriculture the productivity of the farm worker has increased likewise. In Southern Europe the labour input into farming today is about one-quarter of the total for 50 years ago, despite the increased productivity. Put another way, three-quarters of the farming population have left. Obviously the timing of this process has varied, and Greece and Portugal still employ

Figure 13.4 Mechanisation has transformed southern European agriculture, even on small farms run by part-time farmers, such as this owner tending his modern vineyard near Pescara, south-central Italy

three times as much labour per 100 hectares as the average in the EU (Eurostat 1995, p. 141). Older techniques for maintaining soil fertility were normally premised on mixed farming but chemical fertilisers abolish the necessity for this and open the way to greater specialisation, including many areas of monoculture. Mechanisation is most effective on larger fields, providing a spur to increased farm size and land consolidation schemes. Progress is rarely straightforward, and each of these technological advances has had negative consequences: the elimination of much heavy manual labour has led to the dislocation effects of mass out-migration; whilst the increased productivity of the land has been achieved at the cost of the very heavy use of non-renewable energy sources, loss of soil quality (from monocultures, salinisation, etc.), and pollution from chemical fertilisers and pesticides.

Two key trends stand out in this process. First, agriculture in the strict sense now has a much reduced place in the overall economy: it is a small and diminishing sector squeezed between two giant agro-industrial sectors: those supplying farms (machinery, agro-chemicals) and the food processors. Farmers once made their own work implements, bred their own work animals, cut the corn by hand, and processed their crops for consumption. Now they drive tractors and combine harvesters, assembled by urban industrial workers, while their crops go to food industries. Goodman *et al.* (1987) have argued that while capitalist agriculture, organised through wage labour, is much rarer than capitalism in industry (indeed, rarer now than 100 years ago when the latifundia used very labour-intensive technology), these industrial developments have appropriated agricultural labour processes and organised them along capitalist lines.

Second, there has been the impact of markets. The Mediterranean region has always traded in agricultural goods, ranging from specialist luxury items (silk, citrus fruits, fortified wines) for long-distance trade to the export of bulk goods, like wheat from Sicily (see Chapter 5). What has changed is the place of those markets in the overall rural economy. Large sectors of Mediterranean agriculture were organised for subsistence needs, with a small surplus for local markets. There are now far fewer peasants (in fact, they are virtually extinct in most of southern Europe), and where they survive, agriculture does not provide all of their household income. Local markets are also much attenuated: a visit to the grocer in most Mediterranean towns or villages will reveal that an increasing proportion of the goods for sale have been grown or processed in distant regions. A generation ago, the local butcher (if there was one) would have been selling locally reared animals; now in Italy, with the increasing specialisation of agriculture in the European Union, half the meat for sale is imported from Northern Europe, and even the local shepherd is being undercut by lambs from Eastern Europe. Even in relatively remote areas of the Maghreb, village shops supply an increasing proportion of groceries as diet habits change and the cash economy prevails. On the other side of the equation, we can find increasing quantities of fresh vegetables, flowers or strawberries grown under plastic in Salerno, southern Crete, southern Spain or southern Morocco and airfreighted to the shops of Northern Europe. In short, the modernisation of agriculture has created a much more competitive economic environment for the farmer. Those producing wheat or wine using labour-intensive techniques have to sell their surpluses in a market place alongside those using capital-intensive technology, often in much more favourable climatic conditions. The result, as we have seen, has been a steady decline in the number of people employed in agriculture (Table 13.1) and an increasingly polarised agrarian structure for those who remain.

Let us first consider those farming households that have been least able to modernise their operations. These farmers are generally found in the least favoured regions, untouched by effective land-reform programmes; they work small land areas, often on fragmented holdings which are hard to mechanise. Previous experience and local values have generated great resistance to land sales (even by those who have migrated),

TABLE 13.1 Employment in agriculture in selected EU countries, 1983–92

	Absolute number ('000)		% of total employment	
	1983	**1992**	**1983**	**1992**
Spain	2 108	1 257	19	10
Portugal	1 059	517	27	11
Italy	2 466	1 657	12	8
Greece	1 060	804	30	22
UK	587	569	3	2
EU12	12 191	8 385	9	6

Source: Eurostat (1995)

and the average size of agricultural holdings is increasing only very slowly in these districts (see Eurostat 1995, pp. 123–7).

Within Southern Europe, the fate of these small farmers has been heavily conditioned by the intervention policies of the Common Market (now European Union) and by the member states. The Common Agricultural Policy has tended to absorb two-thirds of all EU expenditure, but the main beneficiaries have been the farmers of the northern regions rather than the Mediterranean. There are a number of reasons for this. Subsidies, generally speaking, have been strictly proportional to the size of the farm, and the farms of Northern Europe are larger. Second, market interventions have been made more consistently for the agricultural products of Northern Europe than for the south. Certain other factors have also operated, including the exclusion of many part-time farmers (more numerous in Mediterranean Europe) from eligibility for funds, and the lack of administrative support structures which would allow southern farmers to find their way through the maze of bureaucratic procedures and claim support. Neither the penetration of market forces, nor the main agricultural policies of the European Union (which are often represented in the UK as subsidising inefficient farms) have favoured this sector. The much discussed policy shift indicated in the 1992 MacSharry Report towards income support rather than price support would do far more for these small farmers (Commission of the European Communities 1992).

If the Common Agricultural Policy has not done much for the small-farm sector, other forms of intervention have done more. The European Union has devoted an increasing proportion of its budget towards regional development funds for the more deprived areas, while member states have also enacted regional policies (for everything from the building of heavy industrial plant to the development of rural infrastructures and afforestation schemes), of which the most famous is the Italian *Cassa per il Mezzogiorno*, which has pumped money into the south of the country for most of the post-war period (much of it has gone north again, but that is another story). In addition, state welfare policies (on pensions and various forms of unemployment benefit) have been a form of transfer payment which *de facto* has favoured Italy's southern regions (see Pugliese 1985). Moreover, many of these households combine sources of income through 'pluri-activity', since they are underemployed on the land and able to seek work off the farm, especially in the tourist industry which is a large seasonal employer.

The situation is complex, in that in large parts of rural Southern Europe we find households who are farming the land, but who are in fact gaining more than half their income from either off-farm work or welfare payments. The foodstuffs they produce for home consumption can still provide a significant part of the household needs, but they have to sell products at a price which inevitably represents a very low return to labour. There are various interpretations possible of the long-term trends in this sector. It has been argued that it is important to subsidise household incomes (through policies like the EU regional development funds) in order to maintain rural communities, and the mixed farming systems they practise are at least environmentally friendly. On the other hand, the low level of investment in agriculture is creating an increasingly unbridgeable gap between these farming systems and the modernised sector. Certainly, there are paradoxes in subsidising both this sector and the high-technology/high energy specialised agricultural systems which have made it uncompetitive.

Overall, within Europe, 20 per cent of the farming enterprises produce 80 per cent of the agricultural product, and some of the 20 per cent are to be found in the Mediterranean region. They include some very large estates using wage-labour (for example, in southern Spain and many parts of Italy), but the trend has been towards an increasing dominance of family-based farms. They include small farms practising intensive horticulture, and very large farms growing wheat or olives. What they all share is a high level of investment, whether it is in irrigation systems and pesticides or in labour-saving machinery like combine harvesters, grape harvesters or crop-spraying helicopters. While there are some areas of the Mediterranean which because of favourable soils and irrigation can compete with Northern Europe in crops such as maize or sugar beet, on the whole, investment has gone into regional specialisation. Wine production has been revolutionised, partly in an attempt to produce higher quality wine in a period of declining consumption; olive and citrus groves have been replanted (especially in Spain) to allow mechanisation of the harvest. But perhaps the most general and most striking tendency is the export-led boom in fruit and vegetable production, either of crops which can be got to northern markets earlier in the season than local growers (*primeurs*), or as a result of the creation of new markets for Mediterranean crops in the evolving diet of north Europeans. For details of this process in Spain see Yruela (1995); for Greece, Goussios (1995). In all cases their fate has been shaped by the EU and the creation of an enlarged and unified market linking the Mediterranean areas of Southern Europe to the north.

RURAL DEVELOPMENT IN THE MAGHREB

Within the countries on the southern shore of the Mediterranean, poverty is one of the defining characteristics of many *fellaheen* (rural dwellers). El-Ghonemy (1993) quotes figures for the 1980s showing that between 17 and 30 per cent of rural households in Morocco and Tunisia are classed as poor. Consequently, since independence, agricultural policy in all the Maghreb countries has been built around the need to improve the standard of living of the rural population, to ensure that agriculture contributes to economic growth through exports, and to provide a basic food supply both for the rural population and the burgeoning urban areas. However, the rapid demographic expansion and the even more dramatic shift of the population to urban areas has meant that rural development policy has been unable to sustain levels of output and income commensurate with these goals. Table 13.2 sets out some of the numerical parameters of these problems.

In the first place, each country has had to contend with different structural conditions resulting from independence, in particular the varying legacy of alienated land. Also, there have been various patterns of traditional land tenure which have been subject to reform. For instance, after the *coup d'état* in 1969 which brought Gaddafi to power, Libya swept away the powers of tribal chiefs and thus altered the political basis of power which had resided in land. Initial attention was directed to the redistribution of Italian-owned land but later a programme was developed which introduced socialist-style reforms to all sectors of the economy (Allan 1982). Similarly, in Algeria, which had been part of the French metropolitan economy since the 1840s and experienced a particularly violent war of independence, agricultural

Table 13.2 North Africa: population and food output

	Population 1991 (m)	% employed in agriculture 1991	Popn growth 1961–92 annual ave. %	Growth in food output 1961–92, annual ave. %
Morocco	25.7	35	2.50	3.01
Algeria	25.6	23	2.76	1.66
Tunisia	8.2	23	2.12	3.29
Libya	4.7	14	3.98	4.96

Source: FAO Annual Statistics. Rome: FAO, 1993

reform initially concentrated on the state takeover of former *colon* lands and a programme aimed at socialising agriculture. After the removal of the 'colons', the state then introduced co-operatives and extended this arrangement to the traditional dryland sector, with variable success. By contrast, Morocco, which also had a settler community, made the transformation with relatively little violence. Despite state control of the land repossessed from *colons*, the authorities have been cautious in their attempts to alter the institutional framework of traditional agriculture. Some land reform measures have taken place but there has not been a concerted move to improve the distribution of land in favour of the small farmers.

Second, the modernisation of agriculture in the Maghreb has focused upon the development of irrigated agriculture partly because of the climatic circumstances and also because of its potential as a source of export earnings. Relatively little attention has been given to rainfed production despite the fact that most of the region's grain is produced by small farmers facing the vagaries of dryland production. In all the countries of the Maghreb before independence, the settlers and rich locals developed systems of irrigation which formed the basis of an export-orientated agricultural economy. In Algeria, wine, tobacco, olive oil and vegetables were produced as part of French domestic production. According to Griffin (1976) this sector produced 60 per cent of agricultural income in Algeria, despite the fact that *colons* accounted for only 3 per cent of the population. After independence, alongside the socialist restructuring programme, the irrigated area was expanded but at an insufficient rate to reduce the inherent instability of rainfed production. Similar developments have taken place in Tunisia, where agricultural development policy has included a concentration on the development of irrigated land, the reform of support institutions to facilitate exports and special subsidies.

In Morocco, the lowlands between the Atlas mountains and the Atlantic were visualised as another California, provided sufficient water could be obtained (Swearingen 1988). A programme of dam construction was initiated and water provided mainly, but not exclusively, to French settlers. This programme was continued and enhanced after independence as the '*politique des barrages*', with the object of providing 1 million hectares of irrigated land. Schemes such as the Gharb, the Moulouya and the Haouz have involved both extensive technical investment but also the development of settlement schemes for relatively small-scale farmers (Ait Kadi 1988). These operate alongside large, highly capitalised estates which use 'state-of-the-art' irrigation technology. Although these estates were originally operated by parastatals most are now in private hands. By the late 1980s these irrigated areas had been expanded to about 500 000 hectares. Much of the land under irrigation has been developed for export crops or high-value products for domestic consumption. Vegetables and fruit, particularly citrus fruits, olives and tobacco, have expanded, but also sugar for a rapidly growing domestic market. In the Souss valley of Morocco, for example, the last decade has seen a massive expansion in the production of tomatoes and strawberries. Much of this is carried out 'under plastic' and destined for specific market niches in the EU. Since 1987, the Atlas Fruit Board has assumed the coordinating role for the export of these products formerly controlled by parastatals. Altogether, production under this organisation provides work for about 3 million families.

Third, as noted above, much of the domestic production of cereals (mainly wheat and barley) takes place on rainfed land, mainly by farmers with small plots and limited technology. The rural development programmes targeting this sector have concentrated upon agrarian reform and, in Libya, Algeria and Tunisia, various models of collective agriculture. The post-independence socialisation of agriculture in Algeria, which included the development of production cooperatives alongside land reform, failed to realise a significant increase in the output of food crops despite subsidised inputs. Since 1982, there have been improvements as the 'retreat of the state' promoted more local control of agricultural institutions and more oil revenue has been directed to agriculture for investment. None the less, wheat output remains well below that required by domestic demand levels. The Moroccan case shows similar trends, with state policy fluctuating between a heavy priority on irrigated land and attempts to target rainfed farmers. Numerous programmes for the spread of tractor use have been tried, with some limited local success, but overall cereal output remains highly unstable, primarily as a result of rainfall variations but also because of the limited impact of pricing structures (Kydd and Thoyer 1993). For example, from a relatively high level of cereal production in the late 1980s and for one year, 1994, Moroccan cereal production has plummeted, largely because of drought and the small proportion of production taking place under irrigation.

In the Maghreb, government financial support for agriculture has been no less obvious than in the countries to the north, although clearly the overall level of funds available has been much less. Attempts to stimulate a viable capitalist agriculture were developed during the colonial period and targeted towards the settler population. Since independence, such policies continued, directed to the modernised sector, until the 1980s with the onset of Structural Adjustment policies (not, of course, applicable to Libya). In most cases these policies were designed to stimulate the supply of selected commodities without increasing consumer prices. This strategy was an integral part of the post-independence industrialisation model in which low wage levels formed an essential part of the accumulation process. The budgetary costs of this strategy have become extreme, especially for Morocco and Tunisia which do not have oil wealth as a source of transfer payments to finance such policies. This has been one of the main issues addressed by the World Bank. However, attempts to reduce consumer subsidies have met with considerable political unrest, especially in urban areas. Walton and Seddon (1994) document the pattern of political protests throughout North Africa as governments attempt to juggle budgetary requirements and popular demands. This issue alone indicates the fragility of the state support structure in these countries when compared with those of the north.

Despite some 40 years of independence, the Maghreb countries have become increasingly dependent upon food imports and, in some cases, food aid. Clearly, in the case of Libya, oil wealth provides the basis for an 'oil for food' exchange though political requirements have, in this case, made food self-sufficiency a development goal. In both Morocco and Algeria the self-sufficiency rate of cereals (the percentage of domestic cereal consumption produced locally) fell dramatically between 1970 and 1985 (El-Ghonemy 1993), and more recent data suggest little sustained improvement. Consequently, food imports, especially of soft wheat for bread-making, have risen dramatically. The rising demand for cereals has its origin in population growth but also results from the changing food tastes of an increasingly urbanised population, who now consume products such as manufactured bread where the food technology requires soft wheat.

Agriculture in the Maghreb faces a number of key challenges which are distinctly different from, but closely interconnected with, the trajectory of rural development on the northern shores of the Mediterranean. First, the dramatic population increases of the post-1945 period have inevitably set the broad development agenda into the next millennium. Although oil wealth has managed to ameliorate some of these effects in Libya, and less so Algeria, the problems posed are of a different order of magnitude to those of the northern countries. The basic design of the rural development policies of the Maghreb countries contains many of the ingredients present in the experience of the north, such as use of new crop varieties, irrigation, expansion of credit and the introduction of modern technology, but the structural framework has been very different. Each country has also delivered its own package of incentives, especially associated with subsidised inputs and controlled pricing, but these are being dismantled, in part under pressure from international agencies as the condition for providing further credit.

Second, the rural development goals up to the 1980s, although varied between the Maghrebian states, all contained some elements of 'poverty focus' with its implied target of meeting basic needs and limiting the rural–urban movement of population. These were often of limited success, but the Structural Adjustment programmes largely forced on these countries by the IMF and World Bank (and vigorously supported by the powerful members of the EU) have largely removed direct intervention as a method of achieving these goals. The results so far have suggested some improvement in agricultural production overall but little decline in poverty levels as the policies have failed to create sufficient off-farm employment or support for on-farm activities. It is useful to compare the almost dogmatic implementation of these policies with the protracted negotiations and compromises which have characterised reforms of the CAP during the last decade. Although these EU reforms contain certain similar elements (i.e. reductions in price support, etc.), their impact has been mediated by the fact that the social costs of this readjustment have been met by funds drawn from affluent member states. No such assistance has been available for the North African countries with the possible exception of the use of oil revenue in Libya.

Third, agricultural development in the Maghreb countries is intimately associated with wider socio-political movements in the area (Joffe 1993). There is little doubt that rural poverty and the contrasting visible affluence of some of the urban elites have provided a focus for the demands for a return to basic Islamic principles of social organisation. These include specific forms of inheritance arrangements, different conventions concerning the use of credit, and land ownership systems which can be seen to conflict with models based upon Western ideals (Siddiqi 1981). There is considerable academic debate amongst Islamic scholars about the interpretation of these principles but the sophisticated niceties mean very little in the context of stringent cries for social justice. Whilst some Islamic principles have been incorporated into the legal codes of the Maghreb, their provisions are seen to be increasingly bypassed in the drive for 'efficiency' and 'development', especially when this involves an attack on social provisions promulgated as a result of external pressure. Algeria provides an example of the political instability that results, and this all too quickly is translated into a decline in agricultural output and incomes.

Finally, whereas most of the countries have derived some benefits from an expanding EU market and privileged access under the 'former territories' agreements, the post-1986 enlargement of the EU has meant that competition with Spain, Portugal and Greece is now balanced very much in the north's favour. For Morocco and Tunisia, countries which so far have been unable to exploit oil wealth, the potential loss of export earnings may well have a devastating impact on both national growth and on the large numbers of agricultural workers employed in the export sector. Prior to 1986, the Maghreb countries could export agricultural products to the EU on the same terms as Spain and Portugal with whom they competed directly (Stevens 1990). Now, however, not only has the degree of self-sufficiency in the EU increased for several key products – one estimate suggests 76 per cent for citrus fruits, 100 per cent for vegetables and 100 per cent for olive oil (Ghaussy 1990), but former competitors will be protected by the high EU tariffs as well as being eligible for considerable EU regional development assistance. At the same time, overproduction of key products such as citrus fruits and olive oil will reduce market prices. These issues are readily recognised by both the Maghreb and EU states but a satisfactory policy profile has yet to developed, as we saw in Chapter 10.

CONCLUSIONS

The transformation of agriculture has been profound in the countries on the northern shore of the Mediterranean, especially since the end of the Second World War. The fundamental elements in the restructuring process have been the decline in the proportion of the population who gain a living directly from agriculture and a concomitant increase in the capital input into agriculture, together with the development of specialist production units serving an increasingly global market. Much of the institutional support for this process has been derived from the EU, especially in the last decade with the accession of Spain, Portugal and Greece.

In the Maghreb, changes have been no less significant, but here the historical legacy of colonial occupation set the agenda for subsequent post-independence policies for agricultural development. The context is profoundly different geographically, economically and socially with the over-riding theme being the need to combat the rural poverty which still dominates much of the Maghreb. Development efforts over the last 40 years have followed a similar trajectory to those in the north, in the sense that attempts have been made to modernise agriculture through technical innovation. However, the result has been patchy, with the greatest improvements being seen within the most favourable areas and largely confined to the richer (and larger) farmers.

In both areas there has been an exodus of population from rural areas, but in the Maghreb this movement is driven by population growth rates that remain in excess of 2 per cent per annum. In addition, whereas the out-migration in the north has been largely absorbed by the relatively rapid increase in non-agricultural employment, or funds have been made available to assist in cases of deprivation, in the Maghreb employment outside agriculture remains limited and migrants have sought employment in mainland Europe. This is now generating social and political tensions which are recognised by the EU as needing urgent attention (Chapter 11).

During the colonial period the main thrust of investment within the agricultural sector of the Maghreb was directed to the production of export crops such as wine, olives, wheat and citrus fruits destined for European markets. This market has always been very competitive and, in the post-independence period, special provisions prevailed to ensure that agricultural goods from the former colonies received some tariff respite. However, the inclusion of Spain and Portugal within the EU has meant that the countries of the Maghreb face enormous competition for their products which, in turn, may cause further agricultural decline and increase migration.

In the face of these problems, there is renewed interest in seeking ways in which the countries to the north and south of the Mediterranean can be drawn into a range of cooperative agreements which would significantly enhance the economic development of the south and, equally, contribute to the north's social and political stability. Agreements governing aid to both the agricultural and non-agricultural sectors, arrangements for managing migrant flows and cooperation in many other sectors have been proposed but most await effective implementation. Agreement amongst the Maghreb countries remains elusive and this exacerbates any negotiation with the EU.

REFERENCES

AIT KADI, M. 1988: *Major Features of Moroccan Large Scale Irrigation Projects.* London: ODI/IIMI Management Network Paper 88/1d.

ALLAN, J. (ed.) 1982: *Libya since Independence.* London: Croom Helm.

BERQUE, J. 1978: *Structures Sociales du Haut-Atlas.* Paris: Presses Universitaires de France.

CAMPBELL, J.K. 1964: *Honour, Family and Patronage.* Oxford: Clarendon.

COMMISSION OF THE EUROPEAN COMMUNITIES 1992: *Agriculture in Europe (the MacSharry Report).* Luxembourg: CEC.

DAVIS, J. 1973: *Land and Family in Pisticci.* London: Athlone.

EL-GHOMENY, M.R. 1993: *Land, Food and Rural Development in North Africa.* London: IT Publications.

EUROSTAT 1995: *Agricultural Statistical Yearbook.* Luxembourg: Commission of the European Communities.

FAO 1993: *FAO Annual Statistics.* Rome: FAO.

GHAUSSY, A.G. 1990: *The Impact of the EC Southward–Enlargement on the Islamic Countries of the Mediterranean Basin.* Hamburg: Institut für Wirtchaftspolitik, Universität der Bundeswehr, Discussion Paper No. 9.

GOODMAN, D.E., SORJ, B. and WILKINSON, J. 1987: *From Farming to Bio-technology: A Theory of Agro-industrial Development.* Oxford: Blackwell.

GOUSSIOS, D. 1995: The European and local context of Greek family farming. *Sociologia Ruralis* **35**, 322–34.

GRIFFIN, K. 1976: *Land Concentration and Rural Poverty.* London: Macmillan.

JOFFE, G. 1993: *North Africa: Nation, State and Region.* London: Routledge.

KYDD, J. and THOYER, S. 1993: Agricultural policy reform in Morocco, 1984–91. In Goldin, I. (ed.), *Economic Reform, Trade and Agricultural Development.* London: St Martin's Press, 135–64.

MARTINEZ-ALIER, J. 1971: *Labourers and Landowners in Southern Spain.* London: George Allen and Unwin.

MILLER, J.A. 1984: *Imlil: A Moroccan Mountain Community in Change.* Boulder, CO: Westview.

PRATT, J.C. 1994: *The Rationality of Rural Life: Economic and Cultural Change in Tuscany.* Chur, Switzerland: Harwood Academic Press.

PUGLIESE, E. 1985: Farmworkers in Italy: agricultural working class, landless peasants or clients of the welfare state? In Hudson, R. and Lewis, J. (eds), *Uneven Development in Southern Europe.* London: Methuen, 123–39.

SCHNEIDER, J. and SCHNEIDER, P. (1976): *Culture and Political Economy in Western Sicily.* New York: Academic Press.

SIDDIQI, M. 1981: *Muslim Economic Thinking. A Survey of Contemporary Literature.* Leicester: Leicester University Press.

SNOWDEN, F.M. 1986: *Violence and the Great Estates in the South of Italy.* Cambridge: Cambridge University Press.

STEVENS, C. 1990: The impact of Europe 1992 on the Maghreb and sub-Saharan Africa. *Journal of Common Market Studies* **29**, 217–41.

STEVENS, C.E. 1966: Agriculture and rural life in the Later Roman Empire. In Postan, M.M. (ed.), *Cambridge Economic History, Volume 1.* Cambridge: Cambridge University Press, 92–124.

SWEARINGEN, W.D. 1988: *Moroccan Mirages: Agrarian Dreams and Deceptions, 1912–1986.* London: Tauris.

WALTON, J. and SEDDON, D. 1994: *Free Markets and Food Riots: The Politics of Global Adjustment.* Oxford: Blackwell.

YRUELA, M.P. 1995: Spanish rural society in transition. *Sociologia Ruralis.*, **35**, 276–96.

14

TOURISM AND UNEVEN DEVELOPMENT IN THE MEDITERRANEAN

ALLAN WILLIAMS

INTRODUCTION

The long association of the Mediterranean Basin with travel has meant that the region's name has become synonymous with particular forms of tourism. Dawes and D'Elia (1995, p. 15) express this in terms of 'the cultural weight of the South, and the very place of the South in the mind of the traveller'. The precise destinations within the Mediterranean have changed over time: in Hellenistic times to Egypt, the cradle of divinity; in the Middle Ages to the Holy Land; in the later Middle Ages to Rome.

There was a social reconstruction of the Mediterranean as a tourist destination in the late nineteenth and early twentieth centuries. It became the object of annual seasonal winter visits by Europe's wealthy classes, drawn more by its restorative and climatic attributes than by its cultural heritage. Reynolds-Ball (1914), writing in the second decade of the twentieth century, emphasised that the Mediterranean shores could be considered the world's greatest winter playground. In this way, tourism became established as an activity of the 'periphery', whose dynamic force was the growth of a wealthy upper middle class in the 'core' of Northern Europe and, much later, in North America. As it was reliant on the railway and on passenger ships, its impact was limited to a few narrow coastal strips on the Rivieras of France and Italy, the South of Italy, the Adriatic, Algeria and Egypt. The outstanding example of this winter 'colonisation' was the French Riviera. Reynolds-Ball (1914, p. 23) captures its attraction:

> No doubt the Riviera is commonplace, overcrowded and hackneyed, ... but then what other visitor region offers such a wide choice of pleasant winter quarters, with such a mild and sunny climate, within from a day to a day and a half's journey from London?

After the Second World War, Mediterranean tourism was transformed by mass tourism which was fundamentally different in terms of the social construction of 'the tourist gaze' (Urry 1990), the social access to tourism, and the geographical distribution of the industry. The main concern of this chapter is with the nature, development and impact of mass tourism in the Mediterranean. However, it is important to emphasise that this is only one of the multiple (and overlapping) forms of tourism in the region, and that it coexists with, for example, cultural tourism, rural tourism and urban and mega tourism. There are some Mediterranean regions such as Umbria or the Alentejo where mass tourism is virtually absent and where, instead, there are thriving rural tourism industries which are well integrated with local economies and cultural systems. In recent years there has been rapid growth in such forms of tourism, evident for example in the successful promotion of *Turismo de Habitação* in Portugal. These alternative forms of tourism lie beyond the scope of this chapter, which is mainly about mass tourism.

It is the summer beach and sunshine holiday product which has propelled the Mediterranean into pole position in world tourism, with the countries on its shore accounting for about one-third of the global total of tourist arrivals (Frangialli 1993, p. 15). As a result, Mediterranean countries dominate the world league table for tourism (Fig. 14.1), accounting for three of the top four places and eight of the top 30 positions in 1991. These data also underline the massively uneven distribution of tourism in the region, with the northern shore countries attracting far more tourists than either the eastern or the southern shores. The nature and causes of this uneven distribution are considered in the next section of the chapter.

The massive temporal and spatial polarisation associated with mass tourism has left its imprint on the economies, environments and cultures of the region, and the latter sections of the chapter review such impacts. Of necessity, much of the discussion will be at the level of the Mediterranean Basin as a whole or of individual countries, but it must be stressed that, despite a shared social construction, Mediterranean mass tourism is not a homogeneous phenomenon. Its nature and impact vary by region and, indeed, by resort. Similarly, its impact on local structures is conditional on the levels of development in the destination areas (Shaw and Williams 1994, pp. 189–92), on whether tourism investments are in greenfield sites or in existing settlements (Marchena Gómez and Vera Rebollo 1995, pp. 9–10), and on the stability and sustainability of cultural and environmental systems. However, before such issues can be discussed, we first need to review the evolution of mass tourism in the Mediterranean region.

THE CHANGING MAP OF MEDITERRANEAN TOURISM

The take-off of mass tourism in the Mediterranean region can be traced in the histories of individual resorts, and nowhere exemplifies this better than Benidorm. In 1950 it had 1400 inhabitants who lived mostly from fishing, and even in 1966 it had only 4333 hotel bedspaces. Thereafter, improvements to the transport infrastructure (airport and motorway connections) opened up Benidorm to large-scale charter arrivals and, by the late 1980s, there were more than 31 000 bed spaces in serviced accommodation, and up to 100 000 in apartments (Manrique 1989, p. 350). The result has been the creation of Benidorm's Manhattanesque skyline, one of the best-known symbols of Mediterranean tourism. Similar fishing-villages-to-high-rise transformations took place at the Costa del Sol at Torremolinos, Fuengirola and Marbella (see Fig. 14.2).

The reasons for the growth of resorts such as Benidorm and Torremolinos lie in the conditions of demand and supply. On the demand side, it is linked to the emergence of mass consumption in Europe in the 1950s and 1960s, which in turn is related to aggregate economic growth, the distribution of income and public expenditure, and access to guaranteed paid annual leave in Northern European countries (Shaw and Williams 1994, pp. 75–6). This is not to discount the role of domestic tourism, but international tourism was the motor for growth, given both its rate of expansion and its levels of expenditure. On the supply side, the key elements were reductions in the costs of and barriers to travel, whether by air or road, and the emergence of the highly competitive package-tour industry. The first air-inclusive package holiday was from the UK to Corsica in 1950, but by 1990 charters accounted for 56 per cent of the European air travel market (Jenner and Smith 1993, p. 88).

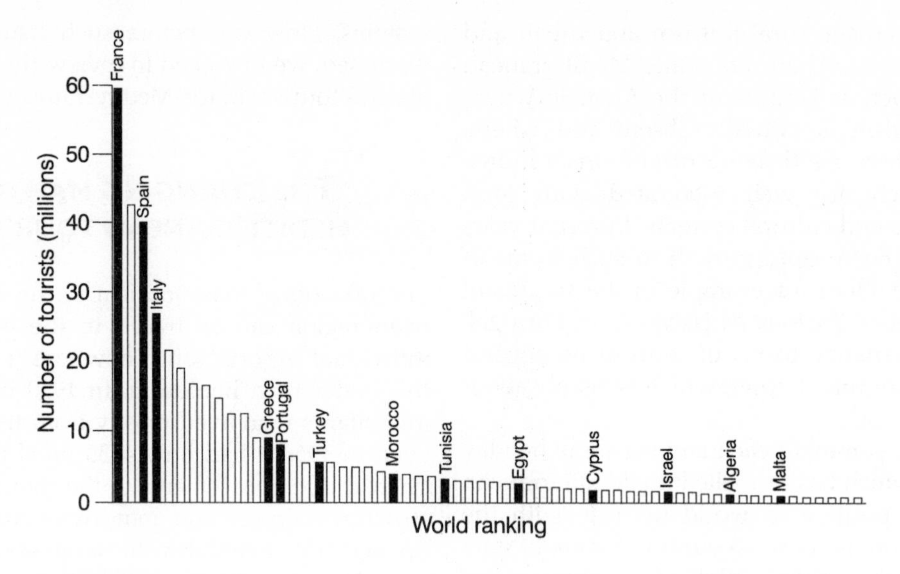

Figure 14.1 Global ranking of the Mediterranean countries in terms of tourist arrivals, 1992

Source: World Tourism Organization (1994)

The advantages of international tour companies, in terms of the volume of passengers carried, together with their 'bundling' of a range of services, enabled them to pressurise downwards the prices paid to sub-contracting enterprises in the destination countries, as well as the real prices charged for holidays (Williams 1996). The result was a virtuous economic circle of falling real costs, rising demand and enhanced economies of scale (Shaw and Williams 1994, p. 178). The virtuous circle was reinforced in most instances by state interventionism. In Egypt, for example, tourism was favoured with special incentives after 1974 as part of Sadat's *al-infitah* (open-door) economic policy and, in the 1980s, was the first Egyptian industry to be subject to economic liberalisation so as to stimulate private investment.

Figure 14.2 High-rise beach-fronts typify many Mediterranean resorts which have mushroomed through servicing mass tourism. The skyline along the coast at Torremolinos in southern Spain.

The diffusion effect

The spatial distribution of mass tourism in the Mediterranean has been highly uneven, corresponding to a diffusion model in some respects. Mass tourism originated in a few countries, regions and resorts and then spread outwards from these nodes. The initial nodes were the French and Italian Rivieras but Spain subsequently emerged as a leading pole of mass tourism. In 1949 it had only 283 000 foreign visitors, rising to 1.3 million in 1951, and 4.2 million in 1959; thereafter, it leapt to 14.1 million in 1964 (Manrique 1989, p. 34). There were also significant developments in this phase in Italy (especially the Adriatic), and in Languedoc-Roussillon (Frangialli 1993, p. 6). By the late 1960s, international tourism had become widely diffused along most of the northern shores of the Mediterranean, with the exception of Turkey and Albania. Greece, Portugal and the former Yugoslavia all experienced considerable growth in this period.

In the late 1970s and early 1980s there was further 'diffusion' of mass tourism, with Morocco, Tunisia and Turkey all experiencing significant increases in international arrivals, along with Malta and Cyprus. While strongly reliant on Northern European markets, they also drew on important Arab markets, both in North Africa and the Middle East. Other non-European countries have also sought to establish themselves as major tourist destinations in the 1980s, especially Israel, Egypt and post-Boumédienne Algeria but, for a variety of political and economic reasons, these countries have had mixed success. Finally, there is a group of countries which has stood aside from the pursuit of mass tourism, either because of ideological choice or the constraint of circumstances, including wars and civil wars; among these are Libya, Syria and the Lebanon. In the 1990s the disintegration of the former Yugoslavia wrecked its tourism industry, although there are signs of revival in Slovenia and parts of the Croatian coast (Istria and Dalmatia).

The outcome of these changes in the distribution of Mediterranean tourism can be quantified using national statistical sources. Coverage is incomplete, and data are lacking for Albania and Lebanon, and very limited for Libya, but all three countries are relatively insignificant as destinations. Difficulties also arise because of differences in definitions of tourists and tourism and in the means of collecting such statistics (Williams and Shaw 1991a). There is also the issue of how the Mediterranean region is to be defined. The simplest approach is to take the countries bordering the Mediterranean Sea, but this excludes Portugal whilst including large parts of France and Algeria which have non-Mediterranean characteristics. A more precise approach is to disaggregate national-level data so as to provide estimates for the specifically Mediterranean regions: this is the approach of Jenner and Smith (1993) who, for example, estimate that 80 per cent in Italy, but only 15 per cent in France, of all tourists visit the Mediterranean regions. Montanari (1995a) considers such an approach to be unduly simplistic because of its mechanistic reliance on the criterion of immediate proximity to the Mediterranean Sea. Instead, he recommends a definition based on cultural and bioclimatic features and which thereby includes southern Portugal, Jordan and the Black Sea regions in Bulgaria, Romania and the Crimea.

These different definitions of the Mediterranean region produce very different estimates for international tourism. In 1991 the Mediterranean countries attracted 187 million foreign tourists, while the Mediterranean regions attracted 113 million (Jenner and Smith 1993, p. 7). If domestic and unregistered tourism are also taken into account, then the Mediterranean regions are estimated to have attracted *ca.* 200 million tourists in 1991. This compares to Montanari's (1995a) estimate of 160 million officially registered foreign tourists in 1990, and to Frangialli's (1993, p. 5) much lower estimate of 90 million.

Whichever definition is employed, there is consensus that tourist numbers stagnated in the late 1980s; for example, the number of arrivals in the Mediterranean regions only increased from 111 million to 113 million between 1987 and 1991 (Jenner and Smith 1993, p. 7). The Mediterranean, therefore, has been losing market share, and its proportion of total world international tourist arrivals fell from 31 per cent to 27 per cent over this same period. To some extent this reflects the changing global geography of production and of consumption, with Japan and the Pacific Rim countries beginning to challenge Europe as the prime motor of international tourism. It also reflects the globalisation of tourism markets, with northern Europeans increasingly undertaking a global scan of trans-continental and more 'exotic' objects of the 'tourist gaze' (Shaw and Williams 1994, pp. 21–6). There has also been a decline in the price competitiveness of the Mediterranean region

(Frangialli 1993). However, to write about the overall growth or decline of tourism in the Mediterranean region is a gross over-simplification for, in reality, it is an aggregate of very diverse national experiences.

The absolute scale of tourism arrivals in the Mediterranean countries has already been noted (Fig. 14.1), particularly the dominance of France, Italy and Spain which account for approximately four-fifths of arrivals in the Mediterranean countries. If only the Mediterranean regions are considered, then Italy and Spain are the dominant market forces accounting for 68 per cent of all arrivals in both 1985 and 1992 (Table 14.1). There was, however, a reversal between 1985 and 1992 in the ranking of the two countries, due to a decline in Italy's market share while Spain made strong advances, boosted by the impact of the Barcelona Olympics, the Seville Expo and Madrid's designation as cultural capital of Europe.

Over time, the statistics show that the distribution of destinations has been changing, reflecting the 'diffusion' pattern noted above. Between 1971 and 1986, the highest growth rates (over 10 per cent per annum) were in the 'middle-age' or maturing markets of Cyprus, Greece, Tunisia and Egypt. Medium growth rates (of 5–10 per cent per annum) were recorded in the emerging markets of Turkey, Israel and Syria and also in Morocco and Malta. The lowest growth rates were in the mature markets of France, Spain and Italy, and also in Yugoslavia (Grenon and Batisse 1989, p. 142).

In the late 1980s there were further shifts in destinations. Only Cyprus maintained its position in the highest growth group, where it was joined by Turkey, Morocco and Portugal. The last of these was excluded from the 1971–86 analysis, but had relatively modest growth in this period (Lewis and Williams 1991). France,

TABLE 14.1 Country shares of the Mediterranean regions' international tourism market, 1985 and 1992

	International tourism arrivals			
	'000		% of total	
	1985	**1992**	**1985**	**1992**
Albania	50	50	–	–
Algeria	535	840	0.5	0.7
Cyprus	922	1 991	0.8	1.7
Egypt	1 245	2 404	1.1	2.0
France	8 812	8 939	7.8	7.5
Greece	8 465	9 331	7.5	7.9
Israel	1 493	1 798	1.3	1.5
Italy	45 269	40 070	40.0	33.9
Lebanon	50	50	–	–
Libya	50	89	–	–
Malta	594	1 002	0.5	0.8
Monaco	260	246	0.2	0.2
Morocco	1 226	1 512	1.1	1.3
Spain	31 607	40 391	28.0	34.1
Syria	650	1 130	0.6	0.9
Tunisia	1 521	2 372	1.3	2.0
Turkey	2 788	5 661	2.5	4.9
Yugoslavia	7 424	616	6.6	0.5
Total	112 961	118 492	100.0	100.0

Notes: Data refer only to Mediterranean regions of certain countries (e.g. Algeria, France, Egypt); 1992 data for Albania, Lebanon and Yugoslavia are estimates
Source: Jenner and Smith (1993); author's calculations based on World Tourism Organization (1994)

Spain and Italy continued to have relatively modest growth rates in the 1980s, whilst the heady expansion of the previous decade fell away sharply in Greece and Tunisia which were becoming more mature markets. Israel and the former Yugoslavia actually experienced major declines as a result of domestic and international conflicts.

Interpreting the changing map of mass tourism

A number of factors are important in explaining the changing geographical distribution of mass tourism, including the product cycle, price–quality competitiveness, and political/military conflicts.

The importance of the product cycle is evident in the slow growth in more mature markets such as France, Spain, Italy, Malta and, more recently, Greece. Italy's market share, for example, has been in slow decline since at least the mid-1970s, whilst in the 1980s it had the lowest growth rates of any of the major destination countries (Table 14.2). This was due to the ageing of the tourist product (a declining cost-quality ratio), the particular circumstances of the Italian economy (including poor quality banking facilities, and an unfavourable exchange rate) and the well-publicised incidence of algae on the Adriatic coast. The effects of an apparently ageing tourism product were most vividly illustrated in Spain where, between 1988 and 1990, there was a 5 per cent decline in international tourist arrivals, although it should also be noted that it succeeded in recovering market share in the 1990s. In contrast to Italy and Spain, other countries, such as Cyprus, Turkey and Morocco, had relatively 'young' tourism industries and were still in the expansion phases of the product cycle.

Price is another important influence on tourism growth rates. One measure of this is provided by the analyses of package tours from Northern Europe to the principal Mediterranean destinations undertaken by Spain's Dirección General de Política Turística (1992). There is considerable variation in prices with the former Yugoslavia at 63.4 and Corsica at 158.4 representing two extremes around a mean of 100 for prices in all destinations (Fig. 14.3). There is some semblance of an inverse relationship between price and growth rates with France, the south of Italy and some Spanish regions having price levels above average and relatively low rates of demand growth. However, Cyprus, Portugal and the Mediterranean coast of Morocco – all of which have high demand growth rates – also have above average prices. At the other extreme, Malta, Greece and Egypt (as well as some Spanish regions) have prices between 80 and 90 per cent of the mean, but only relatively modest growth rates. The lowest price levels of all, excepting Bulgaria, are in the Italian Adriatic and the former Yugoslavia, where prices presumably reflect a collapse in demand following, respectively, environmental and political/military crises.

This brings us to the third major influence in the pattern of growth: the discontinuities associated with domestic or international crises. These can be generalised such as the effects of the 1991 Gulf Crisis which depressed tourism growth throughout the eastern and southern

TABLE 14.2 International tourist arrivals, 1980–91

	1980 ('000)	Average annual growth rate (%) 1980–91
France	30 100	5.76
Spain	23 403	3.82
Italy	22 087	1.79
Portugal	2 730	11.06
Greece	4 796	4.80
Turkey	921	16.96
Morocco	1 425	10.23
Tunisia	1 602	6.56
Egypt	1 253	4.86
Yugoslavia (former)	6 410	−12.59
Cyprus	353	13.23
Algeria	946	2.13
Israel	1 116	−1.52
Malta	729	1.86

Source: Jenner and Smith (1993)

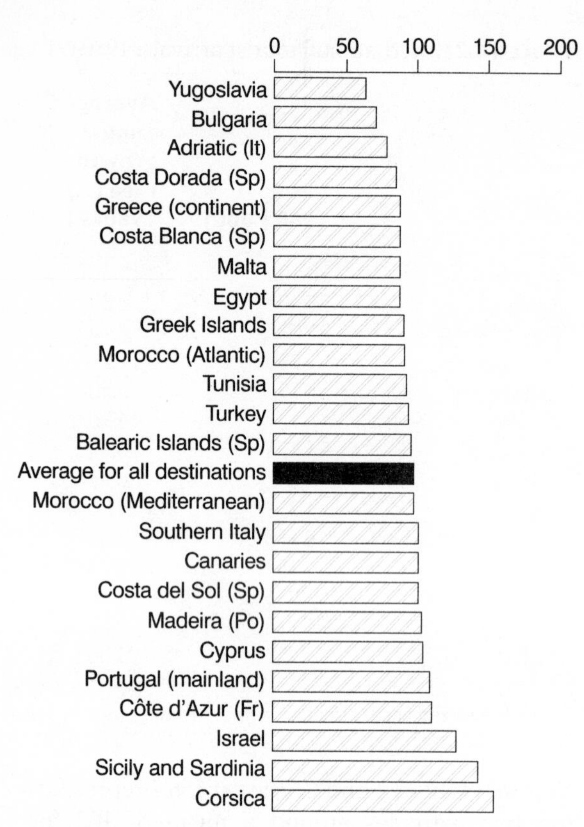

Figure 14.3 Price competitiveness 1992: based on 7-day half-board package holidays in medium–high quality accommodation

Source: After Dirección General de Política Turística (1992, p. 66)

Mediterranean countries. Alternatively, they may be country-specific such as the effects of military coups and the Cyprus conflict on the Greek tourist industry (Buckley and Papadopoulos 1986, p. 87). Violently contested territorial sovereignty can also have a negative impact on tourism; this is evident in the case of ETA campaigns in Spain and PKK activities in South-east Turkey, especially as both have targeted tourists at particular times in their campaigns of violence. In contrast, Egypt illustrates the impact of international conflicts on the tourist industry; there was rapid growth of tourism from the 1970s, partly due to expansion of the Arab market, but also benefiting from the Lebanese crisis, which saw Cairo replace Beirut as the major destination within the region for Arab tourists (Economist Intelligence Unit 1991, pp. 53–8). However, the Gulf War and the rise of violent Islamic fundamentalist movements severely dented prospects in the 1990s. The Lebanon and the former Yugoslavia represent the two most extreme cases of how the highly sensitive and volatile tourist industry can be decimated by violent conflicts. In the former Yugoslavia, for example, there was a 79 per cent decline between 1989 and 1991 in international tourist arrivals (Hall 1995).

MASS TOURISM: SOCIAL CONSTRUCTION AND POLARISATION

Mass international coastal tourism – focused on the Mediterranean – is socially constructed primarily as a sand, sea and sunshine product, with secondary features related to low prices, particular types of entertainment, and the scope for modified individual behaviour. Its main focus is the beach and it is essentially acultural (Shaw and Williams 1994, pp. 180–92). It is fundamentally different to the Mediterranean coastal tourism of the late nineteenth and early twentieth centuries in terms of its mass social character, its wider areal transformation of coastal areas, its focus on the summer rather than the winter season, and the increasing preference for hotter southerly latitudes within the region (Marchena Gómez and Vera Rebollo 1995).

Spatial and temporal polarisation

There are a number of important implications which follow from this social construction. The first of these is massive spatial polarisation of the industry. Not only is it highly regionally uneven, but there is also intra-regional polarisation with mass tourism rarely penetrating more than a few hundred metres inland from the coast (Fig. 14.2). In Spain, for example, the Canaries and Balearics accounted for almost one-half of all overnights and two-thirds of foreign overnights in 1993 (Economist Intelligence Unit 1994a), with most tourist accommodation being on or adjacent to the coast. In Tunisia 80 per cent of total accommodation capacity is to be found in seaside resorts (Poirier 1995, p. 159).

The social construction of tourism as a sunshine and beach product also means that it is subject to strong temporal polarisation. This is reinforced by the mass nature of the industry, and its marked reliance on working- and middle-class individuals and families, the timing of whose holidays is institutionally constrained by the organisation of leisure time in workplaces and in schools. As a result, more than 40 per cent of all tourist arrivals in the Mediterranean countries occur in the three summer months of July, August and September. According to Jenner and Smith (1993, p. 71), seasonality problems are particularly acute in Israel (12 times more business in August than in February), and Greece (nearly six times more business in August than in February). In contrast, those southern Mediterranean countries which can offer reliable winter sunshine are less prone to extreme seasonality. For example, in Morocco only 31 per cent of foreign tourists arrive in the three peak months (Economist Intelligence Unit 1992).

The pattern of arrivals in the southern Mediterranean countries is, however, made complex by the existence of two distinctive major sub-markets. This is illustrated by Egypt; the European sub-market has a relatively even spread throughout the year with only 33 per cent arriving in the three peak months (in the spring and autumn) whereas the Arab market is massively concentrated (54 per cent) in the three main summer months. Climatic differences in Spain also account for the more even temporal spread of tourism in the southern region of the Costa del Sol, compared to, say, the Costa Brava on its colder north-east coast (Valenzuela 1991). The Italian Adriatic Riviera, a low-cost holiday strip mainly patronised by domestic and German tourists, is dormant for most of the year (Figure 14.4).

Figure 14.4 Resorts given over to summer beach tourism remain closed for up to two-thirds of the year. Sea-front hotel at San Benedetto del Tronto on Italy's Adriatic coast. This picture was taken in May and the hotel had yet to open for the season.

Social construction and tourism markets

The social construction of Mediterranean tourism as a mass product based on the 'consumption' of sunshine and beach spaces also has implications in terms of market profiles. Above all, the Mediterranean was constructed as a pleasure periphery for Northern Europeans seeking escape from the climate, the workplace and everyday stresses of their home areas; as a result, the industry is highly dependent on relatively few markets. This is compounded by the economics of the package tour industry, which is based on achieving economies of scale in marketing, transport and accommodation costs (Williams 1996). In turn, this dictates a high degree of market segmentation, with individual destination resorts and regions being highly dependent on particular countries or even tour companies (Pearce 1987).

There are two principal markets for Mediterranean tourism: the European – especially Northern Europe – and the Arab countries (for the southern Mediterranean destinations). Given the social construction of Mediterranean tourism, its history, and the consumption potential of Northern Europe, it is hardly surprising that Europe is still the dominant market. According to Jenner and Smith (1993, p. 46), Europe accounts for more than one-half of the total market, with Germany (17 per cent), France (12 per cent) and the UK (10 per cent) being the principal sources. The leading international flows are strongly concentrated on Italy, Spain and Greece (Fig. 14.5), underlining the fact that Mediterranean tourism continues to be essentially an intra-European phenomenon. The pattern of all-inclusive air holidays is even more polarised as can be seen from Figure 14.6. For both Germany and the UK, there is a very high degree of concentration on Spain and Greece. In the case of the UK, previous colonial ties also contribute to large tourist flows to Malta and Cyprus. In contrast, Turkey is an important German destination, partly as a result of return visits by *Gastarbeiter* and their families. The expatriate market is also important in North Africa, accounting annually for an estimated 1 million visitors to Morocco and half a million to Algeria (Frangialli 1993).

Whilst international mass Mediterranean tourism evolved mainly around Northern European markets, there has also been a growth of intra-Mediterranean movements, reflecting the emergence of new markets in the more prosperous parts of the region. Italy and Spain are, respectively, the fourth and eighth largest sources of visitors to the Mediterranean region (Jenner and Smith 1993, p. 46).

The second most important market for the Mediterranean region is the Arab countries. These are the largest sources of arrivals, if not always of overnights, for Algeria, Egypt, Morocco, Syria and Tunisia. There are two distinct streams: arrivals from neighbouring Arab countries which are often of a short duration, and longer-stay tourists from Saudi Arabia and the Gulf states. The latter flow was confined initially to a wealthy elite, but after the mid-1970s oil price rises, there was a growing stream of middle-class customers from these countries. Cairo has been the favoured destination, especially since the conflicts in the Lebanon, although Cyprus and some Maghreb states

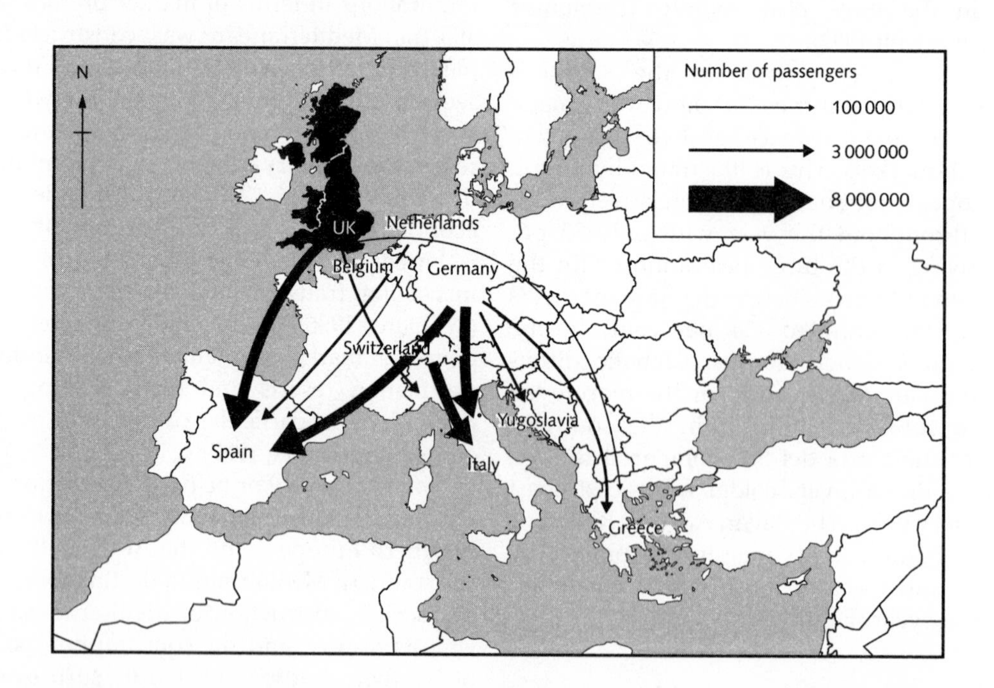

Figure 14.5 Principal tourist flows from northern Europe to Mediterranean destinations, 1991

Source: World Tourism Organization (1994)

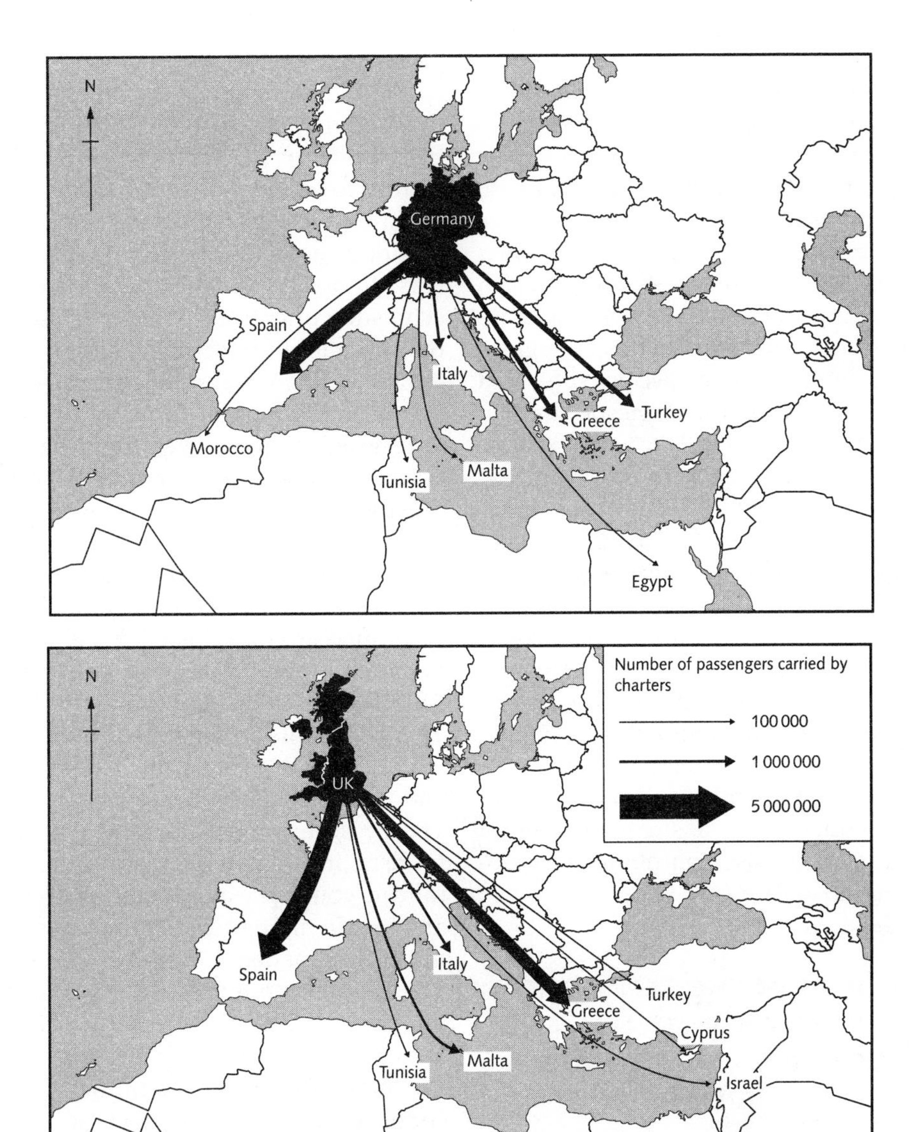

Figure 14.6 Principal charter flights from Germany and the UK to Mediterranean destinations, 1991

Source: Jenner and Smith (1993)

have also benefited. The experiences of Morocco are not untypical of the southern Mediterranean countries as a whole. In 1991 it recorded 3.2 million arrivals, of whom 2 million were from Algeria, and 160 000 were from other Arab countries, including the Gulf (Economist Intelligence Unit 1992, pp. 46–53). However, in terms of overnights, all Arab countries, including Algeria, accounted for only 11 per cent of the market, while the leading countries of origin were France (31 per cent) and Germany (16 per cent).

While the precise form of market segmentation may vary throughout the Mediterranean region, a high degree of reliance on foreign markets and on a small number of countries of origin is common to most of the main destinations. This means, inevitably, that they are particularly vulnerable to external shocks and to cyclical fluctuations. Not surprisingly, reliance on foreign markets is least in the more prosperous northern Mediterranean countries which have relatively large domestic markets. The domestic market share in the Mediterranean regions is particularly high in France (64 per cent) and Italy (43 per cent), while the domestic share in Spain is only 24 per cent but is rapidly expanding (Jenner and Smith 1993, p. 51). Domestic tourism is also important in Turkey on account of its large population (over 50 million) and consumer market, while its international tourism is still in the relatively youthful stage.

In summary, then, the Mediterranean countries have highly polarised – both spatially and temporally – tourism industries. These two forms of polarisation are mutually reinforcing and play a critical role in mediating the economic, environmental and cultural impacts of tourism. Dependency on a small number of international markets also contributes to shaping such impacts.

FOOTPRINTS IN THE SAND: THE ECONOMIC, ENVIRONMENTAL AND CULTURAL IMPACTS OF MASS TOURISM

Tourism and economic development

In the 1960s national governments in the northern Mediterranean countries saw tourism as a potential engine of economic growth, and a similar view was adopted in later decades by governments on the southern shore. Many states took an active role in the development of the industry. In Turkey and Spain, for example, the state provided generous incentives to encourage investments in hotels and other forms of accommodation. Some researchers in the 1960s also enthused about the supposed advantages of tourism for regional economic growth. Formica (1965, p. 68), for example, wrote that tourism in the Costa Blanca 'has slowed emigration from the region and caused the transfer of thousands of workers from agriculture to services thus promoting a better distribution of the population among different economic activities. Per capita income has increased notably.' Few commentators now would echo such uncritical sentiments after three further decades of tourism expansion in the Mediterranean Basin. Nevertheless, tourism has made a major contribution to many of the Mediterranean economies, even if there have been associated dependency effects, seasonal unemployment and other less favourable consequences.

The contribution of international tourism to the Mediterranean countries in 1990 has been quantified as an estimated $67 billion, whilst in the Mediterranean regions it is an estimated $40 billion (Jenner and Smith 1993, pp. 7, 28). This represents an estimated 2 per cent of GNP, which rises to 5 per cent if domestic tourism and indirect multiplier effects are included. There is, however, massive polarisation of tourism income, with the northern Mediterranean countries in general and France, Italy and Spain in particular accounting for most direct international transfers of expenditure. These three countries all have international tourism receipts in excess of $20 billion per annum, and together account for 78 per cent of the total tourism receipts of the Mediterranean countries (Table 14.3). Even if the Mediterranean regions rather than the Mediterranean countries are considered, there is still strong polarisation with Spain accounting for 34 per cent of total expenditure in 1991 (Jenner and Smith 1993, p. 28).

On the basis of the direct contribution of international tourism to GNP, the Mediterranean countries fall into five groups:

1. The island economies of Malta and Cyprus, where tourism accounts for more than one-fifth of GNP, and a very large proportion of export earnings.
2. Tourism accounts for between 5 and 8 per cent of GNP in the leading North African des-

TABLE 14.3 International tourism receipts in the Mediterranean countries, 1992

	GNP ($m)	Tourism receipts ($m)	Receipts as % of	
			GNP	Exports
France	1 311 563	25 000	2	11
Italy	1 205 898	21 577	2	12
Spain	567 231	22 181	4	34
Turkey	108 140	3 639	3	24
Portugal	79 992	3 721	5	21
Greece	78 675	3 268	4	34
Israel	68 478	1 876	3	14
Algeria	43 167	75	0	1
Egypt	34 602	2 730	8	88
Morocco	27 640	1 360	5	34
Libya (1990)	23 016	6	0	0
Syria (1991)	16 224	400	2	13
Tunisia	15 269	1 074	7	27
Slovenia	12 347	670	5	12
Cyprus	7 210	1 539	21	154
Malta (1991)	2 646	574	22	46

Source: World Tourism Organization (1994)

tinations of Egypt, Morocco and Tunisia, and for between 27 and 88 per cent of their export earnings. Although all three countries have faced difficult conditions in the 1990s, after the Gulf War and adverse publicity regarding Islamic fundamentalism, tourism continues to be a major component of their economies.

3. The third group is composed of Portugal, Spain, Greece and Turkey, where receipts of $3–4 billion (Spain $22 billion) are equivalent to around 3–5 per cent of GNP. Tourism has been particularly important to Greece, where the manufacturing sector has performed weakly in recent years. Turkey's tourism industry is still at a relatively early stage of development but it has the potential to catch up with those of Portugal and Greece in terms of its economic contribution. Spain can be considered a member of this group, but the relative importance of tourism has been declining due to the growth, increasing diversity and sophistication of the national economy. Slovenia can also be added to this group although the recent history of its tourism industry is complex given the crisis in the former Yugoslavia. Israel potentially could also join this group, even though its tourism market is distinctive and highly vulnerable to changing political conditions (Economist Intelligence Unit 1994b).
4. The fourth group is composed of France and Italy where enormous tourism receipts account for only 2 per cent of GNP, reflecting the overall size, strength and diversity of these economies. This is also the group with the largest 'loss' of tourism income from the expenditure on foreign holidays by its own nationals.
5. The final group is constituted of those countries where tourism is of minimal importance and includes Algeria, Libya and Syria as well as Albania and Lebanon for which data are not available.

Crude estimates of tourism receipts provide only a broad indicator of the economic impact of tourism. The ability of countries or regions to benefit from tourism is mediated by, among other factors, the intermediary role of tour companies. The strong dependence of particular destinations on a small number of international markets (see Fig. 14.6) is reinforced by the dominant position of a few tour companies within these. As a result, these tour companies can exert oligopsonistic powers and depress the prices paid to sub-contracting hotels, restaurants and other local service firms (Williams 1996). While only one-fifth of the total international tourism market is catered for by tour companies, economies of scale allow them to exert considerable market influence, especially in those countries and regions which are heavily dependent on air transport for international arrivals. This is illustrated by the regional differences in the gross operating profits (GOP) of 4 star hotels in Portugal: a GOP of 39.6 per cent of turnover in Lisbon compares to 30.8 per cent in the Algarve which is dependent on mass tourism, and to only 20.1 per cent in Madeira which is almost exclusively reliant on air travel and the international tour companies (Economist Intelligence Unit 1993, p. 38).

Receipts are only one dimension of the economic impact of tourism. Another equally important aspect is employment. Tourism is a relatively labour-intensive industry and, in the 1960s, one of its attractions to policy-makers was the capacity to absorb inter-sectoral transfers of labour, especially from low labour productivity agriculture. In the 1990s the particular appeal of tourism lies in it being one of the few economic sectors with strong growth prospects. In practice, it is difficult to quantify the employment impact of tourism. Jenner and Smith (1993, p. 33), for example, estimate that direct employment is approximately 3 million, and that direct and indirect employment amounts to 5 million jobs. Such figures have limited value given that definitional problems undermine international comparisons of tourism employment. However, the importance of tourism employment in many countries is underlined by national statistics; in Spain, for example, there were an estimated 822 000 direct and 576 000 indirect jobs in tourism in 1992 (Economist Intelligence Unit 1994a, p. 72).

Even if total employment data were available, these would be of limited value for the economic impact of tourism is conditional upon the divisions of labour in the sector and how these interrelate with the organisation of labour in other sectors and households via multiple, part-time or family employment. There are also important questions related to the social divisions of labour and the way these are linked to gender and age relationships, and to migration systems (see Montanari 1995a; Shaw and Williams 1994, Chapter 7; Williams and Shaw 1991b).

The discussion of economic impacts has thus far focused on the level of individual countries. In practice, however, the impact of tourism is highly polarised inter- and intra-regionally, as indicated in the earlier discussion of tourism arrivals. Not only is tourism polarised, but in some countries its economic impact is becoming even more disproportionately concentrated. In Tunisia, for example, 65 per cent of tourism employment was concentrated in just three out of seven regions in 1988 but, by 1992, this polarisation had increased further. The uneven impact of tourism does not, of course, stop at the intra-regional level; it can have a highly differentiated impact within individual resorts. In Nazaré (Portugal), for example, the ability to prosper from tourism is critically dependent on house-ownership as the principal means of income extraction is rent from accommodation services (Mendonsa 1983).

The economic impact of tourism on the Mediterranean region has been highly uneven and this poses the question of whether it has been a force for overall economic convergence or divergence in recent years. Some insights into this question are provided by comparing the growth of tourism receipts 1987–92 (i.e. including the first year after the Gulf crisis) with the overall level of GNP (Fig. 14.7A). There is in fact no clear relationship between the growth in receipts (measured in $) and levels of GNP per capita. The few countries which recorded a growth in receipts in excess of 80 per cent are ranked first, fifth, ninth and twelfth out of 14, in terms of GNP per capita. Spain and Israel (in third and fourth ranks) had relatively low growth rates below 50 per cent, but so did Algeria and

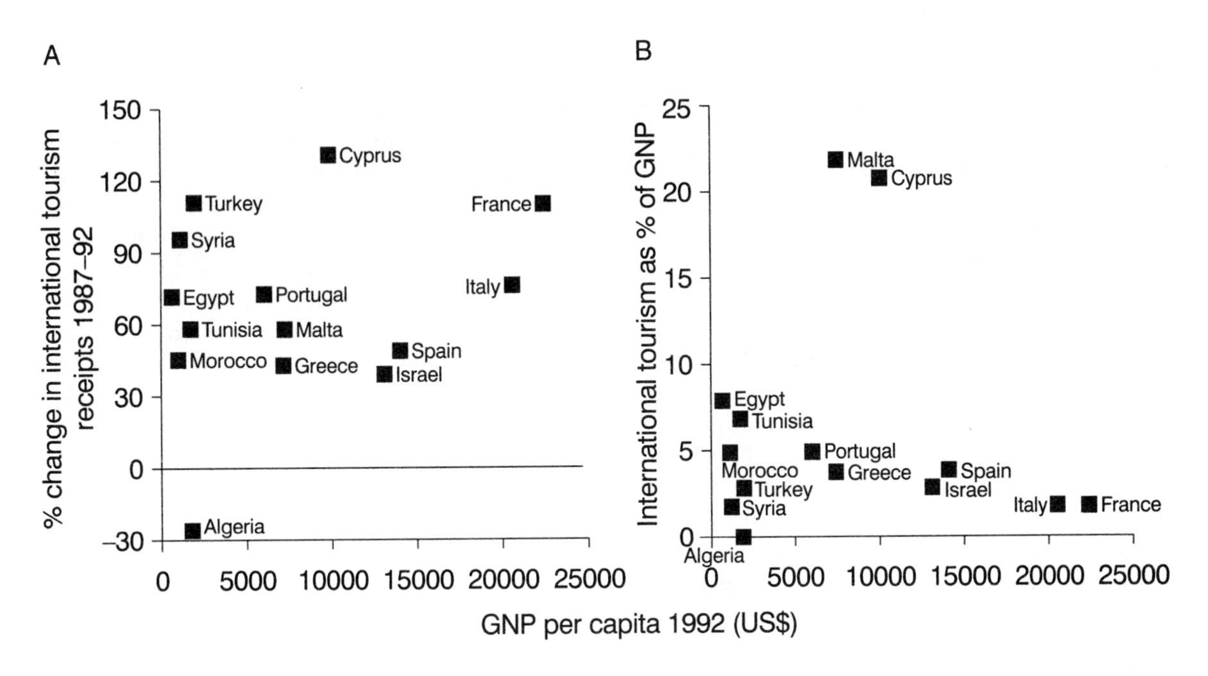

Figure 14.7 (A) Relationship between GNP per capita 1992 and percentage growth in international tourism receipts, 1987-92 (B) Relationship between GNP per capita and share of international tourism receipts in GNP, 1992

Source: World Tourism Organization (1994)

Morocco in the lower rungs of the rankings. A comparison of tourism's contribution to GNP with GNP per capita (Fig. 14.7B) similarly reveals no clear relationship. If the special cases of Malta and Cyprus are excluded, then two features are evident: there is considerable variation in the importance of tourism to the poorer countries, while the share of tourism in GNP declines relative to the level of GNP in the more prosperous countries. There is no unequivocal evidence, then, that tourism has been a force for economic convergence *within* the Mediterranean region, either in the longer or the shorter term.

Footsteps in the landscape: the environmental imprint of tourism

Given its highly polarised nature, tourism has a considerable capacity to influence the environment. Over time the relationship between tourism and the environment in the Mediterranean has changed, mirroring the sequences in the evolution of mass tourism. Dowling (1992) argues that these relationships have passed through four phases: in the 1950s there was harmonious coexistence; in the 1960s there was increasing pressure on the environment from mass tourism; in the 1970s there was increasing environmental awareness as tourism–environment relationships were seen to be in conflict; and in the 1980s, even though ecological concerns increased, there was also realisation that tourism development could be beneficial to the environment.

While Dowling's sequence is highly generalised, it applies particularly well to the Mediterranean region, except that in the fourth phase there has been more concern about the consequences than appreciation of the potential of tourism. This was exemplified by the Blue Plan, which forecast an increase in tourist numbers in the Mediterranean by the year 2025 to between 379 and 758 million (Grenon and Batisse 1989). The impact of any increase in numbers is, of course, amplified by the spatial polarisation of tourism, which means that impacts are cumulative, mutually reinforcing and offer little opportunity for recovery in environmental systems. The coastal concentration of tourism development is the key feature here, and the complex problems of coastal management are the subject of Chapter 17. Not all coasts are equally affected, however. The extent of regional polarisation is underlined by the estimate that tourism developments already occupy 2000–5000 sq. km of surface area in the Mediterranean countries, and that 90 per cent of these developments are in Spain, France and Italy (Montanari 1995a). It has been argued that the over-development of the north-western parts of the Mediterranean coast may have favoured the more recent growth of tourism in the eastern and southern parts of the region (Frangialli 1993, p. 7).

The environmental impacts of tourism are complex and diverse, but can be illustrated by reference to landscape degradation, excessive demands on water supplies and problems of waste management. In terms of landscape change, the imprint of tourism is most clearly evident in the nearly continuous urbanisation of much of the northern coast of the Mediterranean. The precise extent of the consumption of land for tourism purposes is a matter of speculation. Frangialli (1993, p. 11), for example, estimates it is 5000 sq. km while Grenon and Batisse (1989, p. 156) put the figure at no higher than 2200 sq. km. The despoliation of landscapes was compounded by the fact that much of the development in the northern countries occurred in the 1960s and 1970s prior to the establishment of effective urban planning controls in most of the region. The resultant problem is the creation of massive and unsightly tourism settlements. Recent attempts to control such developments, such as Spain's *Ley de las Costas*, which restricts construction within 100 metres of the shoreline, can do little to ameliorate the already extensive landscape degradation in the region. These difficulties are compounded by the very speed of tourist development.

The procurement of adequate water supplies is another problem: tourism requires large volumes of fresh water supplies, but must compete with other users for this very scarce resource and these problems are discussed in Chapter 15. The Blue Plan estimates that local tourism demand for water will increase between 1.7 and 2.3 times between 1984 and 2000 (Grenon and Batisse 1989, p. 156). This poses a dilemma for policy-makers. On the one hand, restricting the supply of water to tourism can damage future demand. However, increasing the supply of water to tourism can have implications for other users. For example, in Spain's Cota de Doñana wetland national park, the water table fell as water was extracted for nearby tourist resorts, resulting in the loss of 85 per cent of the wetland since 1972, as well as the near-destruction of an important feeding ground for hundreds of thousands of birds, including greylag geese, flamingos, spoonbills and egrets (Williams 1993).

There are also complex relationships between tourism development and waste disposal and pollution (see Chapter 17). Tourists are one of the major producers of waste (in concentrated zones over a relatively short period of the year) and yet can be repelled by the clear evidence of pollution, from whatever cause. The various forms of waste generated by tourists can be efficiently managed, as Spain has shown in its determined attempt to improve its treatment of sewage in recent years. However, rapid tourism development still tends to outstrip the already inadequate infrastructure in many Mediterranean countries. In addition, exogenous events and developments can have a major impact on tourism. For example, the spillage from the Haven oil tanker severely affected large areas of the French and Italian coasts, while a 1000-km-long algal bloom (caused by the outwash of nitrate fertilisers from the Po Valley) devastated the Rimini tourist industry.

Managing the tourism–environment relationship poses severe challenges for the national and the local state (Montanari 1995b). Tourism can generate income which funds the conservation of historic monuments or provision of facilities for local populations. However, the financing of tourism infrastructure can be burdensome especially as the demand for tourism is not only spatially but also temporally polarised. The essential contradiction faced by the state is how to reconcile requirements from different fractions of tourism capital so that it permits further investment in tourism facilities, provides adequate infrastructure, and yet somehow manages to conserve an environment which is attractive to tourists. Furthermore, the power of the state *vis-à-vis* international tourism capital is limited; given the highly internationalised markets for mass Mediterranean tourism, the bargaining strength of any one country is severely constrained.

The cultural imprint of tourism

Tourism is frequently held to be an agent of cultural transformation but, while it is a significant social force in many localities, it is often cited as the cause of changes which result from other social processes, including extension of the reach of the mass media.

The cultural impacts of tourism assume many forms, but cultural commoditisation is one of the more widespread. Greenwood (1989, p. 179), for example, states that 'Culture is being packaged, priced and sold like building lots, rights-of-way, fast foods and room service', and argues that 'the commoditization of culture in effect robs people of the very meanings by which they organise their lives'. Norman Lewis in his powerfully written *Voices of the Old Sea* (1984, p. 77), describes the experiences of one ex-fisherman who went to work in the emerging tourist industry of the Spanish village of Farol in the 1950s. Even smiling at the tourists has become commodified:

> You have to smile all the time. It's part of the tariff – that's our joke. We get a peseta a smile, and five pesetas for saying 'Good evening sir, good evening madam. It's nice to see you again.' ... You can safely say that we have sold ourselves.

The relationship between tourism development and cultural change is, of course, a conditional one, depending in particular on the scale of arrivals and on the level of development and social organisation of the host community. It is, therefore, a relationship which has evolved over time. This is typified by the case of the Costa del Sol where, until the 1950s, fishing villages such as Torremolinos absorbed small numbers of tourists with little cultural discontinuity. Then, in the 1960s, mass tourism arrivals had a major impact on the coastal region: the large gap in standards of living between Northern Europe and Andalusia meant there was a strong demonstration effect in terms of material conditions as well as a sharp challenge in terms of values and behaviour at a time when Franco's Spain was still relatively closed culturally. In the 1990s there is a far smaller demonstration gap and the influence of tourism has been swamped by that of the mass media.

Extending the Spanish experience to the rest of the Mediterranean, it is evident that even in mass tourism there are major national and regional differences in the cultural impact of tourism, as is evident in the contrast between the prosperous and well-organised communities of the French Riviera, say, and the still relatively underdeveloped regions of southern Turkey. Such differences can be highly localised as Tsartas (1992) has demonstrated in comparative studies of Greek islands. Even where the cultural impact of tourism is very strong, as in Malta, tourism may not be resisted because of the economic benefits it brings. In other contexts, tourism may also be welcomed as less culturally disruptive than the alternative of mass emigration of young people from what would otherwise be relatively poor rural communities.

The potential clash of host and guest cultures is probably greatest in Islamic societies. Only the most secular of Islamic countries such as Turkey, the Turkish Republic of Northern Cyprus and Bosnia-Herzegovina have been able to absorb significant numbers of tourists without major conflicts with local religious/cultural systems. Other Islamic societies have responded to this potential clash either by discouraging tourism (as in Syria or Libya), or in trying to isolate tourism in enclaves, as in

Tunisia. However, the development of enclaves cannot eliminate the impact of tourism. Poirier (1995, p. 167) writes that, in Tunisia,

> sights that are common in the West, such as scantily clad visitors on the beach or around the hotel pool, and open affection between men and women, offend many Tunisians. Additionally, tourists knowingly or unknowingly violating rules of piety in and around mosques and Islamic religious activities, provide fuel to Islamic fundamentalists who criticise the excessive Westernisation of Tunisian society.

CONCLUSIONS

Tourism in the Mediterranean has had an uneven impact – economically, environmentally and culturally – on different regions and countries. To some extent there is a form of diffusion, with the initial focus of international tourism being the French and Italian Rivieras, but subsequently spreading to Spain, then to Cyprus, Malta, Greece and Portugal, and finally to Turkey and some of the northern African countries. Domestic tourism coexisted with international mass tourism in all these countries but its massification has tended to lag. Only in France, Italy and, more recently, Spain has domestic tourism become a significant force shaping Mediterranean coastal tourism.

Turning to the future, tourism in the Mediterranean region is predicted to continue to expand, even if its share of the global market declines due to distance from the emerging new centres of consumption in the world economy, increased competition from destinations in the earlier stages of the product cycle, and growing concerns about the environmental quality of the Mediterranean. Estimates of future growth are, of course, dependent on the health of the global economy, comparative costs, and such exogenous influences as wars, civil strife and terrorism. Nevertheless, neither of the two principal estimates of future demand seems improbable. Jenner and Smith (1993, p. 160) consider that by the year 2000 there will be 330 million foreign and 205 million domestic tourists in the region. The Blue Plan estimates that there will be between 379 and 758 million tourist arrivals in the region by 2025 (Grenon and Batisse 1989).

More debatable is the likely distribution of tourism within the Mediterranean Basin. There are still relatively 'immature' destinations such as Morocco, Egypt and Turkey which have considerable potential for further development. Whether they will do so depends on the strength of both their Arab and European markets, as well as domestic and international political relations. There is also the potential to build, or rebuild, virtually new tourist industries in Albania, the Lebanon and parts of the former Yugoslavia. While they have the advantages of being relatively new tourism products, with low cost structures, their success will depend on being able to convince highly sensitive foreign tourist markets of their security and stability. Even if they fail to attract foreign tourists, however, most of these countries stand to benefit from expansion of their domestic tourism markets following anticipated economic growth. Their success will also depend on the ability of the established European Mediterranean countries to maintain or enhance their market shares. That, in turn, is dependent on the latter's ability to respond to significant shifts in holiday spending, the pursuit of quality and the demand for more individualised holidays.

Perhaps the greatest challenge for the Mediterranean countries is to secure an effective redistribution of the economic benefits of tourism. This breaks down into three specific issues: the distribution of income between the international tour companies and the host economies; the distribution of tourism expenditure between the coastal and interior zones; and the distribution of tourism income between the state, fractions of tourism capital and the tourism labour force. The outcome of any such redistributions depends fundamentally on the future economic and political organisation in the host countries, but it is also structurally constrained by the globalisation of the tourism industry.

REFERENCES

BUCKLEY, P.J. and PAPADOPOULOS, S.I. 1986: Marketing Greek tourism: the planning process. *Tourism Management* **7**, 86–100.

DAWES, B. and D'ELIA, C. 1995: Towards a history of tourism: Naples and Sorrento (XIX century). *Tijdschrift voor Economische en Sociale Geografie* **86**, 13–20.

DIRECCIÓN GENERAL DE POLÍTICA TURÍSTICA 1992: Los precios de los packages turísticos temporada verano 1992. *Estudios Turísticos* **115**, 55–86.

DOWLING, R.K. 1992: Tourism and environmental integration: the journey from idealism to realism. In Cooper, C.P. and Lockwood, A. (eds), *Progress in Tourism, Recreation and Hospitality Management: Volume Four*. London: Belhaven, 33–46.

ECONOMIST INTELLIGENCE UNIT 1991: Egypt. *EIU International Tourism Reports* **1**, 53–71.

ECONOMIST INTELLIGENCE UNIT 1992: Morocco. *EIU International Tourism Reports* **3**, 41–57.

ECONOMIST INTELLIGENCE UNIT 1993: Portugal. *EIU International Tourism Reports* **1**, 23–42.

ECONOMIST INTELLIGENCE UNIT 1994a: Spain. *EIU International Tourism Reports* **3**, 71–89.

ECONOMIST INTELLIGENCE UNIT 1994b: Israel. *EIU International Tourism Reports* **3**, 25–45.

FORMICA, C. 1965: La Costa Blanca e il suo sviluppo turistico. *Rivista Geografica Italiana* **22**, 42–68.

FRANGIALLI, F. 1993: El turismo en el Mediterraneo: La apuesta del desarrollo sostenible para un gran destino frágil. *Estudios Turísticos* **119–120**, 5–21.

GREENWOOD, D.J. 1989: Culture by the pound: an anthropological perspective on tourism as cultural commoditization. In Smith, V.L. (ed.), *Hosts and Guests: the Anthropology of Tourism*. Philadelphia: University of Pennsylvania Press, 171–86.

GRENON, M. and BATISSE, M. 1989: *Futures for the Mediterranean Basin: The Blue Plan*. Oxford: Oxford University Press.

HALL, D. 1995: Tourism change in Central and Eastern Europe. In Montanari, A. and Williams, A.M. (eds), *European Tourism: Regions, Spaces and Restructuring*. Chichester: Wiley, 221–44.

JENNER, P. and SMITH, C. 1993: *Tourism in the Mediterranean*. London: Economist Intelligence Unit.

LEWIS, J.R. and WILLIAMS, A.M. 1991: Portugal: market segmentation and regional specialization. In Williams, A.M. and Shaw, G. (eds), *Tourism and Economic Development: Western European Experiences*. London: Belhaven, 107–29.

LEWIS, N. 1984: *Voices of the Old Sea*. London: Hamish Hamilton.

MANRIQUE, E.G. 1989: El turismo. In Ory, V.B. de (ed.), *Territorio y Sociedad en España II, Geografía Humana*. Madrid: Taurus, 341–67.

MARCHENA GÓMEZ, M.J. and VERA REBOLLO, F. 1995: Coastal areas: processes, typologies and prospects. In Montanari, A. and Williams, A.M. (eds), *European Tourism: Regions, Spaces and Restructuring*. Chichester: Wiley, 111–26.

MENDONSA, E.L. 1983: Tourism and income strategies in Nazaré, Portugal. *Annals of Tourism Research* **10**, 213–38.

MONTANARI, A. 1995a: The Mediterranean region. In Montanari, A. and Williams, A.M. (eds), *European Tourism: Regions, Spaces and Restructuring*. Chichester: Wiley, 41–65.

MONTANARI, A. 1995b: Tourism and the environment: limitations and contradictions in the EC's Mediterranean region. *Tijdschrift voor Economische en Sociale Geografie* **86**, 32–41.

PEARCE, D.G. 1987: Spatial patterns of package tourism in Europe. *Annals of Tourism Research* **14**, 183–201.

POIRIER, R.A. 1995: Tourism and development in Tunisia. *Annals of Tourism Research* **22**, 157–71.

REYNOLDS-BALL, E. 1914: *Mediterranean Winter Resorts*. London: Kegan Paul, Trench, Trüber and Co.

SHAW, G. and WILLIAMS, A.M. 1994: *Critical Issues in Tourism*. Oxford: Blackwell.

TSARTAS, P. 1992: Socio-economic impacts of tourism on two Greek isles. *Annals of Tourism Research* **19**, 516–33.

URRY, J. 1990: *The Tourist Gaze*. London: Sage.

VALENZUELA, M. 1991: Spain: the phenomenon of mass tourism. In Williams, A.M. and Shaw, G. (eds), *Tourism and Economic Development: Western European Experiences*. London: Belhaven, 40–60.

WILLIAMS, A.M. 1993: Spain and Portugal: coping with rapid changes. In Williams, M. (ed.), *Planet Management*. New York: Oxford University Press, 124–31.

WILLIAMS, A.M. 1996: Mass tourism and international tour companies. In Barke, M., Towner, J. and Newton, M.T. (eds), *Tourism in Spain: Critical Issues*, Wallingford: CAB International, 119–35.

WILLIAMS, A.M. and SHAW, G. 1991a: Tourism and development: introduction. In Williams, A.M. and Shaw, G. (eds), *Tourism and Economic*

Development: Western European Experiences. London: Belhaven, 1–12.

WILLIAMS, A.M. and SHAW, G. 1991b: Western European tourism in perspective. In Williams, A.M. and Shaw, G. (eds), *Tourism and Economic Development: Western European Experiences*, London: Belhaven, 13–39.

WORLD TOURISM ORGANIZATION 1994: *Yearbook of Tourism Statistics*. 2 vols, Madrid: World Tourism Organization.

15

WATER: A CRITICAL RESOURCE

BERNARD SMITH

INTRODUCTION

The Mediterranean Basin is characterised by such extreme variations in the availability of freshwater and in the pattern of water requirement that generalisation is difficult. None the less, there is an underlying seasonal or annual scarcity of rainfall which, set against rapidly increasing demand – typically in the driest part of the Mediterranean Basin – requires careful and imaginative water harvesting, storage, distribution and recycling. Even with careful management, however, it seems inevitable that the scarcity and hence the value of freshwater are set to increase (see Fig. 15.1) and that this will significantly influence patterns of economic and social change. In particular, water shortage increases the scope for conflict where there is competition for resources. This may occur within countries where regional autonomy has stimulated environmental protectionism but also, more importantly, between nation–states where rivers flow across and groundwater basins straddle international borders. It is hardly surprising, therefore, that freshwater is increasingly seen as a constraint upon, rather than as an enabler of Mediterranean development (Grenon and Batisse 1989).

In addition to uncertainties over water availability, much of the Mediterranean Basin is ecologically fragile and as such is endangered by current social and economic trends (Falkenmark and Lindh 1993, pp. 80–91). Water supply and waste disposal are crucial elements in the region's development, but their mismanagement can trigger damaging environmental impacts. Safe and efficient provision is further hampered by the concentration of permanent and tourist populations in coastal areas, which creates an uneven demand in space and time. Many coastal areas are also extremely sensitive to water pollution and their soils prone to erosion. When this sensitivity is combined with rapidly expanding urban populations and the continued growth of irrigated agriculture, it is clear that what is required is 'early, anticipatory water planning rather than waiting for even more serious water shortages and pollution to occur' (Falkenmark and Lindh 1993, p. 88). To be successful, plans have to ensure the long-term maintenance of the balance between supply and demand. This must be done in such a way that the needs and aspirations of populations are met without serious or long-lasting damage to the environment or sources of supply. In this chapter problems of water supply and demand will be examined in the context of their regional characteristics. A series of case studies will then be used to demonstrate how management has been approached under different environmental, economic and political conditions. Finally, future prospects for water availability will be discussed together with the strategies necessary to maintain supplies.

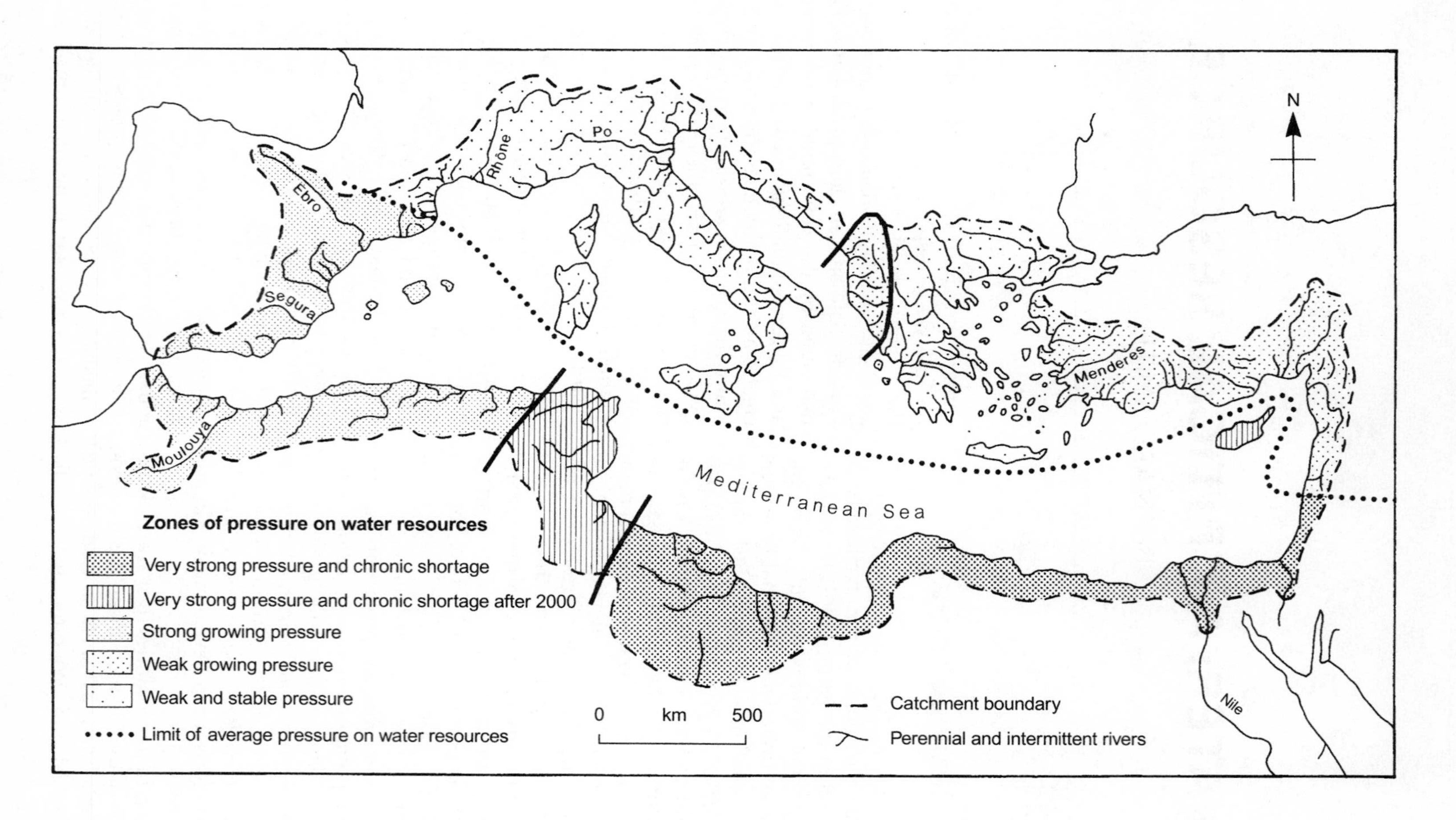

Figure 15.1 The Mediterranean Basin, showing the main rivers draining into the Mediterranean Sea

Source: After Milliman *et al.* (1992) and Ajuntament de Calvià (1995)

WATER SUPPLY

Unless Mediterranean countries have a high proportion of their land area in climatic zones wetter than the Mediterranean (e.g. France, Table 15.1), can import water (e.g. Egypt) or can mine fossil groundwater (e.g. Libya), freshwater supply is primarily dependent upon annual rainfall. Because of this, the Mediterranean Basin can be divided into two broad, natural hydrological regions. The northern shore is relatively well-watered, with annual precipitation of the order of 400–1000 mm. Many of the rivers are perennial, but with a wide seasonal variation in flow, and here Ambroggi (1977) estimated annual potential water resources at approximately 2 000 m^3 per capita, compared to an annual per capita use of approximately 400 m^3. Conversely, the southern and eastern shores of the Mediterranean are essentially arid. Apart from the Maghreb, annual precipitation is often less than 100 mm and annually renewable resources rarely exceed 1000 m^3 per capita. Drainage networks are poorly developed and flows (if they do not derive from outside the Mediterranean climatic zone) are characteristically intermittent or ephemeral.

The proportion of available resources that are usable depends upon a number of factors, not the least of which is the storage potential of seasonal rainfall and runoff. The geology of the

TABLE 15.1 Population and water resources of Mediterranean countries

Region/country	1990 population 10^6 people	Annual internal renewable water resources		Annual river flows	
		Total km^3/yr	1990 (10^3m^3/yr) per capita	From other countries (km^3/yr)	To other countries (km^3/yr)
EUROPE					
Albania	3.25	10.00	3.08	11.30	–
France	56.14	170.00	3.03	15.00	–
Greece	10.05	45.15	4.49	13.50	3.00
Italy	57.06	179.40	3.13	7.60	–
Spain	39.19	110.30	2.80	1.00	17.00
Yugoslavia (1990 boundaries)	23.81	150.00	6.29	115.00	200.00
AFRICA					
Algeria	24.96	18.90	0.75	0.20	0.70
Egypt	52.43	1.80	0.03	56.50	–
Libya	4.55	0.70	0.15	–	–
Malta	0.35	0.03	0.07	–	–
Morocco	25.06	30.00	1.19	–	–
Tunisia	8.18	3.75	0.46	0.60	–
ASIA					
Cyprus	0.70	0.90	1.28	–	–
Israel	4.60	1.70	0.37	0.45	–
Lebanon	2.70	4.80	1.62	–	0.86
Syria	12.53	7.60	0.61	27.90	30.00
Turkey	55.87	199.00	3.52	7.00	69.00

Source: After Gleick (1993)

Mediterranean littoral is predominantly sedimentary and alluvial and comprises a large number of discrete groundwater basins. These form important water stores and explain the traditional reliance in many Mediterranean countries on local aquifers for their freshwater supplies (Fig. 15.2) compared, for example, to the UK and Germany. Unfortunately, because of their coastal location and prolonged over-abstraction, many aquifers are now contaminated by sea water incursion. The rate of groundwater depletion can be rapid and Ben-Asher (1995) describes how in Malaga active wells were found 20 metres from the coast 25 years ago, whereas today no active wells are found within 400 m of the coast. This retreat has been associated with an annual drop in the water table of 20–50 cm and an increase in salinity of 0.1–0.2ds/m/annum. The consequences of salination cannot always be avoided by digging new wells and around Malaga it has reduced vegetable production by 1–3 per cent per annum and citrus output by 2–4 per cent per annum.

Problems of over-abstraction are often most severe in the notionally drier Mediterranean countries such as Israel, where the coastal aquifer has been heavily over-exploited over the last 50 years. However, even in European countries (Fig. 15.3) over-abstraction is widespread and locally severe. One of the largest areas to be affected is in the arid south-east of Spain, where the problem is linked to the rapid growth of horticulture. Other areas include the south-east of Italy and the north-east of Greece. The associated problem of saltwater incursion is significant both in mainland Spain and in the Balearic Islands, where a combination of sensitive coastal aquifers and efficient pumping has severely depleted groundwater. Further consequences of over-abstraction include threats to many wetlands. The Mediterranean is well endowed with wetlands, most of which are nominally protected by the Ramsar Convention. These sites are important locally for their indigenous wildlife and as controls on surface water flows, but many are also vital staging points on bird migration routes. None of this has prevented local encroachment into these areas, especially by tourism, but also by agriculture, pollution, drainage and falling water tables. Possibly the largest single area under

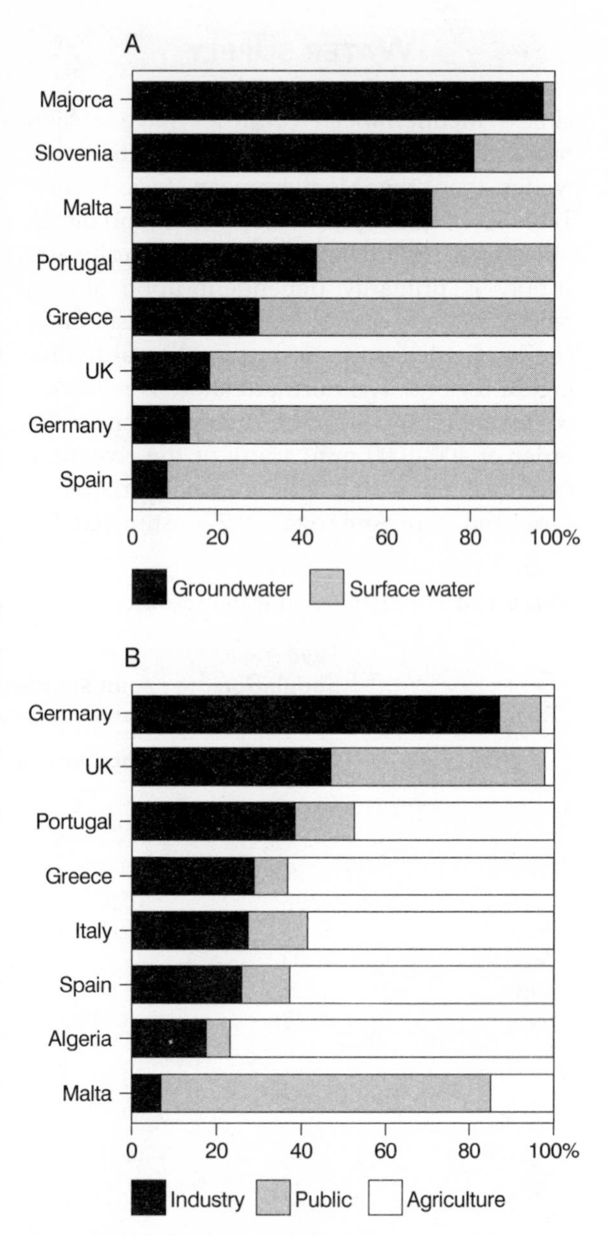

Figure 15.2 Diagrams showing freshwater sources (A) and patterns of water use (B) in selected countries

Source: Adapted from diagrams in Stanners and Bordeau (1995) and from data in Pons (1989) for Majorcan water supply in 1984

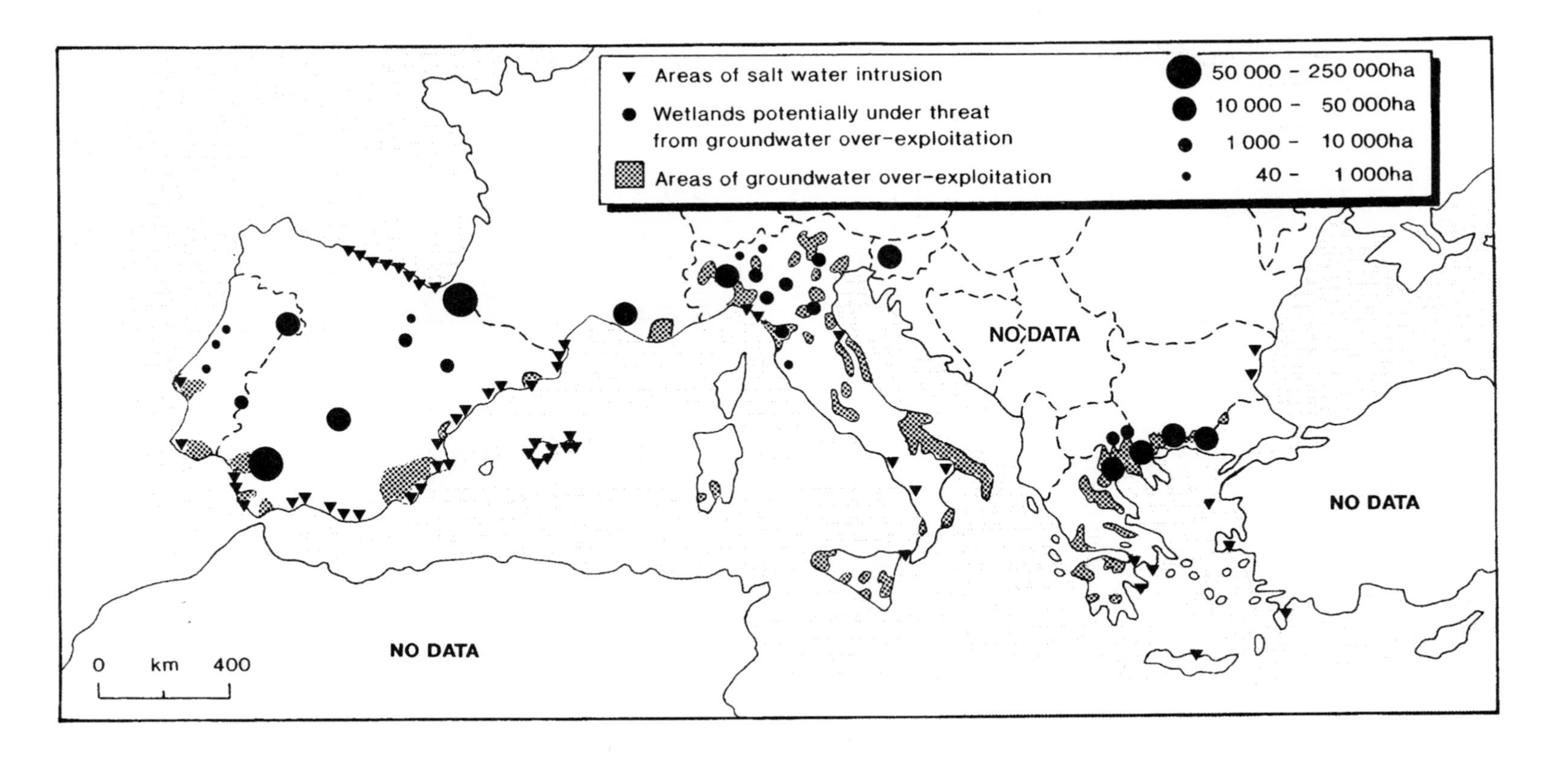

Figure 15.3 Map of the northern shore of the Mediterranean showing areas of saline incursion into aquifers and wetlands threatened by groundwater over-abstraction

Source: Adapted from maps in Stanners and Bordeau (1995)

threat is the marshland of Las Marismas at the mouth of the Guadalquivir between Seville and Cadiz in south-west Spain. This area is threatened not only by the spread of cultivation around its margins but also indirectly through contamination by pesticides and fertilisers as agriculture intensifies.

The complex tectonics that created the numerous groundwater basins inhibit regional transfers of water and restrict water management. A similar compartmentalisation affects surface water catchments and although the accentuated relief of much of the Mediterranean Basin provides many opportunities for reservoir construction, transfer between basins is limited by the same topography. Thus, 'the relative lack of major graded and converging water systems is a factor complicating the water-use conditions and makes local utilizations less interdependent' (Grenon and Batisse 1995, p. 220). Nevertheless, as demand for water increases and groundwater resources decline, the necessity for surface water storage and inter-regional transfers has increased. In response, many governments have begun to invest heavily in reservoir storage and to investigate expensive water transfer projects. One of the first Mediterranean countries to extensively implement this strategy was Spain, and Figure 15.4 shows the rapid rise in surface water storage capability in this country since the 1950s. It also shows how initially this was achieved by exploitation of large sites, but how since approximately 1960 the size of reservoirs has decreased as suitable sites become increasingly rare. García de Jalon (1987) also highlights problems associated with this expansion of supply where, for example, sewage treatment has failed to expand at the same rate and river water quality has sharply deteriorated. The stimulus for this expansion in Spain has been a combination of increasing industrialisation, a rapid growth in urban population and a very uneven distribution of rainfall across the country. Because of the latter there have been long-standing transfers of water into cities such as Madrid, Barcelona, Bilbao and Seville, but these are now complemented by broader, inter-catchment transfers. The largest of these joins the upper Tajo River with the Segura catchment and since 1979 has transferred 100–300 × $10^6 m^3$ per annum. Since joining the European Union the need for such transfers to meet the needs of an increasing commercial agricultural sector has risen dramatically. Unfortunately this demand has coincided with increasing regional autonomy and an unwillingness by politicians to 'give away' their region's water. For example, in 1994, José Bono, the then President of Castilla la Mancha, was reported as proclaiming that not a single drop of water would be transferred southwards from his region when his own people were running short. Similarly, the President of Aragon was rumoured to have remarked that only over his dead body would water be diverted northwards from the Ebro into Catalonia (Smith 1994).

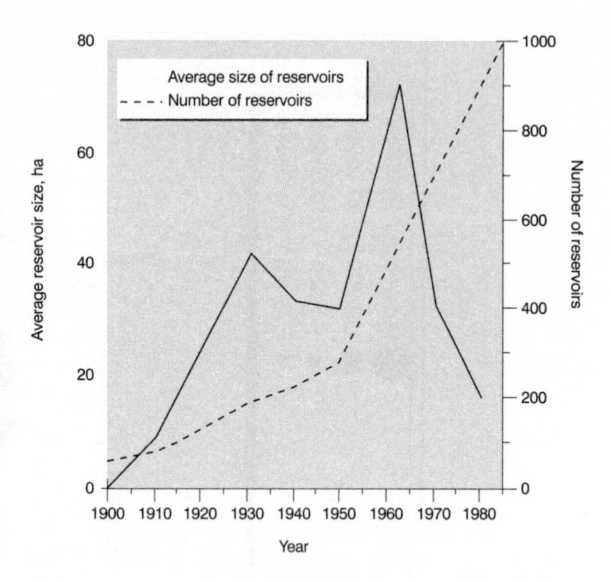

Figure 15.4 Graphs showing the rise in surface water storage capability in Spain, but accompanying decrease in reservoir size

Source: García de Jalon (1987)

Reservoir construction itself is not without hazard. The very tectonics that produced a terrain suitable for dam construction continue to threaten stability through frequent earthquakes, volcanism and the creation of conditions that favour slope instability. Soil erosion rates around the Mediterranean are also high in response to steep slopes, shallow, friable soils, seasonally intense rainfall and a long history of intensive land use, as Chapter 4 has shown. Suspended sediment loads in slopewash and rivers are thus correspondingly high and sediment accumulation behind dams can be very rapid. Grenon and Batisse (1989), for example, pointed out that siltation in Algeria was reducing storage capacity by 2–3 per cent per annum. Likewise, in Tunisia, the Chiba Dam, which was completed in 1963, had lost 35 per cent of its capacity by 1975 and an estimated 70 per cent by 1989.

One way to increase available water may be to harness surface runoff and ephemeral stream flow before they infiltrate. Such flood water harvesting has been practised around the Mediterranean for thousands of years, traditionally in North Africa and the Middle East but increasingly in Southern Europe. In one of the most detailed analyses of water harvesting techniques, Evenari *et al.* (1971) reconstructed a rainfed farm at Avdat in the Negev Desert originally built by the Nabateans from the third century BC. To do this they had to relearn principles of arid lands runoff. This included increasing total runoff by dividing hillsides into small catchments with stone bunds and reducing infiltration by the removal of stones, thus allowing the remaining clay to crust under raindrop impact.

Techniques employed to collect runoff include: roof-top, inter-row, micro-catchment, medium-sized catchment and large catchment harvesting using rock dams, stone or earth contour bunds, cisterns, galleries and trapezoidal bunds around individual trees. The aim of all these methods is to 'concentrate runoff and to make it available for domestic or agricultural purposes, thus preserving other water sources like groundwater or water from wells and streams' (Prinz 1995, p. 136). In addition, harvesting systems can control fluvial erosion, sediment entrainment, transport and deposition and, by trapping soil, increase soil depth and moisture storage (Gilbertson 1986). Within the Mediterranean Basin, Tunisia is the most advanced in its use of water harvesting and conservation. In the arid south of the country some 10 million olive trees are cultivated using water harvesting and in 1990 a national strategy for surface runoff mobilisation was initiated. This includes the building of 203 small earth dams, 1000 ponds, 2000 groundwater recharge schemes, and 2000 flood irrigation schemes by the year 2000 (Prinz 1995).

Where there is no scope for increasing yield from surface runoff, intensifying groundwater abstraction or transferring water from regions of surplus to those of deficit, countries face a limited number of choices. They could introduce expensive technologies such as desalination, which effectively create new sources of supply, or they could negotiate water imports. Ultimately, if supplies are exploited to their limit and neighbours refuse to help, countries may be driven to military conquest of adjacent lands to ensure supplies.

DEMAND

Underlying trends

Against a background of rapid population growth, urbanisation, industrialisation and previous under-provision of domestic supplies, it is hardly surprising that demand for and use of freshwater has dramatically increased around the Mediterranean in recent years. This increase is illustrated in Figure 15.5 for a number of European countries. However, it is likely that in future the greatest increase in demand for water will come from countries of the southern and eastern Mediterranean shores. Most of these states are passing through a demographic transition, in which high birthrates and decreasing mortality produce high rates of population growth (see Chapter 11). These are also the countries that are economically the least able to embark on major water development schemes. Exceptions exist: for example, in Libya oil revenues have been deployed to exploit groundwater resources; but generally these countries already utilise

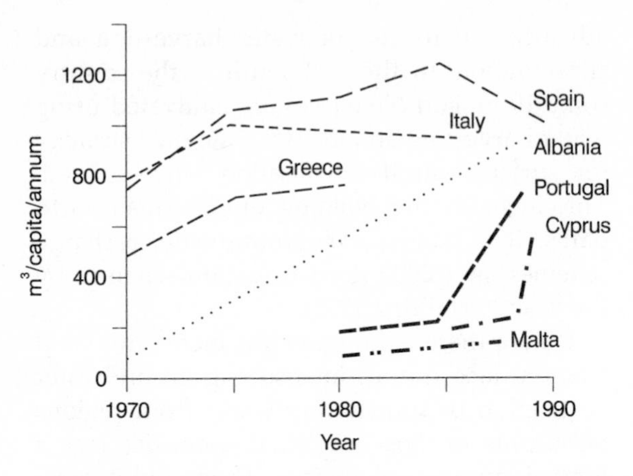

Figure 15.5 Trends in total water abstraction for selected countries

Source: Stanners and Bordeau (1995)

high proportions of their annual renewable water supplies (Fig. 15.6), leaving little scope for exploitation of new, internal sources.

The potential for increased demand is illustrated in Table 15.2, which shows access to safe drinking water around the south and east of the Mediterranean. Most of these states have large rural populations with limited access to clean water, and in some countries (Morocco and Egypt) this actually decreased during the 1970s and early 1980s. Even if there were no increase in rural provision, overall demand will increase because of rural-to-urban migration which will give more people a theoretical access to piped supplies. Other factors that should increase per capita demand are increased education and a growth in the economically active proportion of the population (Grenon and Batisse 1989). Economically, there is likely to be a growth in industrial per capita output and increased demand for irrigation water. As all of these changes occur there is also the possibility of increased losses from the already inefficient water supply and distribution systems that characterise much of the Mediterranean. As an example, in a recent survey of 17 Greek towns and cities water loss varied between 17 and 68 per cent with an average of 45 per cent. If this figure was extended to cover all of Greece it would represent a financial loss equivalent to US$25 million per annum. The recognition of this problem has at least triggered a systematic leak reduction policy in Greece, but this was only possible via financial and technical assistance through the EU's Sprint Programme. Access to such assistance may not always be possible for other Mediterranean countries.

Finally, underlying demands for freshwater are increasingly influenced by the annual in-migration of millions of tourists (see Chapter 14).

Tourist pressures

Tourism has for many years distorted patterns of water need in the European Mediterranean by concentrating maximum demand during summer months of minimum availability. This pattern is, however, changing. First, many older tourist areas in the north-west Mediterranean now actively pursue all-year-round occupancy. Second, the development of longer-haul tourism during the 1980s and 1990s has seen a rapid growth in the numbers of visitors to the drier countries of the eastern Mediterranean such as Cyprus and Israel. For example, between 1970 and 1990 tourist visits to the Mediterranean as a whole tripled from 54 to 157 million, but in Greece the number of visitors grew sixfold and in Turkey it quadrupled (Stanners and Bordeau 1995).

The consequences of increased tourist visits encompass not only the demand for increased quantities of water, but also an expectation that water quality will conform to the extremely high standards demanded at home – most commonly north-west Europe. Such standards may be both technically unachievable, because of local environmental conditions, and also uneconomic, given the present infrastructure and limited financial resources of many countries. Under these circumstances priority is normally given to clean, piped water and greatest stress falls on sewage and effluent treatment and disposal from resorts and other urban areas. Even in the European Mediterranean only approximately 30 per cent of municipal waste water from coastal towns receives any treatment

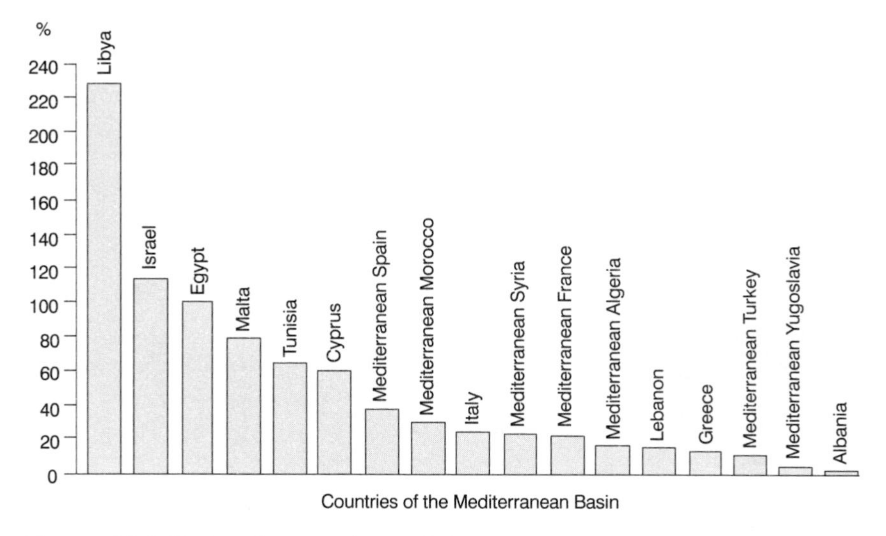

Figure 15.6 Ratio of water demand to resources in selected countries for 1985

Source: Grenon and Batisse (1989)

before discharge. It is estimated (Stanners and Bordeau 1995) that it would take approximately US$10 billion to upgrade sewage disposal facilities within the relevant EU countries alone. Across the Mediterranean Basin as a whole, the solid waste generated by tourists was estimated in the late 1980s as 2.9 million tonnes per annum. This was projected to increase to between 8.7 and 12.1 million tonnes by 2025 and waste water from 0.3 billion m^3 to between 0.9 and 1.5 billion m^3 (Grenon and Batisse 1989).

Nowhere are the pressures on water supplies by tourists greater than on the smaller islands in the Mediterranean. The economies of islands such as those of the Balearics, Aegean, Malta, Crete and Cyprus are now highly dependent upon an annual influx of visitors. Indeed, in many cases communities were only rescued from rural poverty (often dictated by a paucity of exploitable water resources) by the advent of mass tourism in the 1960s and 1970s. Because they are islands there is no scope for the overland importation of water and the very climatic factors that make them attractive to many visitors are also those that limit annual replenishment of water resources. Under these circumstances demand might be met by reducing per capita use, by desalination (e.g. the island of Syros in the Cyclades), deliveries of water by tanker (e.g. Patmos in the Dodecanese) or, for example, by switching water from irrigation to domestic supplies. Such a switch would, however, be counter to the overall trend around the Mediterranean of increasing demand for irrigation water. These strategies are investigated further in case studies of Malta and Majorca later in the chapter.

Irrigation need

A major shift of water resources away from agriculture may be acceptable in small, under-populated, tourist-dominated economies. Generally, however, rapidly growing populations, most of whom still live in rural areas, mean that demand for irrigation water and the political power to insist on its provision are set to continue. This increasing reliance upon irrigated agriculture in many of the drier Mediterranean countries is illustrated in Table 15.3, which also illustrates the problems caused by salination in countries such as Egypt. Supplying further water in the quantities required for irrigation is likely to prove very expensive. Grenon and Batisse (1989), for example, quoted costs as high as US$30000 per hectare for new irrigated land in Syria, whereas to

Table 15.2 Access to safe drinking water and sanitation services in selected Mediterranean countries

	Urban population as percentage of total population		Access to safe drinking water				Access to sanitation services			
			Percentage of urban population		Percentage of rural population		Percentage of urban population		Percentage of rural population	
	1970	1985	1970	1985	1970	1985	1970	1985	1970	1985
AFRICA										
Algeria	39.5	42.6	84	85	61*	55	13	80	6	40
Egypt	43.5	46.4	94	88**	93	64**	ND	ND	ND	10**
Libya	35.8	64.5	100	100**	42	90**	100	100**	54	72**
Morocco	34.6	44.8	92	100	28	25	75	62	4	16
Tunisia	43.5	56.8	92	100	17	31	100	84	34	16
ASIA										
Cyprus	40.8	49.5	100	100	92	100	100	100	92	100
Syria	43.3	49.5	98	98**	50	54**	ND	74**	ND	ND
Turkey	38.4	45.9	ND	95**	ND	62**	ND	56**	ND	ND

Notes: *1975; **1980; ND = no data
Source: After Gleick (1993)

TABLE 15.3 Agricultural population, total cropland and irrigated areas for Mediterranean countries

	Agricultural population 1989 (10^3)	Total cropland (10^3 hectares)	Irrigated area 1974 (10^3 hectares)	Irrigated area 1989 (10^3 hectares)	Percentage of cropland irrigated 1989
EUROPE					
Albania	1 565	707	320	423	59.8
Greece	2 198	3 924	848	1 190	30.3
Italy	3 688	12 033	2 680	3 100	25.8
Spain	4 247	20 345	2 783	3 360	16.5
AFRICA					
Algeria	5 936	7 605	242	336	4.4
Egypt	21 005	2 585	2 843	2 585	100
Libya	616	2 150	195	242	11.3
Malta	14	13	1	1	7.7
Morocco	9 095	9 241	1 032	1 265	13.7
Tunisia	2 015	4 700	115	275	5.9
ASIA					
Cyprus	146	156	30	35	22.4
Israel	199	433	176	214	49.4
Lebanon	247	301	86	86	28.6

Source: After Gleick (1993)

increase the irrigated area in the north-west of the Mediterranean Basin by 3.8–4.0 million ha. would require an estimated US$70 billion at 1985 prices. Where there is no access to increased supplies, any increase in irrigation can only be met by water saving. This could involve more accurate application of water and fertiliser using micro-irrigation techniques, the use of treated waste water and the carefully managed use of saline water.

Any significant increase in water use will adversely affect flow conditions in many rivers. Not only will it concentrate any pollutants, but reduced discharges can also lead to sedimentation, reduction in channel capacity and increased flood risk during high seasonal or storm flows. The clearest example of this increased risk is provided by the Nile. At the beginning of this century the annual discharge to the Mediterranean was 60 billion m^3. By the late 1980s the discharge was reduced to 5 billion m^3, the minimum level required to maintain the channel (Grenon and Batisse 1989), with no safety margin for flow reduction during drought periods.

As available water is reduced, it is inevitable that the need to exploit more costly abstraction and treatment options will result in a rise in unit cost. This particularly affects irrigation water, where real costs have rarely been passed on to the consumer. Either the capital costs of dam construction are subsumed into national development plans, or running costs are subsidised as part of agricultural support and/or as a means of social engineering designed to maintain rural populations by supporting the 'small farmer'. If real costs are passed on, it is likely that this will influence cultivation procedures, crop types and ultimately land tenure. There is already increasing pressure to grow cash-generating crops or crops for import substitution. Furthermore, the technical and financial investment required to improve irrigation efficiency is frequently beyond the capabilities of individual farmers, who are pressurised to sell their land to larger landholders or commercial undertakings with all the benefits of scale and access to capital. The costing of water is not, however, simply a question of economics. Mediterranean peoples are not immune to the cultural traits of other societies that have rebelled against the concept of having to pay for water. Water is perceived by many as a natural resource that belongs to everyone and the pricing of water to promote conservation can be strongly opposed. What may be required to overcome this problem is an explanation of the distinction between paying for water and paying for the service of delivery. So that, 'where strong cultural opposition exists to charging for water itself, charging for the satisfactory provision of water in a distribution system may be possible instead' (Falkenmark and Lindh 1993).

Management of freshwater resources

Within the varied constraints on supply and demand imposed by regional conditions, countries and communities must adopt very different strategies for managing their water resources. Even where the same strategies are employed, local conditions ensure that no two projects are identical. To illustrate some of the problems of water management, this section will examine a number of case studies chosen to reflect three common situations. These are inter-basin transfers within national boundaries, the importation of water across international borders and the particular problems associated with the uneven demands imposed by tourism.

Internal transfers: keeping it in the family

Cyprus

The only means of freshwater replenishment in semi-arid Cyprus is rainfall, most of which falls between October and April on the Kyrenia Mountains in the north and the Troodos Mountains in the south-west. As is the case for much of the Mediterranean, the population and hence demand for freshwater are located primarily in intervening lowlands and surrounding coastal plains. To compensate for this imbalance and the absence of perennial rivers, much of the island has traditionally relied upon groundwater for its water supplies. Springs in the foothills of the mountains are tapped and wells are sunk to exploit the mountain aquifers which run under the plains. The history of using this water goes back over 2000 years, but abstraction accelerated after 1945 with the introduction of pumped boreholes which could extract water at rates of between 25–100 m^3/hour (Konteatis, 1995). Thus, by the 1960s groundwater abstraction was effectively out of control and many private initiatives resulted in aquifer over-pumping, depletion of inland aquifers and saline incursions into coastal aquifers at Morphou, Famagusta and Akrotiri.

The initial response to over-pumping in the 1940s and 1950s was for local irrigation divisions to build small-capacity storage dams on a local demand basis. By independence in 1960, however, it was realised that local initiatives could not solve an island-wide problem and in August 1961 President Makarios, in his inaugural speech to the House of Representatives, set out an integrated water resources policy. This involved: 'a survey of total water resources, the replenishment and conservation of Kyrenian aquifers, the impounding of river water and the supply of piped water to all towns and villages of sufficient quality for domestic use' (Lytras 1993). By 1967 this policy had resulted in the drawing up of a Master Plan for the whole island based on the integrated use of all water resources.

The keystone of the development plan was the construction of two major inter-basin transfer systems in the north and south of the island. The northern system was given priority and was designed to supply irrigation water to the Morphou Plain, which was suffering greatly from over-abstraction of groundwater, and domestic supplies to Nicosia and surrounding villages. The scheme was effectively halted following the Turkish invasion of Northern Cyprus in 1974. Despite the invasion, the impounding of surface water has continued in the south of the island and dam storage capacity rose from 6 million m^3 in 1960 to 197 million m^3 in 1994 – equivalent to approximately 50 per cent of annual runoff (Tsiourtis and Kinler 1995). The reason for this growth in storage has been the success of the second transfer scheme, the Southern Conveyor Project (Fig. 15.7). This provides for the transfer of water from Paphos in the west to Famagusta in the east based upon the construction of a series of dams on the southern slopes of the Troodos Mountains. At the heart of the project is the Kouris Dam with a capacity of 115 million m^3, and the Asprokemos Dam that feeds the Paphos Irrigation Project with a capacity of 51 million m^3, completed in 1983 (Lytras 1993). Water from the Kouris Dam, augmented by the Dhiarizos Diversion Tunnel from the Western Troodos, feeds a 110-km-long main conveyor designed to carry 32 million m^3 for domestic use. At points along the conveyor water is diverted to a number of irrigation projects and conurbations. Originally it was designed to carry water as far as Famagusta. After the partition of the island left Famagusta in the northern Turkish-controlled sector, this part of the plan was 'shelved', but as a consequence of the abandonment of the northern project a spur was added to provide domestic water to Nicosia.

Management of water resources in Cyprus has not only relied upon optimising supplies, successive governments have also pursued strategies of loss reduction and control of per-capita demand. These measures have been summarised by Konteatis (1995) and include:

- Controlling irrigation losses by lining earth channels with concrete, replacing channels with pipes and introducing drip and sprinkler irrigation and water meters.
- Reducing domestic supply losses between reservoirs and consumers from up to 20 per cent down to 8–10 per cent in most cases.
- Groundwater recharge, by constructing storage dams over aquifers and controlled leakage from canals.
- Effluent re-use, which began in 1953 with small plots at Episkopi, Akrotiri and Dhekelia built by the British. Schemes are currently under construction at Limassol, Larnaca,

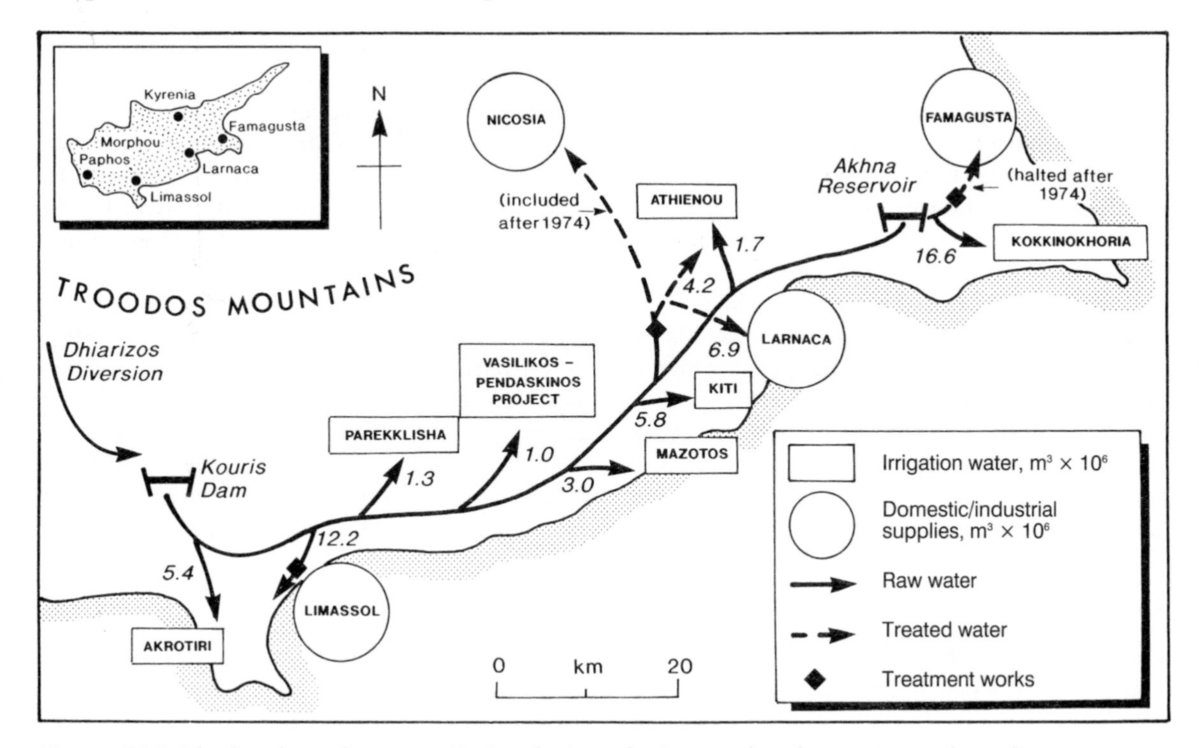

Figure 15.7 The Southern Conveyor Project for inter-basin transfer of water in southern Cyprus
Source: Modified from Pechoux (1989) and Tsiourtis and Kindler (1995)

Paphos, Paralimni and Ayia Napa and may eventually supply up to 40 million m^3 per annum for irrigation.

- Desalination. One plant currently operates on the British Military Base at Dhekelia with a capacity of 1000 m^3/day, but high costs have generally inhibited development although the government has initiated a programme for an additional plant at Dhekelia for domestic use.
- Rain augmentation. This was experimented with in 1972 using cloud seeding with silver iodide.
- Evaporation control. Losses from reservoirs can be up to 15 per cent, but experiments at Yermasoyia Dam using a fatty alcohol as a retarding agent reduced evaporation by 32 per cent at a cost equal to one-third of the water that would have been lost.

Libya: the Great Man-Made River

Perhaps the most ambitious water transfer scheme under development anywhere at present is designed to bring groundwater from the south of Libya to its Mediterranean coast. Unlike the Cypriot schemes, the Libyan transfers do not rely on harnessing renewable resources, but on the effective mining of fossil water. This water lies in a series of seven major aquifers in the Libyan desert that were first identified during exploration for oil in the early 1960s. By 1967 plans were drawn up to exploit the water for irrigated agriculture and when Colonel Gaddafi came to power in 1969 his initial plans were to persuade people to move to farms created around wells at Sarir (Fig. 15.8). However, after a decade of easy living on oil incomes it was not possible to persuade large numbers of people to move from the urban coast to the desert (Bulloch and Darwish 1993). Gaddafi's solution was to shift the water to the people.

The total estimated resources are some 60 000 km^3, most of which accumulated during a pluvial phase between 8000 and 5000 years ago. The plan is to exploit the aquifers in three phases. The first, begun in September 1984 and finished in September 1991, conveys water from 120 wells in the Tazerbo Field (Fig. 15.8) to meet the domestic needs of the coastal towns of eastern Libya and to irrigate 50 000 ha. of farmland (Hillel 1994). The second phase takes a similar quantity of water from a series of well fields in south-west Libya to the western coast, including the capital Tripoli. The final phase will augment supplies from wells at Kufra and allow the extension of the first phase from its terminus at Sirte westwards and from Ajdabiya to Tobruk. In total it is planned to construct 4200 km of pipeline and to irrigate 180 000 ha. near Sirte and 320 000 ha. in the Jabal al Akhdar using an annual abstraction of 2.2 billion m^3. When this yield is compared to estimates of annual recharge that range from 600 000 m^3 to 5 million m^3, it is clear that the project has a finite life. Estimates vary from an official Libyan figure of 50 years to 40 years by the Egyptian government and 100 years by the contractors. As Bulloch and Darwish (1993) point out, however, the truth is that nobody really knows.

In addition to uncertainties over the long-term viability of the project, concern remains over the feasibility of persuading an essentially urban population to become farmers. Disagreement also continues over the ownership of the groundwater. The Kufra aquifer that feeds much of the scheme is believed to extend southwards into Chad and Sudan and eastwards into Egypt. Any depletion must therefore affect water availability in these countries. Such considerations may have been important in Libya's 1973 annexation of the 95-km-wide Azou Strip in northern Chad. Although Libya was eventually forced to withdraw from Chad, the Egyptian border remains a potential area of conflict. Bulloch and Darwish (1993) point out that a particular concern of the Egyptians is that siltation of the Nile behind the Aswan Dam may divert the river towards Libya. These authors claim (pp. 128–9) that if silting does show signs of diverting the Nile, Egypt would be prepared to occupy areas in south-western Libya and/or attempt to divert the flow by construction of massive earth ramparts. Such scenarios must be taken seriously, given the open conflict between the two countries in the 1970s.

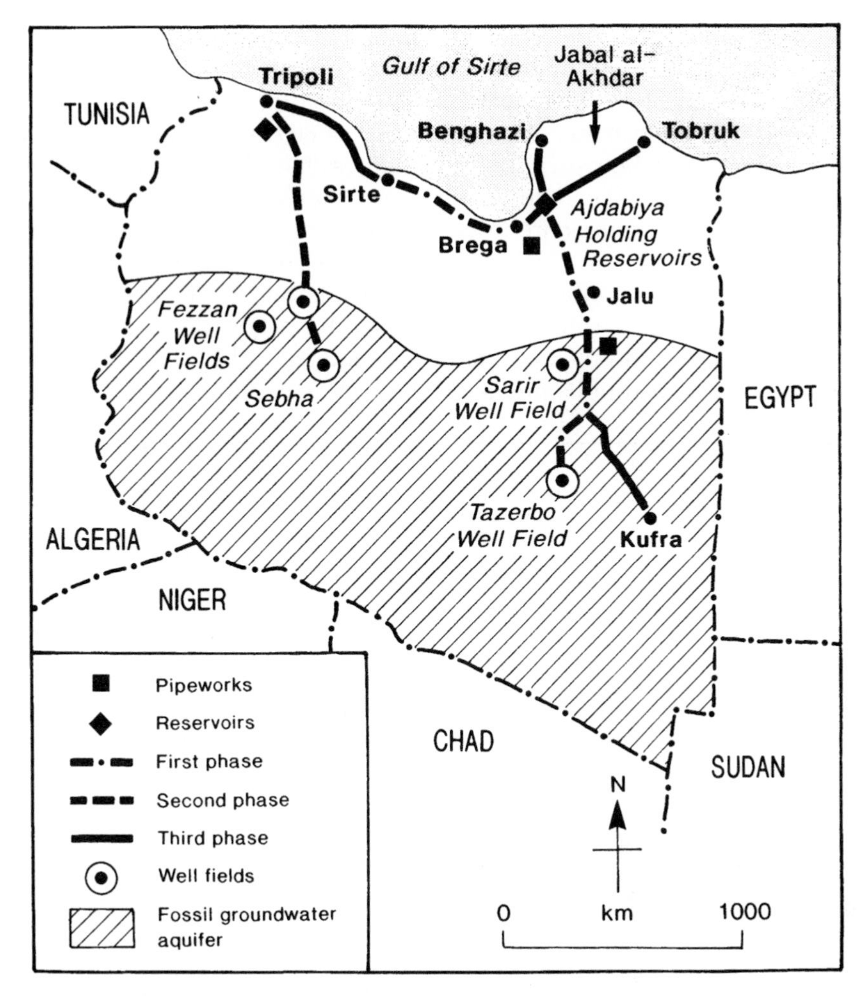

Figure 15.8 The Great Man-Made River Scheme in Libya
Source: After maps in Agnew and Anderson (1992) and Hillel (1994)

Libya and Egypt are not, however, the only countries facing potential conflict triggered by disputes over access to water.

Cross-border transfers: negotiating with neighbours

Many countries of the south-east Mediterranean have less than 500 m^3 of water available per capita per annum and are under severe water stress. Furthermore, it is estimated (Vidal 1995) that by 2025 the amount of water available to each person in the Middle East and North Africa will have dropped by 80 per cent in a single lifetime. To meet these shortfalls countries have consistently looked beyond their recognised boundaries, with far-reaching and sometimes catastrophic consequences. There is a widely held belief, for example, that it is not competition over oil, religion or political ideology that will lead to future conflicts, but disputes over dwindling water resources set against rising populations and economic expectations. This view was first espoused coherently in a volume by Naff and Matson in 1984 – *Water in the Middle East: Conflict or Co-operation?* More recently, it has been recognised at UN conferences, such as that held in Dublin in 1992, that water resources will be critical in determining rates of economic and social development in this area. Specific issues have been investigated in books by Pearce (1992), Bulloch and Darwish (1993) and Hillel (1994), all of whom identify the need for international cooperation and integrated regional planning if demands are to be met and 'water wars' avoided.

Nowhere is the potential for conflict greater than in and around Israel. Ever since its creation in 1948, Israel has intensively exploited available water resources as part of its preoccupation with 'making the desert bloom'. Initially much of this water was drawn from the sandstone Coastal Aquifer (Fig. 15.9), with wells as deep as 800 m. As these wells became increasingly saline during the 1950s, other sources were sought and water was taken from two large springs at Yarkon and Taninim which are fed from the Mountain Aquifer beneath the limestone hills of the West Bank. By the early 1960s abstraction already exceeded recharge for this aquifer and by 1967 an estimated 300 million m^3 per annum were being removed by

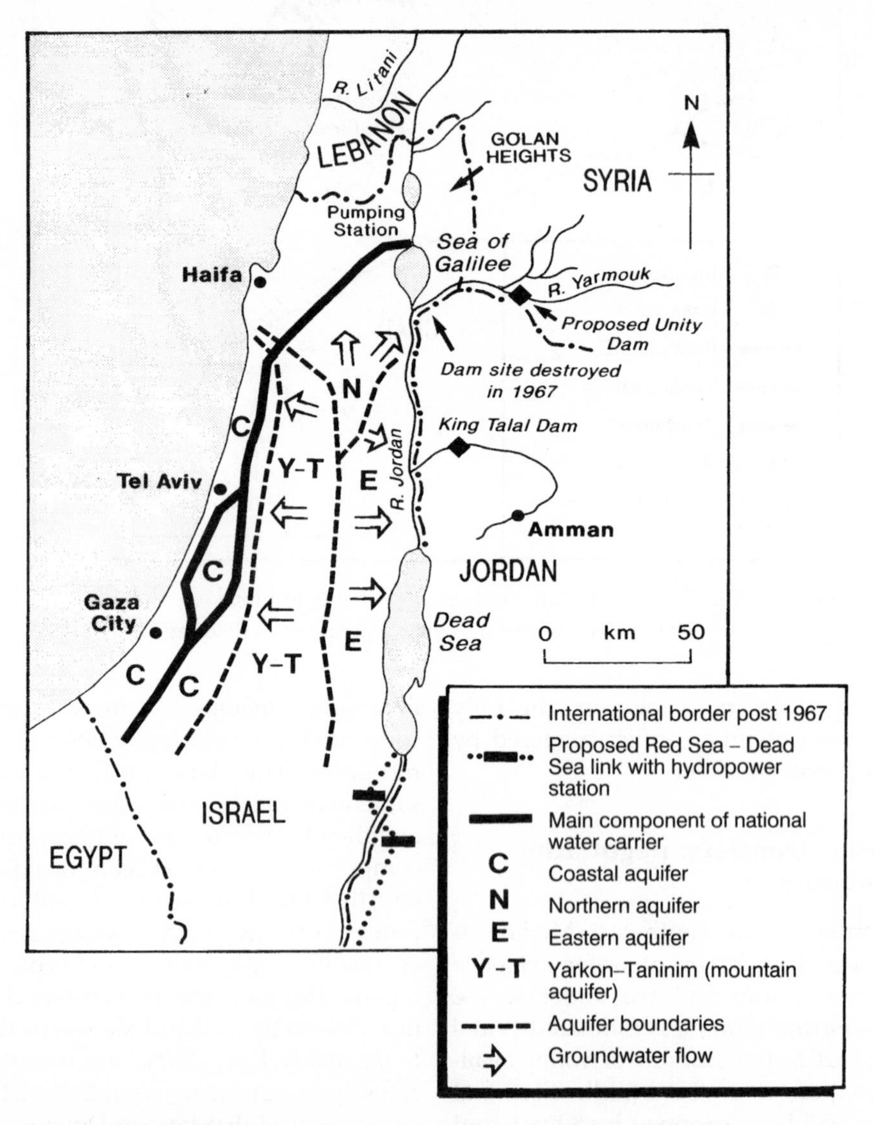

Figure 15.9 Hydrology and water resources in and around Israel

Israel compared to 20 million m^3 per annum by the local population (Pearce 1991a). Israel had, however, already acted to address the imbalance between supply and demand by pumping water 200 m up from the Sea of Galilee to feed the pipeline that runs the length of Israel as the National Water Carrier, designed to carry up to 1 million m^3 per day. Removal of water from the Sea of Galilee on this scale dramatically reduced flow in the River Jordan – the principal source of supply for Jordan. As Pearce (1992, p. 287) noted, 'in 1964, Israel hijacked the waters of the River Jordan ... the seizure happened suddenly and without international agreement'. Since this time the lower Jordan has been nothing more than a saline trickle.

Although Israel agreed that Jordan should continue to have access to 100 million m^3 per annum from the river, this figure was rarely met and Jordan began to look elsewhere for supplies. In addition to aquifers underlying central Jordan, attention was drawn to rivers such as the Zarqa and, especially, to the possibility of constructing a dam on the River Yarmouk which drains into the Jordan just south of the Sea of Galilee. This development was seen as such a threat to flow in the Jordan that it was specifically targeted for destruction by the Israelis during the Six Day War of 1967.

Although the securing of water supplies may not have been the primary reason for prosecuting the 1967 war, it is indisputable that most of the land that Israel annexed continues to be held in part for its hydrological significance (Pearce 1991b). Runoff from the Golan Heights, for example, feeds directly into the headwaters of the River Jordan, while the West Bank is the recharge area for the Mountain Aquifer. After the annexation of the West Bank and Gaza Strip there were severe restrictions on the drilling of new wells by the Palestinian population – ostensibly to protect aquifers already being over-pumped by Israel. Arab organisations such as the Nablus Municipal Water Authority were thus refused permission to tap the deep aquifer beneath the town and at least one hundred West Bank villages are without piped water (Pearce 1991a). In the Gaza Strip, over-pumping of the Coastal Aquifer has reached such a point that the 800 000 population is seriously threatened by salination and pollution of its water supplies, as well as restrictions on quantity that have inhibited agricultural development. The impact of these restrictions is graphically reflected in patterns of water use across the region, whereby annual per capita consumption is 404 m^3 in Israel, 237 m^3 in Jordan and 130 m^3 in the West Bank (Laurance 1993).

Despite all of the restrictions imposed by Israel, the water supply situation remains precarious. By the early 1990s, the National Water Carrier could no longer supply the country's needs and when rainfall fell below average, allocations to farmers were reduced by 50 per cent in 1991 while up to 10 per cent of the Coastal Aquifer had become salinised (Pearce 1991a). Attempts to address this imbalance by, for example, recycling up to one-third of Israel's sewage waste (180 million m^3 per annum) and increased use of saline groundwater beneath the Negev Desert have proved insufficient. At the same time, water quality problems in Gaza grew far worse and new Israeli settlements on the West Bank were extracting more groundwater than ever.

The realisation of the difficulties of achieving a water balance with the resources under their control may be one factor in the change in the political climate in Israel. In the early 1980s, Prime Minister Begin was insistent that an overriding condition for West Bank autonomy was that Israel retained control over water resources. Israel also maintained its threat to bomb any attempt to dam the Yarmouk River by construction of the Unity Dam (Fig. 15.9) proposed by Jordan and Syria. By 1994, however, the Israeli government view had changed to one in which it was agreed that problems of water should be settled in unison, not by haggling over water rights (Kessel 1994). Partly as a response to this change in attitude and the scope for compromise made possible by the peace treaty signed with Jordan, movement finally seems to be taking place towards international cooperation. In October 1994, Israel agreed to allow an extra 200 million m^3 per annum to flow down the River

Jordan and for Jordan to build two new dams to hold this water (Pearce 1995). There is also hope that the Yarmouk scheme may eventually be completed, if peace can be negotiated with Syria, and Israel is now actively promoting cross-border collaboration. One scheme would involve the diversion of water from the Litani River in Lebanon to the headwaters of the Jordan to form part of a Middle East Water Bank. Most ambitious of all the regional projects is a scheme to link the Red and Dead Seas (Fig. 15.9). The World Bank has already conducted a feasibility study (Pearce 1995) that envisaged an initial flow of 1900 million m^3 per annum, settling down to 1200 million m^3 after 10 years. This would be used to generate up to 600 megawatts and, although two-thirds of this would be used to pump water through the system, the remainder would desalinate up to 800 million m^3 annually and be used to trigger an intensive economic development strategy.

All of these developments are symptomatic of an outward-looking cooperative frame of mind that may eventually be the only route to non-military solutions for water resource conflicts in the Middle East and beyond.

Coping with visitors: additional demands imposed by tourism

The rapid growth of international tourism over the last 30 years has posed particular problems for Mediterranean countries, as we saw in the previous chapter. Nowhere are these problems more acute than on small islands with a traditional reliance upon groundwater. Two examples will be examined in this section, Malta and Majorca, each of which has adopted a different strategy to overcome seasonal water shortages.

Malta

Malta is a small (246 km^2), semi-arid, rocky island with a high population density. By 1985 the population was being supplemented annually by over 500 000 visitors who stayed an average of 11 nights, which represented the equivalent of an extra 15 600 residents in addition to the official population of 316 000 (Peplow 1989). By 1991, the number of tourists had risen to 895 000 per annum, set against a rise in the local population to 359 455 (Attard *et al.* 1996). Unfortunately, tourist impact is spread neither evenly in space nor time and the greatest demand is in summer months of least water availability. Nevertheless, water shortage in Malta is not a recent phenomenon. In their review of water provision, Agnew and Anderson (1992, pp. 255–77) noted that Malta developed a 'nationally organised approach to problems of water shortage' early in its history. As early as 1610, a 13-km-long aqueduct was constructed to bring spring water from Rabat to Valetta and from 1866 a comprehensive system of collecting galleries, storage facilities and pumping stations were built to exploit groundwater resources on which the island was almost entirely dependent. The data in Figure 15.10 (A) demonstrate this continued dependence, but also show how the pattern of supply changed between 1966 and 1986. By the early 1960s it was apparent that natural supplies were insufficient to meet the demands of an increasing permanent population and the nascent tourist industry. Thus, initially at a time of cheap fossil fuels, four flash-distillation plants were opened between 1966 and 1968 with a capacity of 20 250 m^3 per day, while a reverse osmosis plant was built at Ghar Lapsi in 1982 to be followed by plants at Marsa, Cirkewwa and Tigne (Agnew and Anderson 1992). The growth of desalination is reflected in the water supply figures for 1986 (Fig. 15.10 (A)), which also show the rapid growth in abstraction from boreholes. This trend followed the implementation of an 'Immediate Action Plan' in the mid-1970s drawn up to counter escalating demand, 'mainly brought about by the ever increasing number of tourists visiting Malta' (Peplow 1989, p. 173). It should be noted that the review of provision occurred after the 1974 rise in oil prices, which made desalination a less economically viable option. None the less, by 1991 desalination accounted for 63 per cent of all potable water on the island (Attard *et al.* 1996).

In addition to providing new sources of supply, Malta has a comprehensive water con-

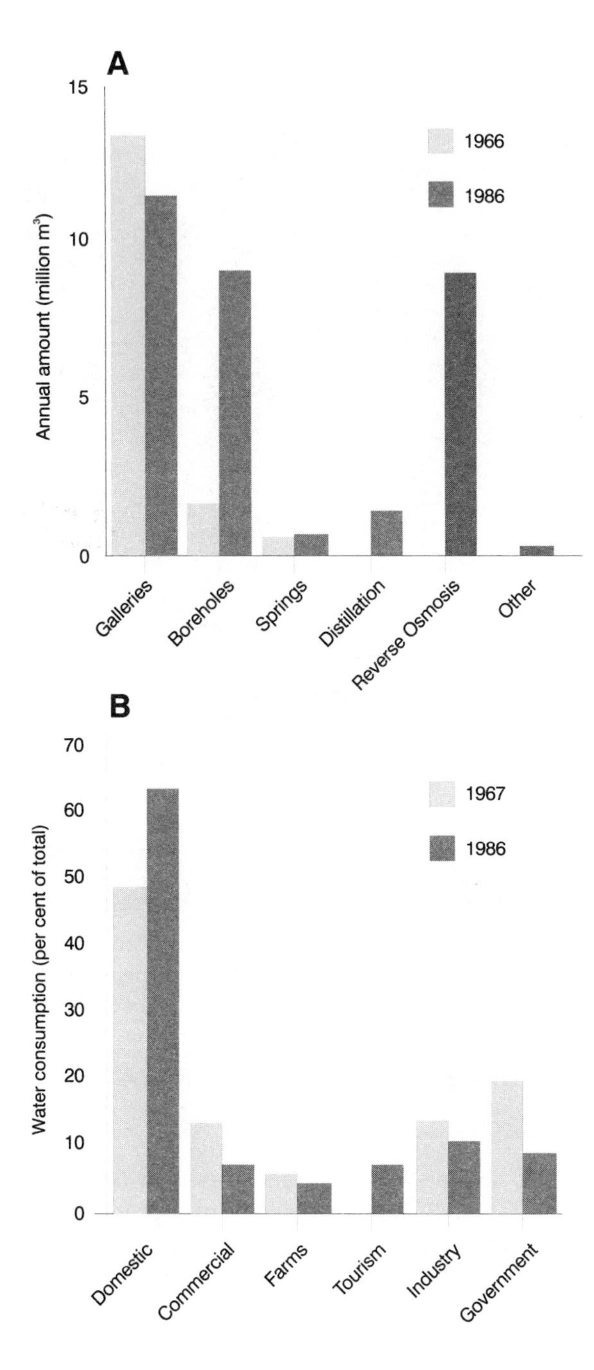

Figure 15.10 (A) Annual water production from different sources in Malta, 1966 and 1986 (B) Water consumption by different sectors in Malta, 1967 and 1986

Source: Data from Agnew and Anderson (1992)

servation strategy. This has included programmes of leak detection and reservoir repair and initiatives such as the sewage treatment and irrigation scheme at Sant' Antrim that can produce 9000 m^3 per day. Despite these measures, leakages continue to exceed 30 per cent and total transmission loss may be as high as 47 per cent (Agnew and Anderson 1992). The need for an efficient distribution network is emphasised in Figure 15.10 (B), which shows that although tourism places an additional, seasonally concentrated burden on supplies, the greatest overall demand and most significant increase has been in domestic supplies to the indigenous population. This indicates a strong underlying trend in increasing demand that questions continued investment in 'technological and non-natural resources ... at a level of expense which must raise questions about the sustainability of the system' (Agnew and Anderson 1992, p. 257).

Majorca

An alternative view of the impact of tourism and government response is provided by the island of Majorca. Although Majorca is for many people the archetypal Mediterranean holiday island it is, firstly, considerably larger than Malta (3650 km^2) and, although much of it is semi-arid, the highest peaks in the Serra de Tramuntana receive an annual precipitation in excess of 1200 mm. The economy of the island has traditionally been dominated by agriculture, and despite a 64 per cent decline in the number of farmers from 72 729 in 1963 to 26 705 in 1983 (Pons 1989), irrigation remains the largest single consumer of freshwater. Most of this demand is met from groundwater (see Fig. 15.2 (A)) and although there are reservoirs in the mountains, supplies are mainly drawn from springs, wells and boreholes which tap limestone aquifers beneath fertile coastal plains.

Like Malta, however, Majorca has reached a point where it can no longer satisfy all-year-round demand from natural resources. Some of this difficulty stems from saline intrusion into coastal aquifers brought about by over-pumping for agriculture and increased local demand for clean piped water. However, it was the rapid growth of tourism that brought water supply to crisis point by the late 1980s (Pons 1989). Since then, the crisis has been exacerbated by the growth of water-demanding activities such as golf (Royle 1996). Thus, whereas in 1981 demand accounted for 92.3 per cent of available supply, by 2010 it is estimated that it will rise to 112.4 per cent (Pons 1989).

The initial response to the worsening situation in the 1980s was to increase abstraction from inland aquifers, construct water treatment plants, campaign for water conservation and search for new aquifers. This has included the exploitation of small groundwater basins from which water is effectively mined. Unlike Malta, however, there has as yet been no introduction of desalination apart from a small plant to the north of Palma, opened in 1995 to treat 30 000 m^3 of contaminated groundwater per day. Instead, it was decided in early 1995 to import water in converted oil tankers from the River Ebro. This water is then transhipped to another tanker moored in Palma harbour from which it is pumped into the island's distribution network. As summer progresses, more water is taken from reservoirs and the effects are amply demonstrated at the island's major reservoir of Gorg Blau (Fig. 15.11). This photograph shows the reservoir in September 1995 when, after summer abstraction, storage had been reduced to approximately 10 per cent of capacity.

FIGURE 15.11 The reservoir at Gorg Blau in the Serra de Tramuntana of Majorca at 10 per cent of capacity in September 1995 after a long hot summer

The policy of importing water is not without its critics. Many in mainland Spain are opposed to exporting scarce resources at a time when they themselves are experiencing water shortage. On Majorca, the expectation is that the cost of water will rise steeply and that increases of up to 90 per cent are predicted once the scheme is fully operational. The success of importation as a strategy thus depends upon the willingness of consumers to pay and upon a relatively nearby partner who is able and willing to sell the water. The drawbacks are the costs, over which the buyer has little influence, and the abdication of strategic control over supplies to the seller in particular and market forces in general.

FUTURE PROSPECTS FOR WATER RESOURCES

There appears to be a consensus that, however serious water supply problems are at present, the situation is set to deteriorate. The important questions are: what are the underlying causes of deterioration; where will pressures be felt most acutely; and what, if any, remedial action can be taken to limit the impacts of any changes?

In the preceding case studies a number of operational solutions to water shortage have been identified. These include:

- Increased and more efficient storage of seasonal surpluses.

- Increased efficiency in water use, especially for irrigation.
- Increased use of treated water.
- Increased use of poor-quality, saline water for irrigation.
- More effective distribution of existing resources.
- Improved water harvesting procedures.
- Leak reduction.

These strategies aim to reduce overall consumption by more efficient supply and/or to retain and utilise maximum quantities of precipitation. However, population and economic growth will ultimately bring many Mediterranean countries to the point where supplies can no longer meet demand. At this point more drastic strategies are required, which may include:

- Importing water.
- Increased use of desalination and other non-conventional sources.
- Rationing of supplies.
- Increased (realistic) charging for water supplies to reduce per capita consumption.
- Economic restructuring to shift resources away from high-demand sectors, especially agriculture, although any reduction in agricultural supplies will limit long-term food production and must be weighed against any future need for food imports (Falkenmark 1986).

Levels of severe water stress have already been reached by some Mediterranean nations (Gleick 1993). Egypt, which currently draws 97 per cent of its water from the Nile which in turn derives 95 per cent of its flow from outside the country, is the prime example. To secure access to this water, Egypt has been active in promoting international treaties that seek to enshrine the rights of downstream nations to flows that originate outside their borders. In pursuit of these rights, Egypt successfully reached an agreement with its immediate neighbour Sudan in 1959, which assumed an annual flow of 84 km^3 and allocated two-thirds of this to Egypt (Gleick 1991). As Pearce (1994) pointed out, however, the agreement left out the seven other upstream countries which contribute most of the flow to the Nile, the only consideration given to them being President Nasser's threat of war if they took any of 'Egypt's water'.

Unfortunately for Egypt, the projected Nile flow has not always been reached since the signing of the treaty and during the 1980s average annual flow was only 76 km^3 and fell to a low of 42 km^3 in 1984 (Pearce 1994). With the deterioration of political relations between Egypt and Sudan during the 1990s, the value of earlier agreements must be called into question, particularly as Sudan strives to improve its own water storage capacity and increase agricultural production. Similarly, Ethiopia has periodically discussed the possibility of taking more water from the Blue Nile (Gleick 1992) and as recently as 1990, Egypt warned Ethiopia of dire consequences when it claimed that Israeli engineers were investigating the damming of Lake Tana at the head of the Blue Nile (Pearce 1994). Although Egypt is an exceptional case, it is only one among several Mediterranean countries that draw a high proportion of their water supplies from outside their borders; Syria derives 79 per cent from its neighbours, Albania 53 per cent, Portugal 48 per cent and Israel 48 per cent (Gleick 1992).

It is obvious that the Nile, like other shared rivers around the Mediterranean, will only be managed effectively and conflict avoided if rights of access are guaranteed by international agreement along its length. Even with international agreement, extra water resources cannot be manufactured, and thus many import-dependent nations will, like those who have to rely on their own resources, see future reductions in per capita water availability.

Figure 15.12 shows projected water availability for 2025 in some of the driest Mediterranean countries given present population trends. These data suggest that in many countries supplies will fall below 1000 m^3 per capita per annum and in some, below 500 m^3. This latter level was suggested by Falkenmark (1986) as a possible minimum in semi-arid regions for the sustainable development of a modern society employing sophisticated water management techniques.

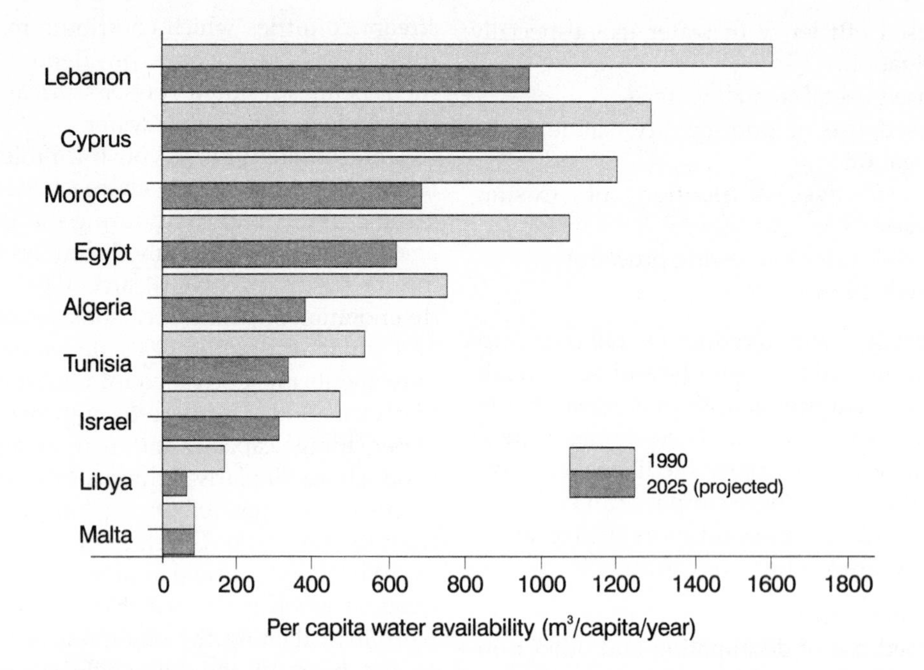

FIGURE 15.12 Per capita water availability in selected countries for 1990 and estimated for 2025

Source: Data from Gleick (1992)

This is highlighted by the case of Israel which, despite stringent and innovative water management, experiences great difficulty in surviving on 470 m³ per capita per annum. These data also identify the increasing polarisation between those countries, mainly in the southern and eastern Mediterranean, in which water shortage is likely to be a constraint on development, and the better-watered countries of southern Europe. In fact, Grenon and Batisse (1989) manage to identify three groups of countries.

1. Countries where water availability will remain adequate until 2025 and beyond. These include countries such as France, Italy and the former Yugoslavia with low rates of population growth; and Albania, Turkey and Lebanon with high population growth but considerable reserves of fresh water.
2. Countries with adequate present supplies which will experience a considerable fall in per capita resources. These include Spain, Morocco, Algeria and Cyprus and they will require development of new resources and internal transfers to maintain supplies. Any increase in per capita demand will place these countries under considerable water stress.
3. Countries where current availability is limited or negligible and where by the year 2000 demand will already have exceeded supply. The imbalance will be compounded by high rates of population growth in countries such as Syria, Egypt and Libya, but will also pertain in countries with low (Malta) and medium (Israel and Tunisia) population growth.

This classification is based principally on projected population change and present-day hydrological conditions. Any predictions should also consider economic, political and environmental changes that will be superimposed upon technological and demographic developments. Within this framework, Grenon and Batisse (1989) recognise three economic scenarios.

1. Slow economic growth, where budget constraints and severely reduced investment make it difficult to meet consumer demand,

especially in North Africa and the Middle East. This will lead to widespread local deterioration in water quality, especially in northern, industrial economies.

2. Stronger economic growth, but with belated concern for the environment, would lead to improved supplies, largely by increasing conventional water management schemes. Wastage would increase and quality decrease as only improvement schemes with immediate advantages to consumers would be implemented. Some non-renewable resources would be depleted in the south and east of the basin and failure to provide sewerage and water protection may lead to deterioration in offshore waters.
3. Medium to strong economic growth, combined with active environmental safeguarding and water management. Supply will be matched to demand through efficiency gains, recycling, re-use and pricing strategies while the environment is protected by sewage treatment and purification which also decreases drinking water costs.

These economic scenarios and their consequences assume that long-term annual precipitation remains constant and that short-term fluctuations in available water resources are limited. Any departure from these assumptions towards an overall reduction in water availability or increased variability in supply must place additional burdens upon already stretched resources. Countries that rely upon fossil groundwater will inevitably experience a reduction in available resources, but the greatest long-term threat across the whole Mediterranean is posed by possible climatic and sea-level changes driven by global warming.

The wider implications of climatic change have been investigated in Chapter 3 of this volume and in a number of recent books (e.g. Jeftic *et al.* 1992; 1996). In terms of water resources, Gleick (1992) pointed out that in studies from Greece a temperature increase of 2–4° is likely to reduce runoff by up to 20 per cent. The greatest eventual threat from any warming is likely to arise from increased evaporative loss and a corresponding rise in demand by humans and natural ecosystems. However, the major immediate threat is the increased uncertainty that warming will bring to rainfall and runoff. Many global climatic models have identified the Middle East as an area where the nature of future changes is particularly uncertain and in which the degree of change could be severe. A review of these models by Gleick (1992) showed, for example, that in the Litani and Jordan catchments predictions of precipitation change varied from –14 to +48 per cent and in his own study of the Nile (Gleick 1991) he identified the possibility of even greater negative fluctuations in runoff.

Any increase in uncertainty over supplies must in turn increase the difficulty of successfully managing water resources. This encompasses problems of accommodating more severe and frequent droughts as well as designing hydrological structures to withstand floods whose magnitudes cannot be predicted from previous runoff patterns. Global warming, through sea-level rise, may also indirectly affect water availability in areas such as the Nile Delta. Because of the deep and wide seaward margin of the Nile Delta aquifer, sea water naturally intrudes beneath the delta under a high potential head. At the base of the aquifer this intrusion already extends some 63 km inland and some increased salinity can be detected as far as 108 km from the coast. This gives a dispersion zone of 45 km which, if sea level were to rise by 50 cm, would increase by an estimated 9 km (Sherif 1995). When this is combined with a 1 km shift in the sea water boundary, it would move the overall saline incursion inland by up to 10 km.

Because of the supra-national nature of the possible causes of change and sources of water supply, the only route by which future conflicts over water resources can be avoided is by negotiation within and between countries. If treaties ensuring the equitable distribution of freshwater are not negotiated, agricultural and industrial development will be curtailed, individual health and environmental quality will be reduced and politicians will be pressured to secure access to supplies by whatever means available. Given the geographical spread of the problems, it is advisable that disputes are settled within the context of international law and with the involvement of intergovernmental organisations such as the United Nations. As Gleick (1992) points out, however, numerous treaties already exist, many of which are under strain from economic and political changes amongst the signatories. Moreover, none of them either explicitly include or are likely to be effective in incorporating the impacts of climatic change. Within this context, what is clear is that

> future climatic changes effectively make obsolete all our old assumptions about the behaviour of water supply. Perhaps the greatest certainty is that the future will not look like the past. We may not know precisely what it will look like, but changes are coming. (Gleick 1992, p. 138)

A similar prognosis can, unfortunately, be applied to political, social and economic changes around the Mediterranean Basin and their likely impact on water resources.

REFERENCES

AGNEW, C. and ANDERSON, E. 1992: *Water Resources in the Arid Realm*. London: Routledge.

AJUNTAMENT DE CALVIÀ 1995: *Calvià 21, local agenda*. Mallorca: Ajuntament de Calvià.

AMBROGGI, R.P. 1977: Freshwater resources of the Mediterranean Basin. *Ambio* **6**, 371–3.

ATTARD, D.J. *et al.* 1996: Implications of expected climatic changes for Malta. In Jeftic, L., Keckes, S. and Pernetta, J.C. (eds), *Climatic Change and the Mediterranean, Vol. 2*. London: Arnold, 322–30.

BEN-ASHER, J. 1995: Soil and water contamination in arid coastal zones. In Tsiourtis, N.X. (ed.), *Water Resources Management under Drought or Water Shortage Conditions*. Rotterdam: Balkema, 235–40.

BULLOCH, J. and DARWISH, A. 1993: *Water Wars: Coming Conflicts in the Middle East*. London: Victor Gollancz.

EVENARI, M., SHANAN, L. and TADMOR, N. 1971: *The Negev*. Cambridge, MA: Harvard University Press.

FALKENMARK, M. 1986: Fresh water: time for a modified approach. *Ambio* **15**, 192–200.

FALKENMARK, M. and LINDH, G. 1993: Water and economic development. In Gleick, P.H. (ed.), *Water in Crisis: A Guide to the World's Fresh Water Resources*. Oxford: Oxford University Press, 80–91.

GARCÍA de JALON, D. 1987: River regulation in Spain. *Regulated Rivers: Research and Management* **1**, 343–8.

GILBERTSON, D.D. 1986: Runoff (floodwater) farming and rural water supply in arid lands. *Applied Geography* **6**, 5–11.

GLEICK, P.H. 1991: The vulnerability of runoff in the Nile Basin to climatic changes. *The Environmental Professional* **13**, 66–73.

GLEICK, P.H. 1992: Effects of climatic change on shared fresh water resources. In Mintzer, I.M. (ed.), *Confronting Climatic Change: Risks, Implications and Responses*. Cambridge: Cambridge University Press, 127–40.

GLEICK, P.H. 1993: Fresh water data. In Gleick, P.H. (ed.), *Water in Crisis: A Guide to the World's Fresh Water Resources*. Oxford: Oxford University Press, 115–453.

GRENON, M. and BATISSE, M. 1989: *Futures for the Mediterranean Basin: The Blue Plan*. Oxford: Oxford University Press.

HILLEL, D. 1994: *Rivers of Eden: The Struggle for Water and Quest for Peace in the Middle East*. Oxford: Oxford University Press.

JEFTIC, L., KECKES, S. and PERNETTA, J.C. (eds) 1996: *Climatic Change and the Mediterranean, Vol. 2*. London: Arnold.

JEFTIC, L., MILLIMAN, J.D. and SESTINI, G. (eds) 1992: *Climatic Change and the Mediterranean, Vol. 1*. London: Edward Arnold.

KESSEL, J. 1994: Realistic new attitudes reduce threat of water wars. *The Guardian*, 25 July.

KONTEATIS, C.A.C. 1995: Integrated water resources management in Cyprus under water shortage conditions. In Tsiourtis, N.X. (ed.), *Water Resources Management under Drought or Water Shortage Conditions*. Rotterdam: Balkema, 79–86.

LAURANCE, B. 1993: Pooled water leaves Palestine parched of success. *The Guardian*, 30 September.

LYTRAS, C. 1993: Developing water resources. In Charalambous, J. and Georghallides, G. (eds), *Focus on Cyprus*. London: University of North London Press, 123–33.

MILLIMAN, J.D., JEFTIC, L. and SESTINI, G. 1992: The Mediterranean Sea and climate change: an overview. In Jeftic, L., Milliman, J.D. and Sestini, G. (eds), *Climatic Change and the Mediterranean, Vol. 1.* London: Edward Arnold, 1–14.

NAFF, T. and MATSON, R.C. 1984: *Water in the Middle East: Conflict or Co-operation?* Boulder, CO: Westview Press.

PEARCE, F. 1991a: Wells of conflict on the West Bank. *New Scientist,* 1 June, 35–9.

PEARCE, F. 1991b: Rivers of blood, waters of hope. *The Guardian,* 6 December.

PEARCE, F. 1991c: Africa at a watershed. *New Scientist,* 23 March, 34–40.

PEARCE, F. 1992: *The Dammed: Rivers, Dams and the Coming World Water Crisis.* London: The Bodley Head.

PEARCE, F. 1994: High and dry in Aswan. *New Scientist,* 7 May, 28–32.

PEARCE, F. 1995: Raising the Dead Sea. *New Scientist,* 22 July, 32–7.

PECHOUX, P.-Y. 1989: Le problème de l'eau à Chypre: aménagement hydraulique intégré ou concurrence entre tourisme et agriculture. In Busuttil, S., Villain-Gandossi, C., Richez, G. and Sivignon, M. (eds), *Water Resources and Tourism on the Mediterranean Islands.* Malta: Foundation for International Studies, 135–53.

PEPLOW, G. 1989: The effects on water production in islands with high population and tourist densities: the case of Malta. In Busuttil, S., Villain-Gandossi, C., Richez, G. and Sivignon, M. (eds), *Water Resources and Tourism on the Mediterranean Islands.* Malta: Foundation for International Studies, 171–80.

PONS, B.B. 1989: Agua y turismo en las Islas Baleares. In Busuttil, S., Villain-Gandossi, C., Richez, G. and Sivignon, M. (eds), *Water Resources and Tourism on the Mediterranean Islands.* Malta: Foundation for International Studies, 17–41.

PRINZ, D. 1995: Water harvesting in the Mediterranean environment: its past role and future prospects. In Tsiourtis, N.X. (ed.), *Water Resources Management under Drought or Water Shortage Conditions.* Rotterdam: Balkema, 135–44.

ROYLE, S.A. 1996: Mallorca: the changing nature of tourism. *Geography Review* **9**(3), 2–6.

SHERIF, M.M. 1995: Global warming and groundwater quality. In Tsiourtis, N.X. (ed.), *Water Resources Management under Drought or Water Shortage Conditions.* Rotterdam: Balkema, 3–9.

SMITH, F. 1994: Water wars threaten to divide Spaniards. *The Guardian,* 4 September.

STANNERS, D. and BORDEAU, P. (eds) 1995: *Europe's Environment: The Dobrís Assessment.* London: Earthscan.

TSIOURTIS, N.X. and KINDLER, J. 1995: Operational water resources allocation under water shortage conditions via linear programming simulation model. In Tsiourtis, N.X. (ed.), *Water Resources Management under Drought or Water Shortage Conditions.* Rotterdam: Balkema, 281–91.

VIDAL, J. 1995: The water bomb. *The Guardian,* 8 August.

16

FORESTS, SOILS AND THE THREAT OF DESERTIFICATION

HAZEL FAULKNER AND ALAN HILL

INTRODUCTION

Grenon and Batisse (1989) state: 'Of all the forest systems in the world, those round the Mediterranean ... have been the most degraded by human action' (p. 208). In this chapter, the history, context and implications of this statement are explored, initially by outlining the main types of vegetation and of soil, and the history of soil and forest use. We then focus on the threat of desertification in the basin from two sets of influences. First, 'anthropogenic influences' are discussed, including the threat to forests and soils from changing patterns of land-use. Second, the effects of global warming on forest and soil management will be considered under the heading 'climatic influences'. Next, Mediterranean soil degradation and the mitigation of erosion will be discussed. The threat to Mediterranean soils and forests from fire risk is considered separately in the final section of the chapter.

PATTERNS OF VEGETATION DISTRIBUTION

The present-day vegetation in the Mediterranean basin has evolved in response to an interplay between climatic, edaphic, floristic and anthropogenic factors (Le Houérou 1990). The governing climatic controls are mean annual temperature and its range, precipitation total and its distribution, and the ratio between precipitation and potential evapotranspiration (called the P/PET ratio). These factors combine to influence both drought stress and cold(frost) stress. Le Houérou (1992) argues that cold stress is 'a potent discriminating factor of vegetation distribution patterns, crop selection and land-use in the Mediterranean', exerting a significant restriction in areas of high altitude. Edaphic factors such as topography can influence vegetation response by exerting an altitudinal control, but geology and topography also influence soil development, affecting nutrient availability and plant response indirectly. Floristic factors include those paleoclimatic and paleogeographic patterns that have both enhanced and constrained biological evolution – Quezel and Barbaro (1982) suggest that 15 000 species of flowering plants are now present in the Mediterranean, a response to a long and complex history of species migration and diversification. Finally, it could be argued that the intensive activities of man and animals (anthropogenic activities) in the Mediterranean are likely to influence ecosystems more than in most other climates.

NATIVE VEGETATION TYPES

Mediterranean vegetation types were originally classified on a climatic and edaphic basis by Emberger (1930). The many subsequent variations on his theme are reviewed by Daget (1977) and Le Houérou (1990; 1992). We use a method here developed by Mazzoleni *et al.* (1992) who identify seven broad vegetation types on the basis of two principal components. The display of these types against their components is illustrated in Figure 16.1. The first (component I) distributes vegetation zones along a precipitation gradient, with positive scores for low-rainfall deserts and negative scores for cold, high-rainfall regions. Reflection shows that the extent to which the winter precipitation maximum is 'pinched out' across the rainfall regime diagram is associated with the prevalence of semi-arid conditions and xerophytic desert species. The second axis of this diagram (component II) reflects large seasonal differences in rainfall and temperature regime, with high scores on Figure 16.1 indicating a high degree of variability. In many texts this is referred to as 'Mediterraneity', defined as climatic variability sufficient to maintain a winter precipitation maximum and good spring regeneration, yet with summer temperatures which risk summer drought stress and winter temperatures which risk cold (frost) stress.

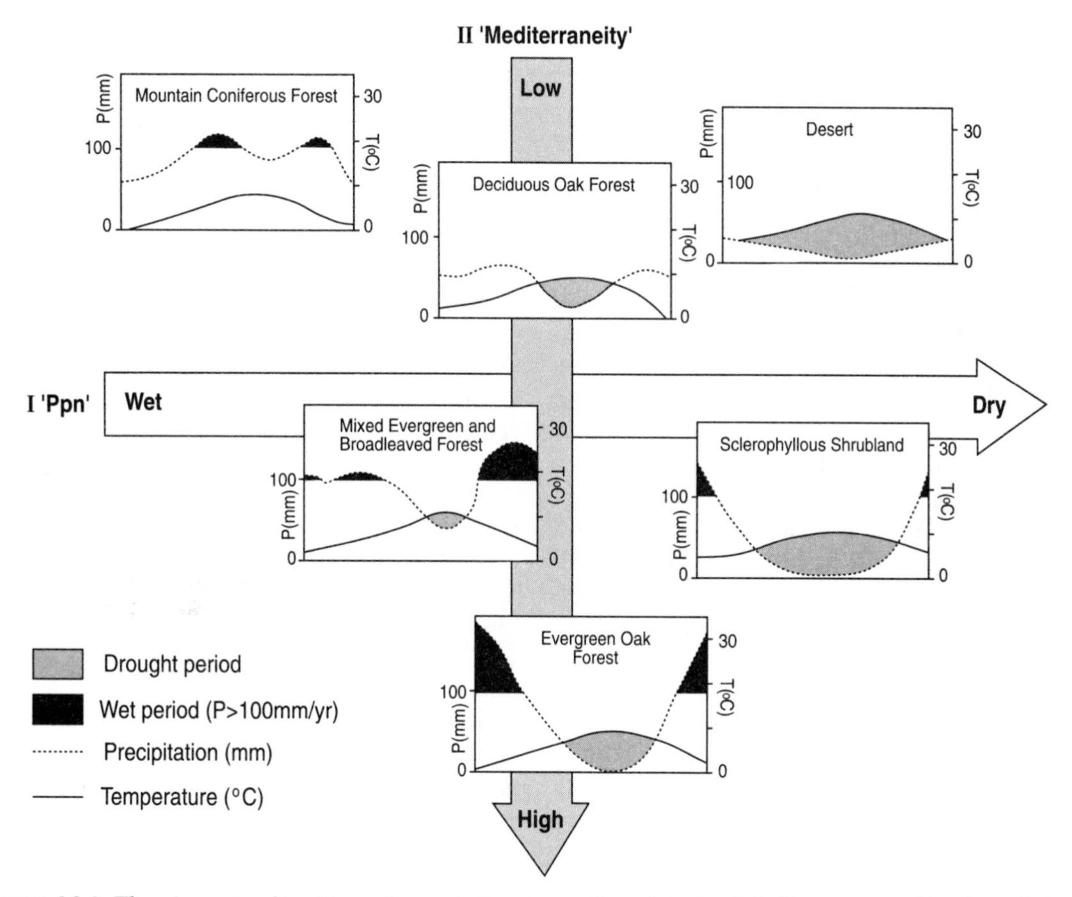

FIGURE 16.1 The six main climatic and vegetation types found in the Mediterranean, distributed along two principal components. *Component 1* = precipitation gradient; *Component* 2 = 'Mediterraneity'.

Source: Modified from Mazzoleni *et al.* (1992)

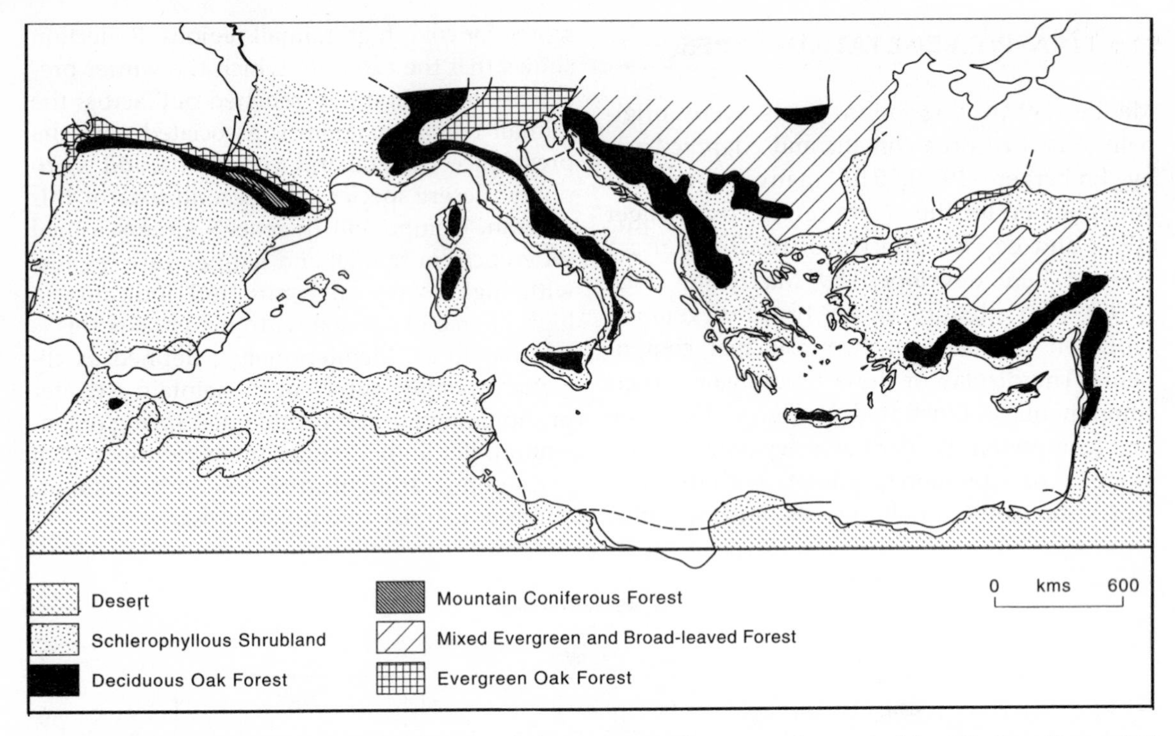

FIGURE 16.2 The geographical distribution of the six vegetation types that were identified on Figure 16.1 across the Mediterranean Basin

The geographical distribution of Mazzoleni's (1992) six main Mediterranean vegetation zones is illustrated on Figure 16.2, which shows that general vegetation belts are laid out along latitudinal as well as altitudinal gradients. Mountain Coniferous Forest occurs in areas with rainfall between 800 and 1500 mm, very low winter temperatures and three months of hard frost. Apart from occasional chestnut and beech, the vegetation in this zone is predominantly coniferous. Deciduous Oak Forest dominates in areas which also have rather high rainfall, (600–2500 mm), but since winters are milder than in the first category, the natural deciduous forests are predominantly oak. Evergreen Oak Forest also has relatively mild winters, but the occasional light frosts in early January mean that evergreens dominate over deciduous species. Rainfall varies from less than 400 to over 1500 mm, meaning that sclerophyllous shrubs and trees are also present. In the Broad-leaved Forests of lowland, low latitude zones, the rainfall regime is Mediterranean but the temperature regime is tropical, and since frost never occurs some tropical species can be found. The Desert zone is made up of steppes, which may be partly grassed, dominated by dwarf xerophytic shrubs and possibly some succulents.

The final zone is Sclerophyllous Shrubland, which Grenon and Batisse (1989) and Le Houérou (1992) describe as the most typical Mediterranean vegetation zone. Figure 16.2 shows its occurrence around large areas of the basin. In this broad zone, rainfall can vary from 350 to 1500 mm and winter temperatures are cool to cold. Depending on which rainfall total applies, the vegetation can vary from semi-arid to hyperhumid, i.e. dominant species are either sclerophyllous or acicular. Four species of oak (*Quercus ilex*, *Q. coccifera*, *Q. suber* and *Q. calliprinos*) are found most commonly alongside a large variety of shrubs, and two species of pines (*Pinus halepensis*, *P. brutus*). The vegetation of this zone is often referred to as *maquis*, defined by Tomaselli (1977) as 'a high plant stand (>2m), generally thick, of evergreen sclerophyllous woody plants, the part above ground not being clearly differentiated as between trunk and foliage, but whose foliage usually extends well down to the base' (p. 42). Figure 16.3 is a photograph of typical maquis vegetation.

Maquis is traditionally viewed as a stage in the degradation (by cutting, burning and grazing) of the broad-leaved evergreen forest, or sometimes as a transitional stage towards that climax. The term *garrigue* is used in an approximately equivalent way by French authors, although Bridges (1978) suggests the usage is restricted to scrub on calcareous soils. *Matorral* is the widely adopted equivalent Spanish term, now commonly applied to all situations where degradation of Mediterranean maquis is well advanced (Tomaselli 1977). Several texts outline the ecological variety of the maquis (Di Castri and Mooney 1973; Di Castri *et al.* 1980; Kruger *et al.* 1983). In the next section, its conservation status is considered more fully.

FIGURE 16.3 Sclerophyllous shrubland or maquis is regarded by many as the most typical Mediterranean vegetation type. It consists of a densely packed mantle of plants up to 2–3 metres in height, as shown in this picture from Majorca.

THE INFLUENCE OF HUMAN ACTIVITY ON MEDITERRANEAN VEGETATION: ANTHROPOGENIC INFLUENCES

The historical use of Mediterranean forests

This huge diversity in native vegetation types has been modified historically by differing land-use strategies, which reflect many centuries of changing socio-economic conditions across the Mediterranean Basin. Delano-Smith (1979) explains that most of Mediterranean Europe was closely forested until the fourteenth or fifteenth century AD, but suggests that Neolithic, Bronze Age and then Roman forest clearance for grazing ensured constant pressure preventing forest regeneration to climax. Early and Classical use of the forests for charcoal and other industrial uses stripped many forests to maquis or heath by the sixteenth century. However, the clearance was piecemeal. Where the deciduous Holm Oak forests were not subject to clearance or degradation, oak, elm, poplar, ash and sycamore forests could still be found in closed canopy stands, particularly in moist valley-bottoms away from mountainous locations. Delano-Smith (1979) vividly describes the 'fierce forests' of medieval Tuscany in such terms, but also recounts in contrast the deleterious consequences of excessive mineral extraction in the Barbagia and Iglesiente districts of Sardinia during Roman and early medieval times, these once-verdant areas now 'wild and desolate'. Other examples of early industrial degradation of the forests include land clearance in Corsica, Languedoc, Provence, and on the Greek island of Delos (Thirgood 1981). Agropastoralism has also been generally accredited with vegetational degradation and soil erosion, especially in the southern Mediterranean Basin, the argument being that in all these circumstances the pervasive maquis and high slope sediment yield act as testimony to the degraded state (see Chapter 4).

Anthropogenic degradation of contemporary forests

Based on the present pattern of residual forestation (Figure 16.2), Le Houérou (1992) distinguishes between the situation in the underpopulated northern (Euro-Mediterranean) basin, and that in the over-populated southern (Afro-Asian Mediterranean) basin. In the northern basin, farmland represents some 36 per cent of the overall land area, rangelands 22 per cent, waste and non-agricultural land 13 per cent, and forests and shrubland 29 per cent. In these seven Euro-Mediterranean countries, there has been a 14 per cent increase in the area of forests between 1865 and 1985 (0.7 per cent per annum), although these afforested stands are sometimes marginalised and neglected due to lack of labour and financial resources (Grenon and Batisse 1989). By contrast, the southern Mediterranean countries experienced a reduction of 13 per cent in the area in forests and shrublands over the same period, cropland increasing by 5 per cent. Le Houérou (1992) argues that the surface area in crops is now 3.5 times what it was in the 1950s, an over-exploitation by poor and growing populations. Stands which cannot regenerate are under threat here from total extinction, particularly in coastal areas because of urbanisation, industrialisation, fire, and too many visitors (Grenon and Batisse 1989, p. 206). The arid steppe rangelands are also badly affected. In the 14 Afro-Asian countries forest cover is now only 6.6 per cent by area, with quite wide variability around this average. This 'resource depletion' has been documented in depth by Thirgood (1981), and the hand of man in the resulting degradation is more or less confirmed by the several contemporary case studies in Fantechi and Margaris (1986). Blumler (1993) observes that 'today, the (southern) Mediterranean and Near East ... have the appearance of used-up land' (p. 287).

In 1975, the Environment Programme of the UN/EEC's 'Mediterranean Action Plan' considered Mediterranean forest degradation under a series of economic and population growth projections up to 2025 (Grenon and Batisse 1989). A chain of forest degradation is suggested in response to these projections (Fig. 16.4), from natural closed forest to managed forest, open forest, forest fallow, bushy woodland and maquis. Grenon and Batisse argue that apart from those areas in the northern basin where planned reafforestation has taken place, native forests in the Mediterranean rarely exist today as mature closed stands, and are therefore degrading and unstable. Even in the northern Mediterranean Basin where some percentage increase in overall afforestation has begun, these rather pessimistic models do not hold out much hope for the preservation of the *status quo*. For instance, under a scenario of slow economic growth, the Blue Plan predicts 50 per cent forest removal of present stands by 2025, and in the southern part of the Mediterranean Basin, only the speed of degradation towards 100 per cent maquis rangelands is open to debate.

Agricultural change and the sustainable ecosystem

Since the dry spell between 1880 and 1920, during which extreme stress brought many ecosystems close to the threshold which supports any sort of normal plant life, the fragility of Mediterranean ecosystems has been appreciated as a fact of normal agricultural life. But the parameters of this agricultural lifestyle are rapidly changing. Once again, it is convenient to divide the Mediterranean region into northern and southern zones when discussing agricultural land use, as we saw in both Chapters 9 and 13. Much of the northern Mediterranean is rationalising the amount of land used for agricultural purposes as efficiency in production improves. As European agricultural policies change, there has been a general migration of the rural population of the northern Mediterranean towards urban centres. This has resulted in the abandonment of traditional, less accessible smallholdings. The ecological damage associated with urban development, deforestation and rural exodus from agricultural smallholdings is further exacerbated by climatic changes. To give an example, abandonment of traditional bench terracing in southern Spain has caused a loss of organic status, reduced soil moisture storage and structural properties and a lowering of infiltration rates (Douglas *et al.* 1994; Faulkner 1995). Although some studies suggest that maquis soon re-establishes itself on abandoned terraces, where terrace walls are no

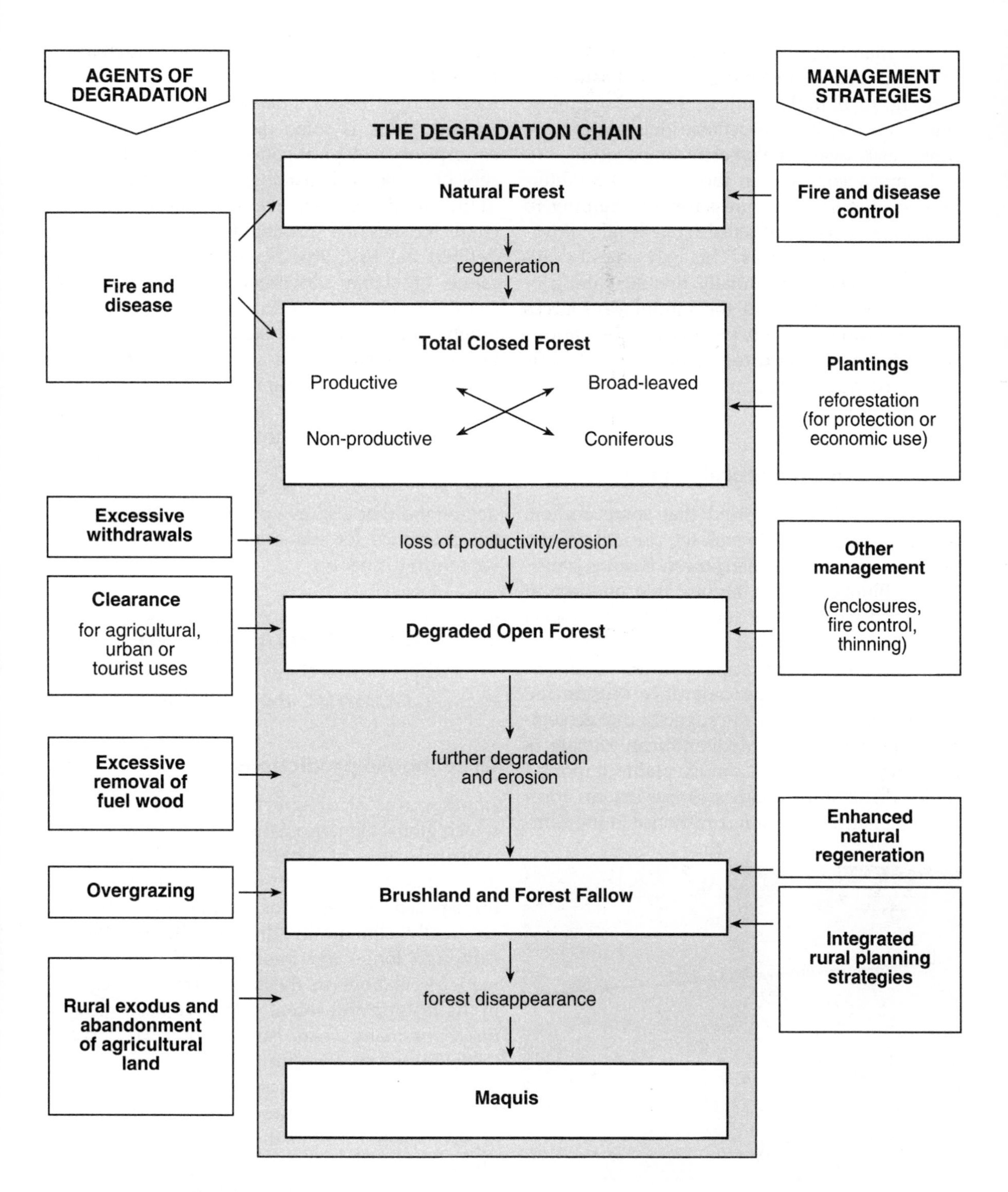

FIGURE 16.4 The classical view of Mediterranean forest degradation. The points in the degradation chain most at risk from various threats, and those points most suitable for a range of remedial management strategies are identified.

longer maintained abandonment produces soil loss and bench breaching during flashfloods (see Fig. 16.5). In the context of global warming, the thresholds for irreversible degradation are more easily overcome on abandoned land.

In many countries in the southern Mediterranean, demographic pressures are helping to fuel an expansion of subsistence-level cultivation into marginal areas. This may often be cultivation which is essentially non-sustaining of soil organic status, and the minimal resources there are can be quickly exhausted. The various ways in which anthropogenic and climatic causes of degradation interact are dealt with later in this chapter.

Other views on 'natural vegetation'

It needs to be emphasised that some authors question the extent to which the present vegetation mosaic in the Mediterranean Basin is transitional. Blumler (1993) argues that models in summer-dry regions are 'topsy-turvy with respect to reality', pointing out that these views are based on an uncritical application of succession models that have better relevance outside the Mediterranean climate. He suggests that ecological instability in the Mediterranean climate is inherent; some of the annual plants imported under disturbance regimes in the maquis form part of a permanent climax response to the summer drought stress (Mediterraneity) that characterises the climatic regime, because seeds are the most drought-resistant plant organ.

FIGURE 16.5 Breached terrace in the Sorbas Basin, Almeria province, south-east Spain, showing the evacuation of sediment from behind the terrace

Since there is some dissent concerning the concept of 'ecological climax', perhaps the concept of 'ecological sustainability' is more useful in the Mediterranean context. The diagnostic characteristics of a sustainable ecosystem can be defined as, first, the persistence of a varied range of stable convergent eco-forms, sufficiently resilient to return to an equilibrium when subjected to extreme climatic or human-induced pressures; and second, the ability to regenerate sufficiently rapidly to meet the needs of the present usage and future demands. Blumler (1993) argues that from this point of view, much of the area under maquis is sustainable, and should not be viewed as evidence of degradation, or a source of concern. This debate has relevance for erosion management, which we return to later on.

GLOBAL WARMING AND MEDITERRANEAN VEGETATION: CLIMATIC INFLUENCES

Greenhouse predictions

Another area of disagreement is the extent to which global climatic change can be viewed as a principal cause of what is commonly referred to as 'desertification'; in other words, how far undisputed patterns of global degradation are really climatically driven (Thomas 1993). Although long- and mid-term patterns of climatic oscillations in the Mediterranean region in the historic and recent past have been recognised for many years (see Chapter 3), Berger (1986) argues that within the next 100 years we are likely to see an increase in aridity within the Mediterranean Basin due mainly to a doubling of atmospheric carbon dioxide associated with human industrial activities in the northern hemisphere – the so-called 'greenhouse effect'. As the Mediterranean is a boundary region, lying between the dry climate of the high pressure subtropical belt of North Africa and the sub-oceanic climate of the cool temperate zone in Northern Europe, it is likely to be severely affected by any migration of the global climatic belts (Francis and Thornes 1990a).

By the middle of the next century some Global Climatic Models (GCMs) forecast a temperature increase of between 3 and 4°C, with a resulting increase in evapotranspiration of around 200 mm (Imeson and Emmer 1992), and a loss of productivity in the Mediterranean's semi-arid zones of around 150–300 kg DM/ha/yr, or 10 per cent. Le Houérou (1993) argues that feedback loops are likely to occur between permanent plant cover, biomass and productivity as a result of the worsening of an already critical soil water budget. As well as an increased soil-moisture deficit between late summer and early winter, increased evapotranspiration could lead to an increase in the areas affected by saline or sodic conditions, particularly in Spain and Italy. Future increases in aridity would probably cause those species of vegetation living on the outer margins of their habitats to be pushed back, mimicking the migration of the climatic bands.

In some of the more arid regions, such as south-eastern Spain and much of North Africa, the existing winter rains are relatively unreliable. Were there to be a change in precipitation patterns, then it is likely that this band of inadequate rainfall would spread northwards. As water is the main limiting factor to plant growth in the Mediterranean, many indigenous species will have to adapt to a lengthening of summer drought, and the marginal moist periods between summer and winter.

Effects of global warming on agricultural crops

On agricultural or fallow land, any permanent reduction in winter precipitation and/or an increase in evapotranspiration and aridity can lead to poor spring regrowth, and the risk of loss of a coherent mat of vegetation to survive the summer months. Inevitably, this would alter vegetational structures and soil and water balances. The concomitant lower rates of organic matter production would cause a general deterioration in the structure of many soil types and alter their ability to partition rainfall between infiltration and runoff. Lower infiltration would further decrease soil moisture availability for late summer growth and have a deleterious effect on crop yield and soil stability. Changes in circulation patterns are also increasing convective storm activity in the late summer months; these must inevitably increase erosion (Blumler 1993; Imeson and Emmer 1992).

Effects of global warming on forest stability and implications for forest management

Le Houérou (1990) has detailed the environmental constraints on particular tree types in the vegetation zones that were mapped on Figure 16.2, predicting in particular the geographical shift in the boundaries of zones in the context of various scenarios of global warming and population growth into the twenty-first century. He suggests that the change in winter temperature would have a significant effect where this factor is restricting plant growth, such as the continental areas of North Africa and the Near East. This would allow an upward and northward shift in cold-sensitive crops such as citrus fruits and leave North Africa open to the development of specialisms in tropical crops, but accelerating the sort of degradational spiral suggested by Figure 16.4 across the deciduous wooded areas (especially the over-used cork-oak forests, such as the Mamora forest in Morocco).

Some 'greenhouse' predictions suggest that many tree species currently used in land reclamation programmes within the semi-arid regions of the Mediterranean are likely to die back and re-expose areas to soil erosion. Such alarm has given rise to international pressure for solutions to this problem. The target is a sustainable forest, but remedial action is the first priority in many cases. The degradational model in Figure 16.4 suggests that the appropriateness of various management strategies (plantings, enclosures, thinning of stands, fire control, re-use of abandoned agricultural land) as well as the relative impacts of differing agents of degradation (fires, global warming, tree diseases, human clearance, use of wood for fuel), vary with the stage of degradation. Thus, intervention must be undertaken with some sensitivity to the regenerative capabilities of the forest in its existing state. Additionally, present functions of the forest for human uses are often only marginally economic and certainly incompatible with any regeneration policy, as well as being in competition with each other. Functions can 'vary in relative importance and priority within the same management plan, whose balance sheet is virtually always in deficit' (Grenon and Batisse 1989, p. 206).

Commercial production figures suggest the gross product from forests in the Mediterranean remains generally low and under threat. However, if biomass production is taken in a broader sense, i.e. in respect of all woody-based products including the use of maquis, the forest is arguably more productive, >5t/ha from mature forest, and <5t/ha from maquis outside of semi-arid areas (Grenon and Batisse 1989). In this way, woody biomass still has a significant role to play in several aspects of rural economies, where residual forests remain locally as ribbon plantings, as fuel-wood lots and fodder stands in farms and villages. Alternative scenarios for sustainable management must take on board such new ways of looking at forest economy. First, planning should be less centralised, fully involving local populations. Second, research and training which foster greater technical and financial cooperation between the northern and southern countries are important. The Blue Plan emphasises that the role of the EU, UNESCO and the FAO is vital here as well as in the dissemination of information.

MEDITERRANEAN SOILS

The zonal soils of the Mediterranean are Brown Earths, Brown and Red Mediterranean Soils, and 'Cinnamon' Soils (classified in Table 16.1; and see the soil map inset into Figure 16.6). The length of the summer drought is an important control on soil development, as is the alternate wetting and drying associated with the 'Mediterraneity' of the annual climatic regime. Most soils have cambic or argillic B horizons, but sequences can range from Brown Earths developing in a leaching environment with less than one month of summer drought, to the Cinnamon Soils where five or six months results in calcification, the latter soil being 'transitional to that of the continental interiors' (Bridges 1978, p. 68).

Parent materials in the southern Mediterranean lands commonly consist of sedimentary rocks (sandstone, shales, and very frequently limestones and calcareous marls). However, the considerable redistribution of parent materials in many parts of the Mediterranean in the Holocene has produced landscapes often devoid of soils on the rocky interfluves interspersed with thick colluvial accumulations in the valleys and hollows. This leads to a complicated soil pattern: 'old soils lying next to immature ones in the same landscape' (Bridges 1978, p. 68).

Brown Mediterranean Soils are similar to those of more humid climates. When present on non-

Table 16.1 Synonyms for soils of the Mediterranean climatic zone

Common names	US soil taxonomy	FAO/UNESCO world map legend
Brown Mediterranean Soils	Hapludalfs	Orthic Luvisols
Red Mediterranean Soils/Red-Brown Earths: 'Terra Rossa'	Rhodustalfs	Chromic Luvisols
Non-Calcic Brown Soils	Haploxeralfs	Orthic Luvisols
Cinnamon Soils	Ustochrepts	Chromic Luvisols

Source: Bridges (1978)

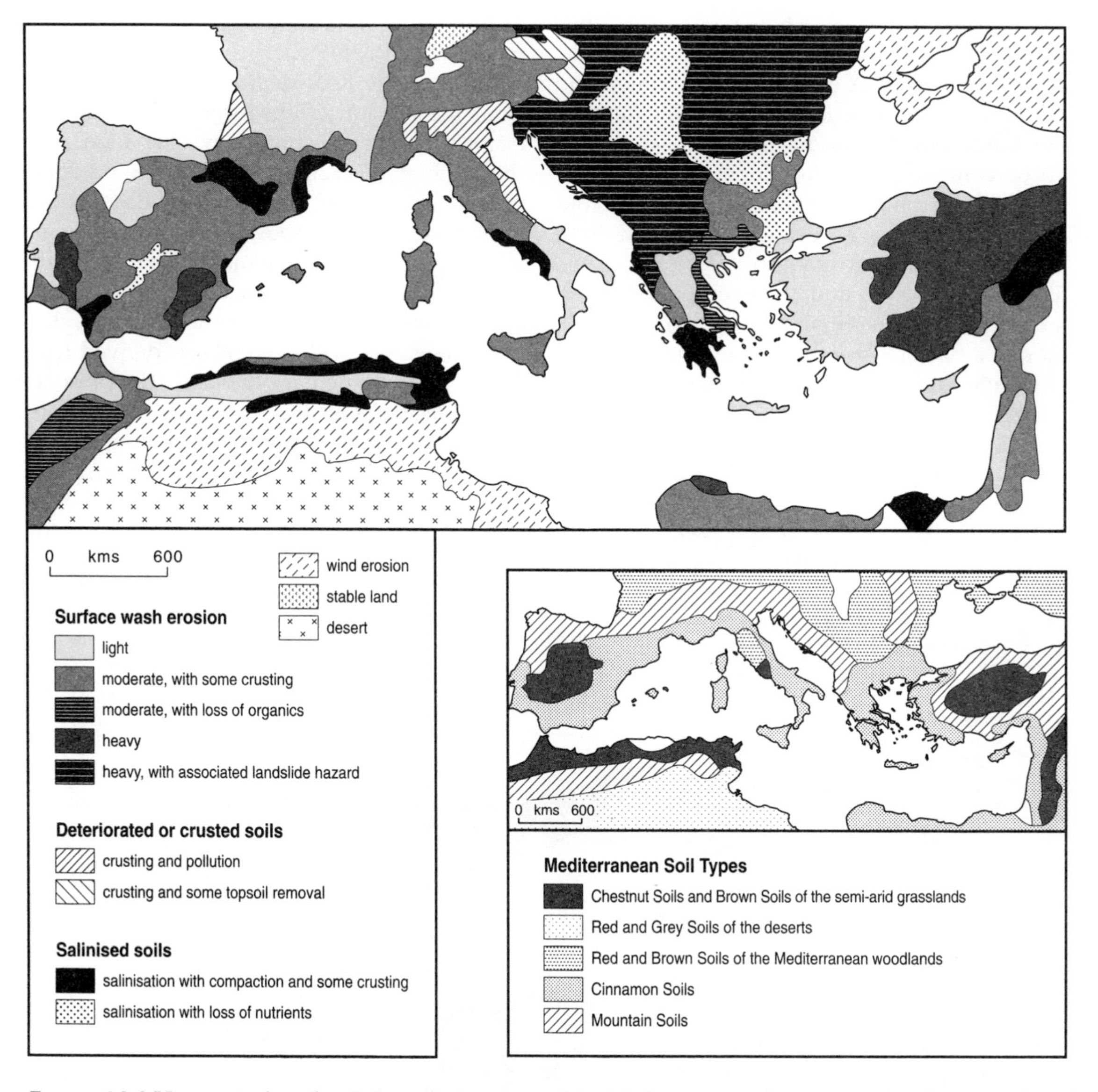

FIGURE 16.6 Human-induced soil degradation map of the Mediterranean Basin. *Inset:* Mediterranean soil type distribution map.

Sources: World Map on Human-Induced Degradation. Rome: FAO (1978); inset map after Bridges (1978)

calcareous or decalcified parent material, Chromic Luvisols form. These have a brown, friable humus-rich A horizon and a denser and less friable argillic B horizon. Where present on calcareous parent material, Calcic Luvisols form. In these situations the upper horizons are decalcified and clay movement is noticeable, redeposited in the lower part of the illuvial horizon and in the fissures of the C horizon, and iron is simultaneously precipitated during the long hot summers to give red colour to the lower horizons (rubefaction).

The deep red clays of southern Europe are often termed '*Terra Rossa*'. These soils are commonly shallow, slightly calcareous and occupy discontinuous pockets surrounded by rocky outcrops (Fig. 16.7). Characteristically, they occur as a dark red clay, organic and friable, above a lower layer of blocky clay. Nihlen and Mattson (1989) believe that the *Terra Rossa* of the Mediterranean has Fe (iron) values too high to be consistent with development from limestone under a hot dry climatic regime, and consider that it may be aeolian, deposited from North African winds in the Holocene. However, Boero and Schwertmann (1989) explain the high Fe values by arguing that under a Mediterranean climate, high internal drainage (due to the Karst nature of hard limestone) can produce neutral pH conditions and Fe substitution for Al (Aluminium), leading to the surprisingly high levels of Fe usually found in the soil. A local origin seems to be the view favoured by Pye (1992).

Some of the Red Mediterranean Soils result from the erosion of the Brown Soils and are deeper and demonstrate clay eluviation from the top horizons. Cinnamon Soils are predominantly associated with parts of the Basin with a prolonged summer drought. They can be found in association with maquis in Spain, Turkey and North Africa, but are actually more atypical of semi-arid areas. Soils are blocky and clay-rich, sometimes with concretions derived from underlying sedimentary deposits. The presence of a surface organic layer depends on land-use, but organics rarely exceed 7 per cent by weight. In very dry locations across the Mediterranean

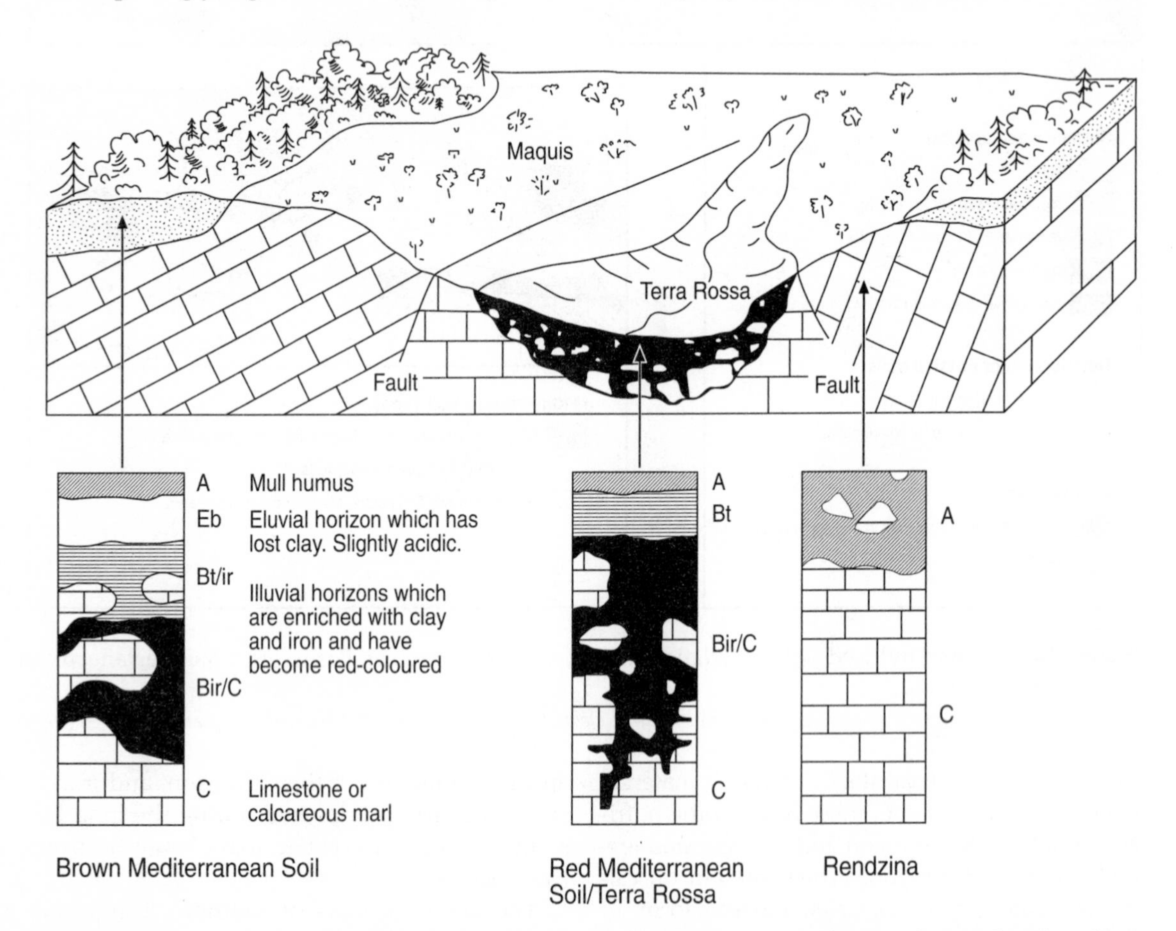

FIGURE 16.7 The relationship between Mediterranean soil types and topography. 'Typical' soil profiles are suggested.

Source: After Bridges (1978)

Basin the term 'soil' is difficult to justify, as the residual material is no more than unstructured, physical weathered rock fragments.

MEDITERRANEAN SOIL DEGRADATION

Historic patterns of soil degradation

Grenon and Batisse (1989) identify the problems threatening Mediterranean soil stability following climatic and anthropogenic change and loss of agricultural space as being twofold. First, there is the degradation of both chemical and physical properties through the excessive intensification of agriculture (in particular, the salinisation of lowland soils). Second, agricultural abandonment can lead to degradation, loss of organic status and soil erosion.

Figure 16.6 displays the 1978 FAO soil degradation map for the Mediterranean Basin. It suggests that even outside the semi-arid belt, high-intensity rain will leave any unprotected soils crusted and at risk from soil erosion by surface wash. Such storms are an increasing possibility bearing in mind the desertification pressures discussed above. Additionally, some alluvial soils affected by increasing rates of evapotranspiration are prone to salinisation, especially where brackish irrigation water has been used, even though poor drainage has not normally been considered a characteristic of Mediterranean soils. The map shows that interior sections of Spain, Italy and Morocco are most at risk from salination.

Saline soils

Salination is widely regarded as a serious threat to contemporary agriculture in the Mediterranean (Fig. 16.6). Productivity is suppressed, due to the inhibitory effects of high amounts of accumulated salts, and the problem is made more severe by anthropogenic activities such as irrigation, deforestation and overgrazing. A typical scenario might be as follows. An initial decrease in precipitation increases salt concentration in soils. Farmers think that adding water will drain these away in solution, and so irrigate the land. The irrigation supply may be from brackish groundwater, and may contain sodium chloride, calcium carbonate or other mobile salts. The irrigated water mobilises otherwise immobile salts within the soil profile and as the PET/P ratio approaches the positive, salts move up the profile and are deposited at or near the surface.

High saltwater concentrations may make water supplies unusable and prohibit a wide range of rural land uses. Since the reclamation of salt-affected soils is difficult and expensive, prevention is the best management strategy. Awareness of the problem is increasing and better managed irrigation systems are in operation in many countries (Ghassemi *et al.* 1995).

Surface wash and thresholds for accelerated soil erosion

Many Mediterranean soils, especially in dry areas, are highly erodible because of their salt status. For example, calcareous salts found in soils in parts of Greece, Spain and France comprise small, unstable soil aggregates, prone to crust formation (see Figure 16.6; also Imeson and Verstraten 1985). In other areas, high levels of exchangeable sodium make clays dispersive and prone to piping (e.g. Baillie *et al.* 1986). But surface wash erosion is likely simply because of vegetation density changes. In the context of a warming climate, semi-arid areas (such as parts of North Africa and south-east Spain) may act as models of what threatens the south of France, parts of Portugal and Italy, and south-west and interior Spain. Semi-arid areas experience no precipitation maximum in the winter, merely occasional but often catastrophic events concentrated into a few weeks in the late summer. Often between one-fifth to one-third of the total annual rainfall can fall within a few hours during such events. For example, two exceptional storms in October 1986 over the Segura basin in south-east Spain accounted for 77.4 per cent of the total annual sediment production (López-Bermúdez 1990). Such landscapes are not agriculturally productive, but are characterised by the chronically gullied terrain described in Chapter 4.

The consensus would appear to be that even potentially erodible soils are only at risk once the vegetation mat has been breached, making this the single most important control on erosion risk. However, there is a view that the critical threshold parameter in determining irreversible deterioration is the structural stability provided by soil organic matter. Additionally, partly decomposed organic matter at the surface can form a mulch which acts as a barrier between erosive rainfall and the vulnerable bare surface and hence inhibits desiccation and crusting. In soils with low organic matter contents, as occur over much of the Mediterranean, particle size is the most important factor in determining resistance to splash detachment following raindrop impact (De Ploey and Poesen 1985). In particular, many fine-textured soils are prone to crusting and resultant low infiltration rates generate rapid runoff, especially during high intensity storms (Garg and Harrison 1992). Very coarse soils may retain high infiltration rates, but their poor structure renders them liable to seepage failures (Faulkner 1995).

Much of the Mediterranean area consequently straddles a threshold which can be defined in terms of deteriorating soil and vegetation resistance, and beyond which lies irreversible erosion. Eventually, the deterioration (ultimately associated with anthropogenic influences) is compounded by the climatic effects of a reduction in the winter precipitation maximum and the increasing intensity of summer storms. Figure 16.8 summarises the ways in which these two effects (increase in power of runoff events, and reduction in surface resistance) combine to define this threshold. On this

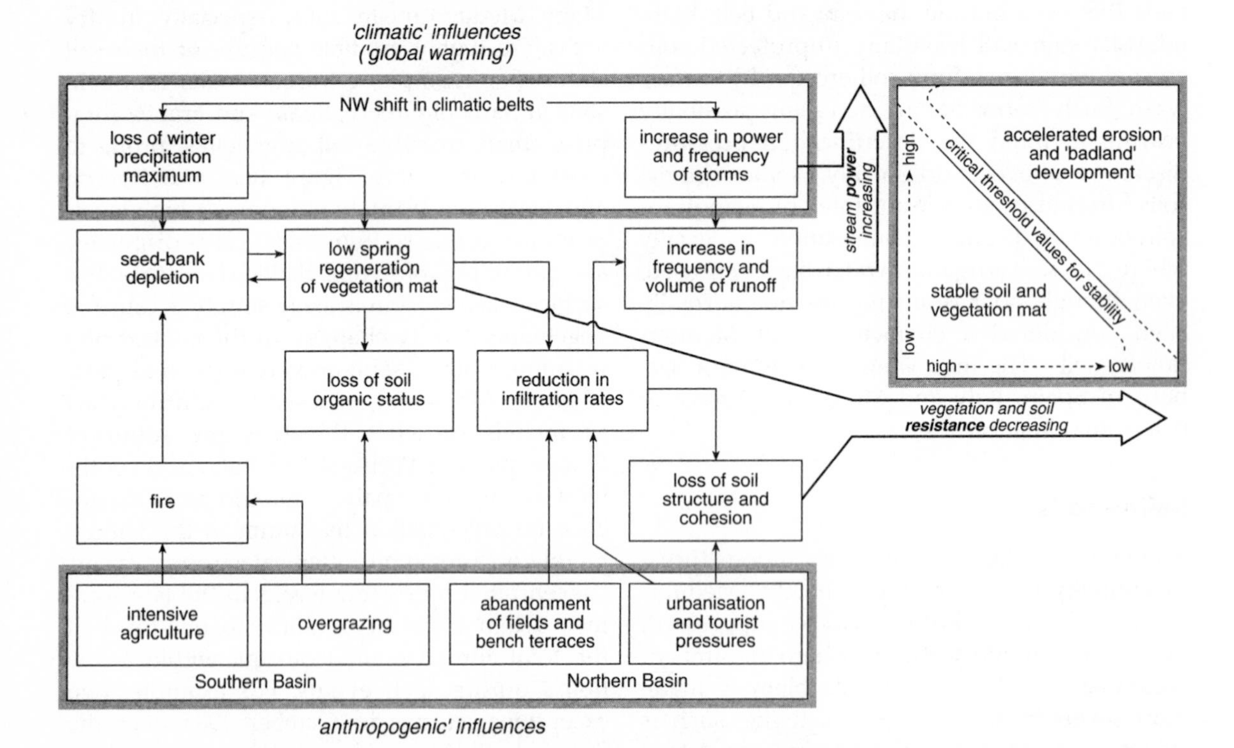

FIGURE 16.8 The manner in which anthropogenic and climatic causes of degradation combine to decrease surface resistance to erosion, as well as increasing the power and frequency of erosive events. The relationship between power and resistance defines a 'stability threshold'.

diagram, the stability threshold is defined by an as yet unquantified discriminant function. This diagram suggests that the changes in either force or resistance (or both) can be linked back via a chain of causal factors to anthropogenic and climatic influences.

MANAGING EROSION

One of the tragedies of the Mediterranean is that many past management practices, such as dry stone wall terracing (the *secano* of the Spanish drylands) and the irrigation systems of Moorish times (the Spanish *regadio* methods), have been neglected in favour of more easily accessible sources of a quick income. This includes the cultivation of unterraced orchards entirely or partly with EU subsidies in mind (Faulkner 1995), or the abandonment of farming completely for the attractions of the tourist industry. The enormous challenge facing the agricultural managers of the future of the Mediterranean would therefore seem to be to resist the shift from sustainable agricultural systems towards irreversible soil loss associated with, for example, semi-arid badlands. This shift can be thought of as crossing the threshold suggested on Figure 16.8 and quantifying it has become the focus for much contemporary Mediterranean erosion research (e.g. the EU-funded MEDALUS projects outlined in Brandt and Thornes 1996) as well as the starting-point for future management stategies.

Revegetation strategies

Inevitably, revegetation is the first thought when considering erosion mitigation in the Mediterranean. The long-term presence of vegetation will gradually alter the structure and fertility of its host soil. In highly calcareous, sandy and silty soils, organic matter plays a vital role in terms of soil fertility and structure, as well as having a very high exchange capacity and retaining nutrients in the soil. It also promotes favourable micro-aggregation and water retention (Imeson and Emmer 1992), which can decrease overland flow and reduce erodibility. A protective litter layer can also increase the irregularity of runoff and erosion mechanisms, making runoff systems less integrated and thus less efficient at soil removal (Sala and Calvo 1990). Finally, in semi-arid areas plant cover shades and protects the ground (Brandt and Thornes 1987); this lowers soil temperatures and decreases water loss by evaporation (Thornes 1989).

So, will replanting mitigate the effects of erosion? The process of post-erosion vegetation recovery is complicated and difficult to model, partly because the constantly changing balance of erosive agents and degrading pressures necessitates a multivariate view of change (as suggested by Figure 16.8). Additionally, event-related erosion is difficult to predict and quantify, and the processes are difficult to model. Seasonal variations in the quantity and quality of the vegetation also make assessment of its protective role difficult. Since Mediterranean-type shrublands are convergent ecosystems, (meaning that different species have adopted similar defensive strategies against the detrimental aspects of the Mediterranean climate), drought-resistant deciduous shrubs might be expected to shed their foliage at similar times, triggered by the seasonal onset of soil-moisture stress. In reality, however, small spatial variations in environmental parameters, such as geology, topography, soils and climate, can influence plant phonology. As a result, variations can occur over time and space and between species (Francis and Thornes 1990b; Kummerow 1983).

Theoretical models of vegetation reclamation

Thornes (1985) observed that it is not realistic to view vegetation recolonisation simply under the heading of 'biomass recovery', as regrowth is constrained by, but also constrains, continual erosion on adjacent sites. Instead, recolonisation must be seen as a spatially variable interaction between plant succession and the 'predatory' effects of soil erosion. In other words,

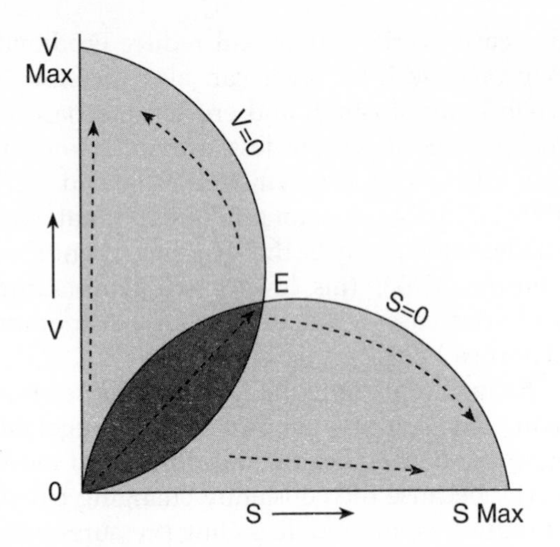

FIGURE 16.9 The interaction of vegetation (V) and erosion (S) isolines and the way in which they define behaviour around the unstable equilibrium 'E'

Source: After Thornes (1989)

erosion and recolonisation can be viewed as competitive, defining more precisely the stability threshold suggested on Figure 16.8. This idea is developed in Figure 16.9, which shows the trajectories that recolonised systems may take. The area of unstable equilibrium ('E') is equivalent to the threshold on Figure 16.8. Successfully competing vegetation moves the system to lower values of erosion and better cover ('V max'). Unsuccessfully recolonising systems move towards a state where vegetation is minimised, 'S max'. Consequently, Thornes (1989) suggests that conservation programmes must attempt to redirect the vegetation-erosion system back towards a vegetation levelling-off point ('V max'), but that the process of land reclamation is likely to be slow because up to 'E' the system is effectively pushed in the opposite direction to that in which it is naturally predisposed. However, if existing erosion-reinforcing feedback loops can be counteracted, then it may be possible to reclaim the degraded area. Once past the central area of unstable equilibrium, the system should switch towards a state of reinforcing vegetational growth, progressing until the levelling-off point ('V max') is reached.

Thus, as one may expect, it is the early stages of vegetation growth that appear to be the most important in the recovery of degraded areas, as here the trajectory is recoverable once the destructive agent is removed. At the other end of the scale, landscapes which have experienced a long history of misuse will have become so damaged and have so few resources left upon which to base a recovery, that it would take an unrealistic amount of effort to force the vegetation-erosion system back to a self-sustaining state. Interestingly, and despite the evidence of impressive badlands in many parts of the Mediterranean Basin, this theoretical model has led Thornes (1989) to argue that some areas within the semi-arid regions of the Mediterranean may have reached a levelling-off point, whereby although they have been severely degraded by erosion in the past, they now resist further change.

Afforestation, terracing, and erosion prevention

Afforestation is a common and often effective means of soil conservation. When successful, it can restrict water movement, reduce runoff and flooding, and lower soil erosion – especially when combined with bench terracing. Much of the northern Mediterranean has the potential capacity to support tree growth and it is only deforestation in historic times which accounts for the lack of trees. It has therefore been argued that trees are the natural climax vegetation in these areas and as such would be preferable in any land reclamation programme.

There are, however, a number of problems associated with afforestation programmes. First, they are long-term projects which are both initially expensive and offer little early protection against erosion. They also require regular maintenance during their first years, including irrigation and fertilisers. Finally, if exotic species are chosen, perhaps for economic reasons, they can interact unfavourably with the surrounding ecosystem. An example is the Eucalyptus which can significantly lower groundwater and is able to deter competitors through the deployment of toxins. Unfortunately these toxins also inhibit understorey growth which plays a vital role in protecting the soil from rainsplash erosion (Francis and Thornes 1990a).

An important limitation on afforestation is that trees generally require non-extreme temperature ranges and adequate supplies of water. Many have a minimum temperature requirement of 10°C in the hottest month which does not prove a problem within the Mediterranean, except at high altitudes. In terms of aridity the 350 mm precipitation isohyet is generally considered the limit to tree growth (Francis and Thornes 1990a). However, in many of the semi-arid regions of the Mediterranean there is insufficient rainfall to maintain a significant tree cover. Unfortunately, it tends to be precisely these areas which suffer the most from land degradation and soil erosion. A few specially adapted species manage to survive in these regions (e.g. *Pinus halepensis*) where they can grow on calcareous, saline and even skeletal soils. Blumler (1993), however, suggests that large-seeded annuals are an important part of the climax vegetation on the more fertile Mediterranean soils and dense shrubby vegetation, where soil is poor or rocky, may be preferred for reclamation outside the northern belt.

For initial reclamation, a mixture of indigenous shrubs may be the best option as they combine relatively thick coverage, minimal seasonal variations and a strong tolerance of the harsh conditions associated with degraded land. Tree species are likely to struggle in these marginal areas and will not provide the initial protection which appears to be so important when attempting to redirect the vegetation-erosion system towards a point of vegetated equilibrium. From their study in Murcia, Francis and Thornes (1990a) conclude that it is good ground cover and not trees that is essential to soil conservation in the Mediterranean, since it is the combination of undergrowth and litter which is the most effective in protecting against erosion (e.g. in *matorral* shrubland communities) rather than the type of forest canopy. They surmise that in the reclamation of large areas of degraded land an effective ground cover is required which is self-generating, cheap and which will eventually lead to natural cover, rather than expensive reforestation schemes.

This view would seem to be confirmed by recent estimations of erosion rates on *matorral* (maquis) as compared to those on reforested terraces in Guadalajara, central Spain. It is important to note, however, that traditional bench terraces, some of which date back to the Moorish period, are still the preferred method of soil erosion prevention where surface wash is the erosion agent, although they are labour-intensive and becoming increasingly unpopular or abandoned. They are less successful if seepage processes or piping are involved (Baillie *et al.* 1986; Faulkner 1995).

Hazard mapping approaches

A broader approach to erosion management is to create a Geographical Information System (GIS), which combines different 'layers' of information, such as land use, soil properties, climatic characteristics and topography to build up an erosion-risk map. Remotely sensed data, either from aerial photographs or satellite images, can be used to identify land cover type, which is a primary factor in determining soil susceptibility to erosion, and to pinpoint major erosional features such as gullies.

Garg and Harrison (1992) in their risk assessment study of the Albudeite catchment in south-east Spain, compared aerial photographs from 1957 and 1986, and isolated changes in land-use over this 29-year period. Topographic parameters, such as slope and aspect, were also extracted from a Digital Elevation Model (DEM). These variables were then combined and analysed using a Geographical Information System (GIS) to highlight regions undergoing land-use change and those areas prone to varying levels of erosion risk. For example, they found that, in general, north-facing slopes were mainly vegetated, whereas south-facing slopes were eroded and often virtually unvegetated. It was suggested that this lack of a protective vegetation layer on south-facing slopes reflects a greater soil moisture deficit, associated with a greater exposure to solar radiation throughout the day. The gully density, slope and land use layers of the GIS were analysed together to produce an erosion-risk map of the Albudeite catchment, based on a compound scoring approach. From this, Garg and Harrison (1992) found that about 50 per cent of the cultivated land was threatened by moderate to very severe erosion risks and concluded that the situation would worsen in the near future unless suitable protective measures were adopted. It is important to note, however, that the resolution at which aerial photographs and satellite images are taken can often be rather crude when compared to the geomorphological features under study. There may also be problems in dealing with seasonal changes in cultivated plants and shadows thrown by height relief.

FIGURE 16.10 Fire is a particular hazard for Mediterranean vegetation; dry, woody maquis is highly flammable. Corsica is one Mediterranean region which suffers greatly from summer fires, as this scene in the north of the island shows.

FIRE: A PARTICULAR MEDITERRANEAN HAZARD

Mutch (1970) suggested that in many parts of the Mediterranean, particularly the sclerophyllous shrublands, plants have evolved physiological mechanisms which increase their inflammability, effectively encouraging fire. Whether or not this is true, there is no doubt that fire is a naturally occurring phenomenon in the Mediterranean forest and many native eco-systems have developed strategies to survive periodic burning, some going so far as to now be dependent on it (Trabaud 1980). In the past, all removal of vegetation by wildfire was assumed to be damaging to the environment, resulting in degraded ecosystems and significant increases in soil erosion. Recent research, however, suggests that, depending on the intensity of the fire and the sensitivity of the affected system, the long-term risk may be negligible or even beneficial to the ecosystem (Moreno and Oechel 1995). On the other hand, rural depopulation and agricultural land abandonment, combined with increased recreational use of the countryside, including many careless visitors, have increased the severity and frequency of fires in some years (see Fig. 16.10).

Post-fire floral diversity in Mediterranean ecosystems

There is controversy over whether species diversity increases or decreases after fire. Naveh (1975), for example, considers that fire may have played an important role in the development of the composition, structure and niche differentiation of maquis-type vegetation in the Mediterranean. Trabaud (1990), working in a *garrigue* (maquis) ecosystem in the French Mediterranean region, also found that floristic richness increases between the first and third year after fire, but then tails off to fairly constant values. In general, less intense fires seem to stimulate regrowth in vegetation by disposing of woody litter, killing less vigorous shrubs and allowing secondary succession to occur (Kutiel 1994). Removal of vegetation cover by burning can also alter runoff and sediment regimes and the nutrient cycle and the vegetation composition in affected areas, causing ecosystems to adjust to new conditions (Kutiel and Inbar 1993; Kutiel and Naveh 1987).

Fire intensity and erosional impact of fires

Fire intensity is largely controlled by biomass,

which provides the fuel. Intense conflagrations generate enormous heat which can burn the organic topsoil as well as the fuel overburden. Spectacular results can occur when fire is followed by intense rainfall. In such circumstances, as well as a decrease in vegetation cover, there may be significant changes to soil hydrological properties, largely due to the destruction of soil aggregates in the exposed upper horizons. Intense burns alter several chemical and physical properties of the soil, reducing infiltration, increasing runoff and promoting erosion (Kutiel and Inbar 1993).

Over recent decades, however, overgrazing has meant that biomass levels are rarely sufficient to initiate intense natural fire cycles in many parts of the Mediterranean; whilst controlled burning to expand grazing areas has the benefit of reducing fuel build-up (Brown 1990). Studies on Mount Carmel, Israel, found that changes in soil aggregate stability did not occur in low intensity fires, and that organic matter remained largely intact. Regarding erosion, low to moderate intensity fires tend to produce a complex mix of rough and smooth surface patches. Smooth patches produce discontinuous runoff and sediment removal, but on rough areas, made up of partially burnt vegetation acting as a mulch and depressions formed by uprooted trees, there was virtually no chance for overland flow to build up over any distance. However, high intensity fires consumed all available vegetation, left only a uniform, smooth ash layer, and led to irreversible erosion (Kutiel and Inbar 1993).

Sensitivity and environmental response after fire

The results from Mount Carmel show that different effects occur after fire depending on microtopography and suggest that environmental sensitivity is a crucial determinant of fire effects. Variations in lithology, slope steepness, aspect, soil and vegetation types all operate to create diverse fire effects. For instance, Marques and Mora (1992) found that residual post-fire debris and protection were influenced strongly by aspect. Post-fire erosion was found to be six times greater on a south-facing slope. The extreme runoff coefficients were 73.3 per cent for this slope and 28 per cent for a similar north-facing one. Rills were seen to develop extensively on the south-facing slope but not at all on the north-facing slope. Naveh (1974) found that after a fire of dense maquis shrubland on a shallow, rocky, brown Rendzina, there was no evidence of overland flow, erosion or soil movement on quite steep slopes (30–40 per cent). This was probably because the soil had a well-developed, humus-rich profile. Naveh suggested that the dangers of post-fire erosion are at their greatest where vegetation cover is minimal (i.e. low shrub cover) and where the soil is less fertile and inherently more erodible. The risk can be further exacerbated if soils are disturbed and compacted by uncontrolled grazing prior to burning.

In areas of tourist pressure, fires commonly occur frequently on the same site and, due to insufficient recovery time, degradation of local ecosystems can result. On the other hand, attempts to exclude natural wildfires allow a build-up of phytomass. Additionally, post-fire erosion can become a problem where the soil structure has been damaged by trampling and machinery and when plant biomass is greatly reduced by intensive grazing. Anthropogenic pressures generally serve to increase the sensitivity of the systems to the rages of fire by reducing surface resistance (see Figure 16.8).

The significance and nature of most post-fire geomorphic effects therefore depend on numerous geographical and anthropogenic factors as well as landscape's sensitivity to change and the fire regime. In general, fierce conflagrations in highly erodible areas will tend to dominate sediment production and may well drastically alter the local ecosystem, whereas the same fire in less sensitive areas may encounter a fairly resilient system response.

Fire and plant recovery strategies

Fires caused by human use of the land are now considered to be a serious problem facing those involved in resource management in the Mediterranean Basin, and 'quick-response' fire management teams are a familiar sight now in Mediterranean forest areas. Post-fire regeneration strategies can be based upon the observation that sclerophyllous plants regenerate vegetatively after fire from buds buried in the soil. Alternative methods involve enhanced germination of seeds, either buried in the soil or invading from neighbouring unburnt areas. According to Kutiel *et al.* (1990), recovery depends on the number of perennial herbaceous plants that survived in the dense tree and shrub cover as shade-tolerant relicts, the availability of seeds from these and other invading species, and the climatic conditions that predominate in the first and second rainy seasons after the fire. The impact of global warming may well be to make these issues even more relevant to future forest managers.

Conclusions

This chapter has explored considerations relevant to the management of the variety of soil and forest types in the Mediterranean Basin, arguing that despite modified views on the stability of the maquis, and the continuing debate as to whether it represents a degradational stage, the evidence suggests that the region is becoming increasingly 'desertified'. The contextual problems contributing to this process have been outlined, and points of focus for future research into the resilience and sensitivity of existing ecosystems have been indicated. The debate as to whether anthropogenic activities or climatic influences are the principal cause of this degradation is left with Le Houérou (1992), who argues that too much emphasis has been placed on predicting the effects of global warming. He stresses that 'the changes would be slight compared to the change induced by the exponential population growth in the south and east of the Mediterranean basin ... man has become a major geological agent in this part of the world' (p. 175). Undoubtably this debate is set to continue for some years.

References

BAILLIE, A.C., FAULKNER, H., ESPIN, G.D., LEVETT, M.J. and NICHOLSON, B. 1986: Problems of protection against piping and surface erosion in central Tunisia. *Environmental Conservation* **13**, 27–40.

BERGER, A. 1986: Desertification in a changing climate with particular attention to the Mediterranean countries. In Fantechi, R. and Margaris, N.S. (eds), *Desertification in Europe*. Brussels: D. Reidel, 15–34.

BLUMLER, M.A. 1993: Successional pattern and landscape sensitivity in the Mediterranean and Near East. In Thomas, D.S.G. and Allison, R.J. (eds), *Landscape Sensitivity*. Chichester: Wiley, 287–308.

BOERO, V. and SCHWERTMANN, U. 1989: Iron Oxide mineralogy of Terra Rossa and its genetic implications. *Geoderma* **44**, 319–27.

BRANDT, C.J. and THORNES, J.B. 1987: Erosional energetics. In Gregory, K.J. (ed.), *Energetics of the Physical Environment*. Chichester: Wiley, 51–88.

BRANDT, C.J. and THORNES, J.B. 1996: *Mediterranean Desertification and Landuse*. Chichester: Wiley.

BRIDGES, E.M. 1978: *World Soils*. Cambridge: Cambridge University Press.

BROWN, A.G. 1990: Soil erosion and fire in areas of Mediterranean type vegetation: results from chaparral in Southern California, USA and matorral in Andalucia, Southern Spain. In Thornes, J.B. (ed.), *Vegetation and Erosion*. Chichester: Wiley, 269–87.

DAGET, P. 1977: Le bioclimat méditerranéen: charactères généraux, modes de characterisation. *Vegetatio* **34**, 1–20.

DELANO-SMITH, C. 1979: *Western Mediterranean Europe*. London: Academic Press.

DE PLOEY, J. and POESEN, J. 1985: Aggregate stability, runoff generation and interill erosion. In Richards, K.S., Arnett, R.R. and Ellis, S. (eds), *Geomorphology and Soils*. London: Allen and Unwin, 99–120.

DI CASTRI, F., GOODALL, D.W. and SPECHT, R.L. 1980: *Mediterranean-type Shrublands*. Amsterdam: Elsevier.

DI CASTRI, F. and MOONEY, H.A. 1973: *Mediterranean-type Ecosystems: Origin and Structure*. Berlin: Springer-Verlag.

DOUGLAS, T., KIRKBY, S.J., CRITCHLEY, R.W. and PARK, G.J. 1994: Agricultural terrace abandonment in the Alpujarra, Andalucia, Spain. *Land Degradation and Rehabilitation* **5**, 280–91.

EMBERGER, I. 1930: La végétation méditerranéenne. Essai de classification des groupements végétaux. *Revue de Botanique* **42**, 641–22; 705–21.

FANTECHI, R. and MARGARIS, N.S. (eds) 1986: *Desertification in Europe.* Brussels: D. Reidel.

FAULKNER, H. 1995: Gully erosion associated with the expansion of unterraced almond cultivation in the coastal Sierra de Lujar, S. Spain. *Land Degradation and Rehabilitation* **6**, 179–200.

FRANCIS, C.F. and THORNES, J.B. 1990a: Matorral: erosion and reclamation. In Albaladejo, J., Stocking, M.A. and Díaz, E. (eds), *Soil Degradation and Rehabilitation in Mediterranean Environmental Conditions.* Madrid: CSIC, 87–115.

FRANCIS, C.F. and THORNES, J.B. 1990b: Runoff hydrographs from three Mediterranean vegetation cover types. In Thornes, J.B. (ed.), *Vegetation and Erosion.* Chichester: Wiley, 362–84.

GARG, P.K. and HARRISON, A.R. 1992: Land degradation and erosion risk analysis in S.E. Spain: a Geographic Information System approach. *Catena* **19**, 411–25.

GHASSEMI, F., JAKEMAN, A.J. and NIX, H.A. 1995: *Salinisation of Land and Water Resources.* Harpenden: Commonwealth Agricultural Bureaux.

GRENON, M. and BATISSE, M. 1989: *Futures for the Mediterranean Basin: The Blue Plan.* Oxford: Oxford University Press.

IMESON, A.C. and EMMER, I.M. 1992: Implications of climatic change on land degradation in the Mediterranean. In Jeftic, L., Milliman, J.D. and Sestini, G. (eds), *Climatic Change in the Mediterranean, Vol. 1.* London: Edward Arnold, 175–227.

IMESON, A.C. and VERSTRATEN, J.M. 1985: The erodibility of highly calcareous soil material from southern Spain. *Catena* **12**, 291–306.

KRUGER, F.J., MITCHELL, D.T. and JARVIS, J.U.M. (eds) 1983: *Mediterranean-Type Ecosystems.* Berlin: Springer-Verlag.

KUMMEROW, J. 1983: Comparative phenology of Mediterranean-type plant communities. In Kruger, F.J., Mitchell, D.T. and Jarvis, J.U.M. (eds), *Mediterranean-Type Ecosystems.* Berlin: Springer-Verlag, 300–21.

KUTIEL, P. 1994: Fire and ecosystem heterogeneity: a Mediterranean case study. *Earth Surface Processes and Landforms* **19**, 187–94.

KUTIEL, P. and INBAR, M. 1993: Fire impact on soil nutrients and soil erosion in a Mediterranean pine forest. *Catena* **20**, 129–39.

KUTIEL, P. and NAVEH, Z. 1987: Soil properties beneath Pinus halepensis and Quercus calliprinos trees on burnt and unburnt mixed forest on Mt. Carmel, Israel. *Forest Ecology and Management* **20**, 269–74.

KUTIEL, P. NAVEH, Z. and KUTIEL, H. 1990: The effect of wildfire on soil nutrients and vegetation in an Aleppo forest on Mt. Carmel, Israel. In Goldammer, J.G. and Jenkins, M.J. (eds), *Fire in Ecosystem Dynamics: Mediterranean and Northern Perspectives.* The Hague: SPB Academic Publishing, 85–94.

LE HOUÉROU, H.N. 1990: Global change: population, land-use and vegetation in the Mediterranean basin by the mid-21st century. In Paepe, R., Fairbridge, R.W. and Jelgersma, S. (eds), *Greenhouse Effect, Sea Level and Drought.* Dordrecht: Kluwer Academic Publishers, 301–67.

LE HOUÉROU, H.N. 1992: Vegetation and land-use in the Mediterranean Basin by the year 2050: a prospective study. In Jeftic, L., Milliman, J.D. and Sestini, G. (eds), *Climatic Change in the Mediterranean, Vol. 1.* London: Edward Arnold, 175–227.

LE HOUÉROU, H.N. 1993: Land degradation in Mediterranean Europe: can agroforestry be a part of the solution? A prospective review. *Agroforestry Systems* **21**, 43–61.

LÓPEZ-BERMÚDEZ, F. 1990: Soil erosion by water and the desertification of a semi-arid Mediterranean fluvial basin: the Segura basin, Spain. *Agriculture, Ecosystems and Environment* **33**, 129–45.

MARQUÉS, M.A. and MORA, E. 1992: The influence of aspect on runoff and soil loss in a Mediterranean burnt forest (Spain). *Catena* **19**, 333–44.

MAZZOLENI, A., LO PORTO, A. and BLASI, C. 1992: Multivariate analysis of climatic patterns of the Mediterranean basin. *Vegetatio* **98**, 1–12.

MORENO, J.C. and OECHEL, W.C. 1995: *The Role of Fire in Mediterranean-Type Ecosystems.* Berlin: Springer-Verlag.

MUTCH, R.W. 1970: Wildland fires and ecosystems: a hypothesis. *Ecology* **51**, 1046–51.

NAVEH, Z. 1974: Effects of fire in the Mediterranean region. In Kozlowski, T.T. and Ahlgren, C.E. (eds), *Fire and Ecosystems.* New York: Academic Press, 401–34.

NAVEH, Z. 1975: The evolutionary significance of fire in the Mediterranean region. *Vegetatio* **29**, 199–208.

NIHLEN, T. and MATTSON, J.O. 1989: Studies on eolian dust in Greece. *Geografiska Annaler* **71A**, 269–74.

PYE, K. 1992: Aeolian dust transport and deposition over Crete and adjacent parts of the Mediterranean Sea. *Earth Surface Processes and Landforms* **17**, 271–88.

QUEZEL, P. and BARBARO, M. 1982: Definition and characterization of Mediterrannean-type ecosystems. *Ecologia Mediterranea* **7**, 15–27.

SALA, M. and CALVO, A. 1990: Response of four different Mediterranean vegetation types to runoff and erosion. In Thornes, J.B. (ed.), *Vegetation and Erosion*. Chichester: Wiley, 347–62.

THIRGOOD, J.V. 1981: *Man and the Mediterranean Forest*. London: Academic Press.

THOMAS, D. 1993: Sandstorm in a teacup? Understanding desertification. *Geographical Journal* **159**, 318–31.

THORNES, J.B. 1985: The ecology of erosion. *Geography* **70**, 222–36.

THORNES, J.B. 1989: Solution to soil erosion. *New Scientist*, 3 June, 45–9.

TOMASELLI, R. 1977: *Degradation of the Mediterranean Maquis*. Paris: UNESCO, MAB Technical Note 2.

TRABAUD, L. 1980: Man and fire: impacts on Mediterranean vegetation. In Di Castri, F., Goodall, D. and Specht, R.L. (eds), *Mediterranean-Type Shrublands*. Amsterdam: Elsevier, 523–37.

TRABAUD, L. 1990: Fire resistance of *Quercus coccifera* L. garrigue. In Goldammer, J.G. and Jenkins, M.J. (eds), *Fire in Ecosystem Dynamics: Mediterranean and Northern Perspectives*. The Hague: SPB Academic Publishing, 21–32.

17

COASTAL ZONE MANAGEMENT

GEORGE DARDIS AND BERNARD SMITH

INTRODUCTION

Although people have lived on the shores of the Mediterranean Sea for thousands of years, over the past 100 years or so pollution levels have been rising at alarming rates, propelled by rapid population increase. By the 1970s the Mediterranean Sea had become a huge waste-bin, a receptacle for millions of tonnes of human and industrial waste, posing severe environmental problems in coastal and marine areas as well as a potential direct threat to inhabitants. Now, no longer is it possible to view the Mediterranean as an infinite resource – rather, it is necessary to look for ways and means to conserve and rehabilitate the sea.

Hindrichsen (1990) rightly argues that people and their needs are both at the centre of these problems and at the centre of the solutions. However, the process of finding solutions to environmental problems in the Mediterranean Basin presents an interesting and difficult task for its inhabitants who represent a hotchpotch of cultures, political traditions, languages and religions, as we saw in Chapter 8. In spite of this, it has been recognised in the past few decades that there is a need to adopt an holistic approach to coastal zone management in the Mediterranean Basin.

This approach is dictated primarily by the geography of the Mediterranean Sea (Fig. 17.1). It is one of the largest semi-enclosed seas on the earth's surface (2 542 000 km^2; Margalef 1985), surrounded by 20 countries, on three continents, with a combined population of 145 million people, sharing a coastline of 46 000 km (Grenon and Batisse 1989). With this combination of geography and political fabric, there is a compelling case to suggest that coastal and marine problems within the Mediterranean Basin are shared problems; this is unlike most other coastal areas world-wide, where coastal problems tend to be relatively localised manifestations of environmental problems (Viles and Spencer 1995).

This chapter examines in some detail the nature of the main coastal problems, and the main factors which contribute to these problems. In the concluding section, a range of management strategies at inter-regional, regional and local levels are evaluated to exemplify some of the difficulties in finding solutions to the problems of the coastal zone of the Mediterranean.

THE MEDITERRANEAN SEA

The Mediterranean Sea is a tectonically active area which straddles the collision zone between Africa and Europe (see Chapter 2). It has an average depth of 1.5 km, although 20 per cent of the total submarine area is less than 200 m deep (UNEP 1989). The coastline contains two

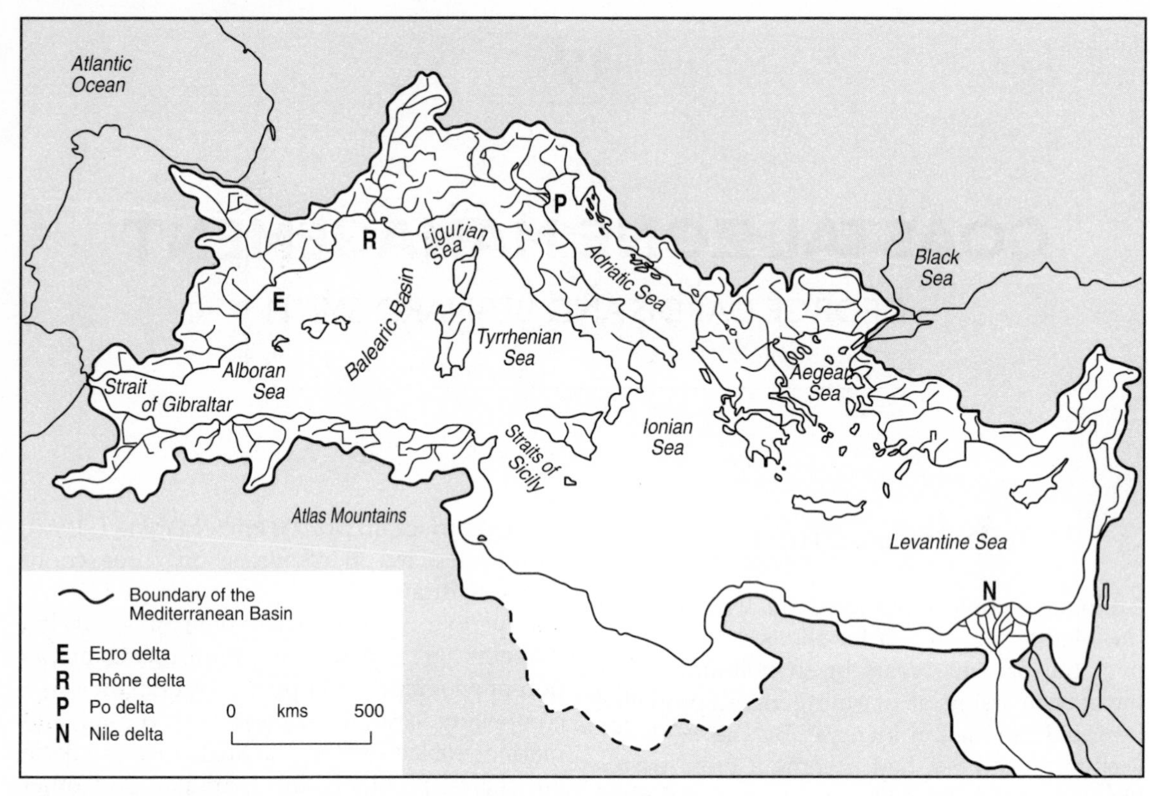

FIGURE 17.1 General physiography of the Mediterranean Sea

Source: Adapted from Grenon and Batisse (1989)

major basins (the eastern and western) as well as a number of smaller regional seas (the Adriatic, Aegean, Alboran, Ionian, Levantine, Ligurian, Tyrrhenian) (Fig. 17.1). The western Mediterranean covers *ca.* 860 000 km^2 and is separated from the eastern Mediterranean by a submarine sill about 400 m in depth extending from Sicily to Tunis.

The Mediterranean is linked to the Atlantic Ocean by the Strait of Gibraltar, through which the flow of water in and out is highly complex. This flow pattern is regulated by salinity and temperature differences within the water masses. Deep Mediterranean water, with 38.5 g of solid matter per kg of water and a relatively constant temperature of 13°C (Margalef 1985), spills out into the Atlantic Ocean, across the Gibraltar–Morocco sill at around 320 m below the surface. At the same time, a slightly higher amount of Atlantic water of relatively lower salinity enters the Strait of Gibraltar on the surface. Relatively fresh surface water also enters the eastern Mediterranean from the Black Sea (Fig. 17.2). These surface waters have a wider temperature range than deep water (about 13°C in winter and as high as 26°C in summer).

This slight surplus of less saline water input compensates for water lost by evaporation, as evaporation generally exceeds both direct rainfall and water discharge from fluvial systems within the Mediterranean Basin. This regulatory process is self-perpetuating as water of relatively high salinity is formed by evaporative concentration and, being more dense, sinks to form the deep, highly saline water layer characteristic of the Mediterranean. This excess of evaporation in the Mediterranean Basin not only draws water from the Atlantic, but also generates a horizontal exchange of water across the Mediterranean, reflected in surface current flow patterns. The slight surplus of inflowing, relatively fresh, surface water also helps to

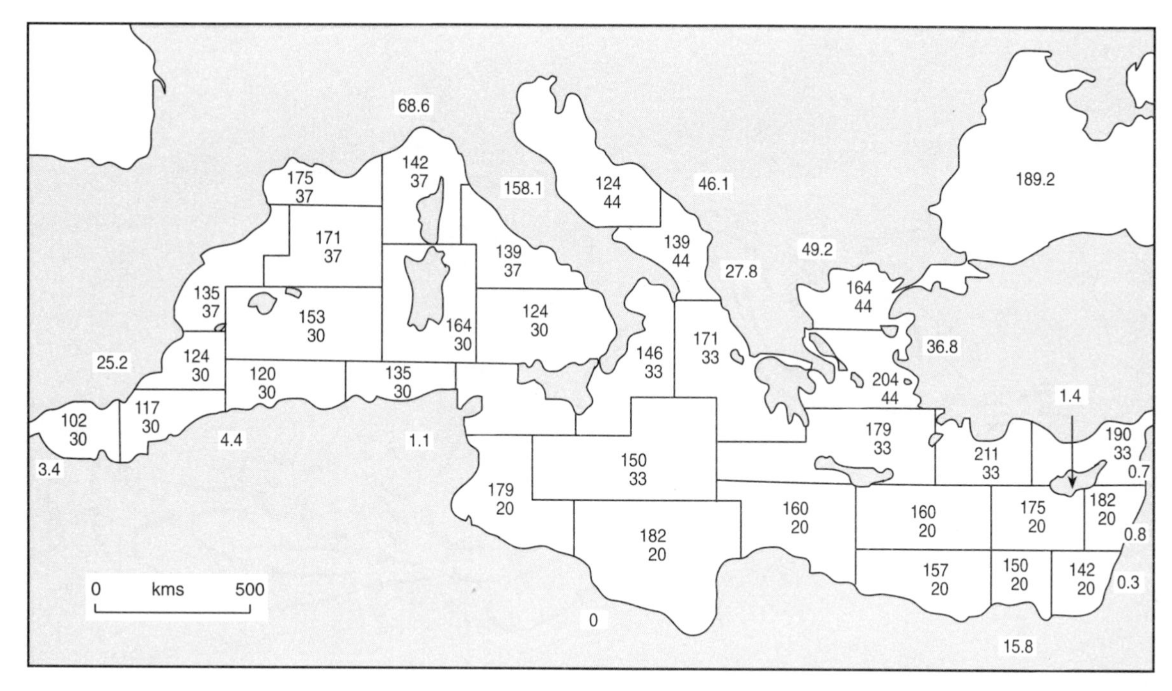

FIGURE 17.2 The estimated values for evaporation (cm/yr cm^{-2}), precipitation (cm/yr cm^{-2}) and runoff (km^3/yr) or the Mediterranean broken into subregions. The values given on the land indicate river runoff and agricultural discharges, while those over the sea represent evaporation (upper) and precipitation (lower).

Source: After Hopkins (1985)

counteract loss by evaporation and thereby maintains the Mediterranean sea level.

Although the Mediterranean has the potential to generate tides independent of the world ocean, they tend to be relatively small. The existence of narrow continental shelves within the Basin results in very little tidal amplification along the coastal margin, to the extent that the Mediterranean is often described as tideless. Strictly speaking, this is not true. Harmonic constants can be relatively high at Gibraltar (38.3 cm) but decrease rapidly eastward (e.g. 1.8 cm at Alicante) into the western Mediterranean (Hopkins 1985). Furthermore, although there is little tidal elevation there is considerable tidal power, which may locally exceed mean current flow (Defant 1961). Because of the peculiar physiognomy of the Mediterranean Basin there is also an additional tide, which in the Western Mediterranean forms by co-oscillations with the Atlantic through the Strait of Gibraltar and with the Eastern Mediterranean through Sicily (Hopkins 1985). This additional tide is often of higher magnitude than the independent tide. These tidal characteristics have considerable implications for coastal land use and general physiography.

THE MEDITERRANEAN COASTAL MARGIN

The tectonic framework of the Mediterranean Basin, coupled with large-scale current flow patterns (Fig. 17.3), have exerted considerable influence on the physiography of the coastal margin, which, in general, consists of a ring of mountains surrounding the sea (Rios 1978). Consequently, the Mediterranean watershed occupies a relatively small land area relative to the sea (Fig. 17.1). This high relief in proximity to the coastal margin has had a significant effect in terms of the

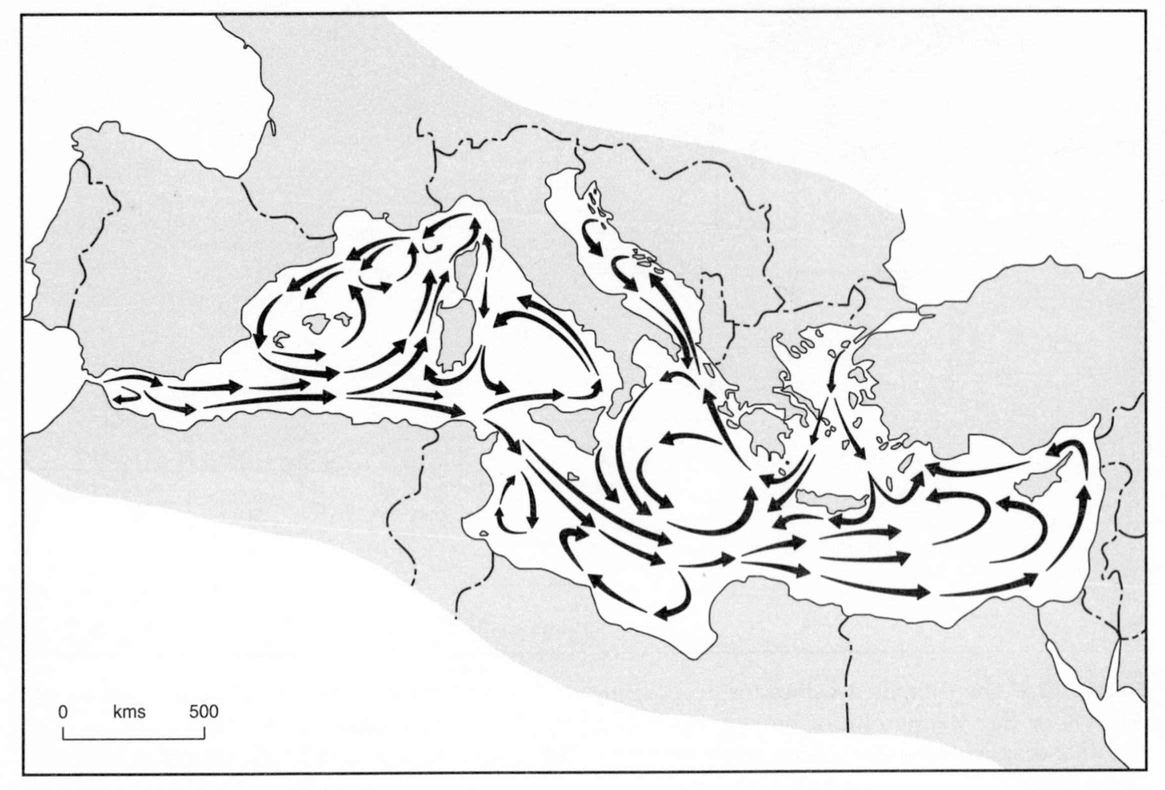

FIGURE 17.3 General trends of surface currents in summer in the Mediterranean

Source: After Grenon and Batisse (1989)

nature of the coastal margin. First, it has led to the development of a predominantly rocky coast; and second, because of the general aridity of the Mediterranean Basin as a whole, it provides ideal conditions for the development of extensive coastal alluvial fan systems. Both of these geomorphological systems present a range of management challenges.

Rocky coasts of the Mediterranean

Between 70 and 75 per cent of the Mediterranean coastline is rocky (Emery and Kuhn 1982), characterised either by cliffs (Fig. 17.4) or extensive shore platforms. The cliffs occur either as cliff lines at the back of large expanses of beaches, or as headlands between bay beaches. Many cliff lines are fault-controlled and may extend for many kilometres. For example, much of the rocky coastline in south-east Spain follows the strong alignment of major Neogene/Quaternary strike-slip fault lines (Bousquet 1977), reflecting widespread differential and intermittent epeirogenic uplift since the Pliocene (Harvey 1990). In most instances hard rock cliff lines show little activity in terms of weathering, mass movements, bioerosion or wave erosion (Viles and Spencer 1995). However, widespread epeirogenic uplift has exposed relatively soft rocks along the Mediterranean coastal margin and increased the potential for instability due to erosion and mass movements. This potential is enhanced by neotectonic activity (Chapter 2).

Shore platforms (Fig. 17.5) are found throughout the Mediterranean and provide very important habitats for marine flora and fauna. They are particularly well developed along the coast near Marseilles (Le Campion-Alsumard 1979) and in north-east Majorca (Kelletat 1985). Shore platforms tend to be geomorphologically complex, with a diverse micro-

FIGURE 17.4 Steep marine cliffs with an abandoned cliff line and frontal beach system in the foreground, Curium, Cyprus

topography, and may be subject to extensive biological erosion in the intertidal zone (Moses and Smith 1994).

Alluvial fans

Alluvial fan systems are commonly found on the coastal margin throughout much of the Mediterranean, particularly in arid and semi-arid areas where vegetation cover is sparse (Harvey 1990). They also occur in tectonically active areas where sediment supply is abundant. These systems normally consist of series of coalescent fans (Fig. 17.6), with surfaces which approximate segments of a cone that radiate down from an apex, where the stream leaves a mountainous area. They are formed by different combinations of fluvial and debris-flow processes, but much of their genesis can be attributed to the abrupt change from confined to unconfined flow within one or a series of distributary conduits. The limited resistance of the depositional apron of the fan encourages a high degree of channel avulsion (lateral migration), which promotes the construction of a low-gradient prograding (clinoform) surface emanating from the fan. The coastal alluvial fans are often fairly extensive. For example, the Campo de Dalías, located west of Almeria in south-east Spain, has an area of 340 km^2, making it comparable in size to the Ebro delta (see below).

FIGURE 17.5 Fifteen-metre raised shore platform in Miocene limestone, south coast of Majorca

The low relief of fans makes them particularly useful for intensive modern agriculture. Indeed, a significant proportion of Mediterranean vegetable production is now found on coastal alluvial fans, making these areas particularly valuable contributors to local and national economies. For example, agricultural production from the Campo de Dalías and the neighbouring Campo de Níjar account for a large proportion of vegetable production in Almeria province.

Deltas

Deltas are probably the most important land systems on the Mediterranean coastal margin, at least in terms of their importance to human activity and sensitivity to environmental change such as sea-level rise. Because they are flat-lying, they provide some of the best agricultural land and, consequently, host some of the highest concentrations of people in the whole of the Mediterranean (cf. Panzac 1983). The deltas, and particularly pro-delta areas, are also extremely important habitats; they contain significant wetlands, made up of complexes of beach-dune barriers, lagoons, and saltwater and freshwater marshes. These provide important sites for enhanced biodiversity and represent essential nesting and resting areas for large numbers of migratory aquatic birds. For example, in the

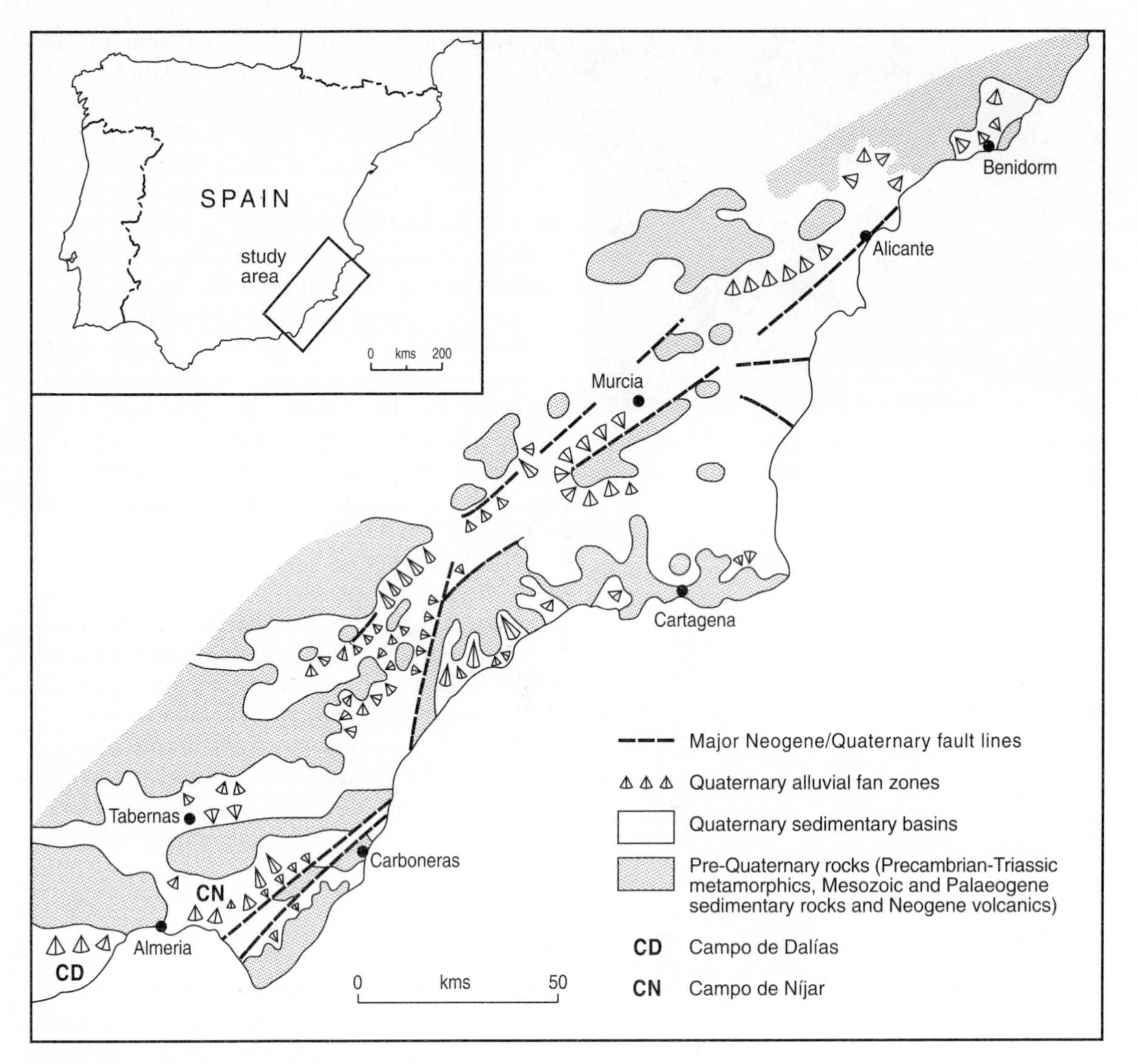

FIGURE 17.6 Coastal alluvial fans in south-east Spain in relation to generalised geology

Source: After Harvey (1990)

reedbeds, brackish lagoons, salt marshes and riverside woodlands of the Ebro delta, there are more than 250 bird types, which alone represent over 60 per cent of the total number of European bird species (Smith 1991).

Despite their importance, and although some of the deltas are large in area, they are not all that abundant within the basin. Four major river systems enter the Mediterranean along the northern coastal margin (the Ebro, Rhône, Po and Tiber) and each has constructed an extensive delta system. In addition, there are numerous smaller river systems, such as the Llobregat, which have extensive deltas. Apart from the Nile (20 000 km^2) and a few rivers draining the Atlas Mountains, relatively few river systems feed onto the southern coastal margin of the Mediterranean (Fig. 17.1). This primarily reflects the aridity of much of the southern coastal margin, which is dominated by dry coastal ecosystems.

Before construction of the Aswan Dams, the

Nile carried well over 120 million tonnes of sediment onto the delta plain and foreslope (Fig. 17.7A) each year (Stanley and Warne 1993a). Sediment delivery onto the delta plain is important. Stanley and Warne (1993b) suggest that initiation of farming settlements in the Nile delta was closely related to deceleration in eustatic sea-level rise at around 6500–5500 BC, which promoted the accumulation of Nile silt and the creation of the widespread and fertile delta plain. The broadening, seasonal flooding provided a setting that was conducive to evolving agricultural activity and was therefore instrumental in the development of Pre-dynastic communities in the Nile delta.

The Ebro delta, located on the coast of north-east Spain (Fig. 17.7B), is the fourth largest delta in the Mediterranean Sea. It drains a catchment of 85 000 km² and it is a triangular-lobate delta, with a surface area of about 285 km² (Mariño 1992). The delta is storm-wave-dominated and micro-tidal (Guillén and Palanques 1993) with two major lagoons along its southern margins, marking the positions of former shorelines. Primary economic activities in the Ebro delta are agriculture and fisheries. About 20 per cent of Spanish rice production and up to 25 per cent of the fish and 40 per cent of the molluscs obtained in Catalonia come from this area (Mariño 1992).

Most of the deltas are prograding clinoform types, characterised either by a deep offshore ramp or shallow shoal water profile. Consequently, they tend to be liable to offshore mass movements and/or on-shore subsidence due to sediment overloading, which can significantly enhance the susceptibility of the delta plain to environmental change. For example, studies of the Nile delta coast by Frihy (1992) show local

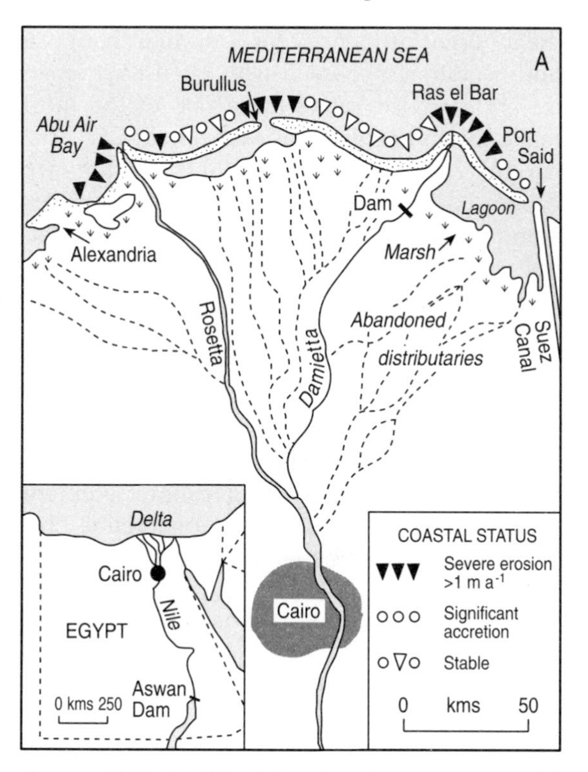

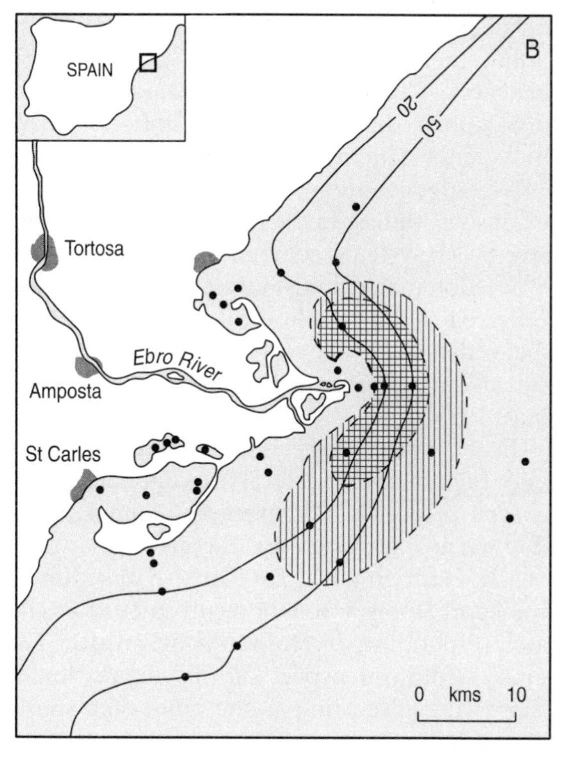

FIGURE 17.7 (A) Nile delta, showing erosion problems *Source:* After Sestini (1992)
(B) Ebro delta, Spain. The offshore dark and light shaded zones indicate, respectively, distribution of allochthonous n-alkanoic acids and hydrocarbons in offshore sediments.

Source: After Albaigés *et al.* (1985)

subsidence ranging from 0.4 to 5.0 mm/yr, while tide gauge records at Alexandria (1944–89) and Port Said (1926–87), north of the Nile delta coast, indicate a submergence of the land and/or a rise of the sea level of 2 and 2.4 mm/yr, respectively. These sorts of changes are mirrored on many of the coastal deltas in the Mediterranean. For example, the Rhône delta, with an area of 1740 km^2, is also slowly sinking under the weight of sediment being delivered onto the delta plain.

Beach systems

Although the Mediterranean is microtidal, much of the coastal margin reflects the direct influence of current flow. This is manifest in the predominance of offset coasts throughout much of the basin. These comprise asymmetrically curved bays joining one headland to the next – termed 'zeta curve', 'logarithmic spiral coastlines', 'headland bay beaches' or 'crescentic beaches' – which reflect coastline evolution determined directly by alongshore sediment movement generated by surface flow (Komar 1973). Alongshore movement is often quite extensive, leading to the development of extensive beach systems, sometimes 50–100 km long. Wave-dominated longshore movement is also reflected in the coastline configurations of the major deltas (see above), which in the main are storm-wave dominated (Guillén and Palanques 1993).

Beach morphology is a consequence of a very high degree of lateral interdependency of coastal processes. For example, Komar (1973) derived an expression for the rate of change of beach width, in which an increase in sediment transport alongshore will result in erosion and a corresponding decrease in beach width. The most significant aspect of this expression is that change occurring at one point on a shoreline may bring about change elsewhere along the shoreline. This interdependency of beach-forming processes has tremendous implications for coastline use and management (see below).

COASTAL PROBLEMS IN THE MEDITERRANEAN

Most coastal problems in the Mediterranean stem directly from stresses on coastal and marine areas due to the way people interact with and use these areas, both past and present (Viles and Spencer 1995). Coastal problems can be considered by distinguishing between factors which promote environmental stress (e.g. people, intrinsic environmental change [subsidence, erosion], accelerated erosion, sea-level change and flooding) and the manifestions of this stress (e.g. water and air pollution, environmental deterioration, and environmental habitat degradation and loss of biological productivity).

Sandy coastlines

The inherent propensity for continual change in sandy coastlines, coupled with intensive human use and activity, make these one of the most fragile environments of the Mediterranean coastal margin. However, this fragility is often not recognised, even though beach environments are now, economically, the most important natural resource in much of the Mediterranean Basin because of their significance for tourism.

The main problems with beaches stem from a variety of anthropogenic processes – dredging, beach sand and gravel extraction, dam construction, reduction of sediment supply from naturally eroding cliffs because of basal protection, and interference with longshore sediment transport due to construction of shoreline obstacles such as piers, jetties and breakwaters (Viles and Spencer 1995). The main effect of these processes is to deplete sediment supplies to the coast or to interrupt longshore transfer of sediments. The effects of these processes can be very significant. For example, approximately 35 per cent of the beach systems of Italy, which comprise around half of the 7500-km-long coastline, are eroding because of extraction of beach materials (Caputo *et al.* 1991). Depletion may also be aggravated by rising sea levels (see below), but the potential impacts in the Mediterranean have not yet been fully evaluated (Milliman *et al.* 1992).

Relative sea-level change

The sea is a large body of water whose level is not static or constant. We can look at sea level as something which is changing over time, subject to transgression (relative movement onshore) or regression (relative movement offshore), and (relative to the land) liable to get higher or lower. There is considerable geological evidence to indicate that sea level has changed markedly within the Mediterranean within the last few million years, even to the point of the complete drying out of the basin during the Messinian Salinity Crisis (see Chapter 2). So, if major changes have occurred in the past, we have to be concerned about possible future changes, not least because most of the people of the Mediterranean live within a few kilometres of the coastline. Furthermore, if sea level rises further in the future, we need to identify the areas most likely to be affected, and the main factors that are forcing relative sea-level change.

Over the past 100 years, tide gauge records suggest that Mediterranean sea level has risen between 10–20 cm (Milliman 1992). Although these types of records are notoriously difficult to interpret, and causitive factors which contribute to sea-level change are often difficult to isolate from one another, it seems highly likely that sea level will continue to rise in many areas. For example, at Alexandria and Port Said on the Nile delta margin, local future sea-level rise is estimated to be about 37.9 and 44.2 cm by the year 2100 (Stanley 1988). These magnitudes of sea-level rise may be highly localised and may not necessarily affect all areas in the Mediterranean – indeed, it is conceivable that relative sea level may fall in some places (Milliman 1992). However, what does seem certain is that the effects of relative sea-level rise will be particularly evident in areas where global eustatic sea-level rise is coupled with local subsidence. This means in effect that all of the coastal deltas of the Mediterranean are particularly at risk from sea-level rise.

Even though relationships between shoreline retreat and sea-level trends indicate that sea-level rise has, by itself, a relatively minor effect on coastal erosion, the anticipitated sea-level rise, combined with other factors, could accelerate coastal erosion, inundate wetlands and lowlands, and increase the salinity of lakes and aquifers (Frihy 1992; Jelgersma and Sestini 1992). Such changes have other significant 'knock-on' effects. For example, many of the low-lying areas and lagoons that form the 'Natural Park' in the Ebro delta are threatened by sea-level change, which in turn could have important ramifications in terms of biodiversity as this area contains the greatest concentration and variety of flora and fauna within the whole of the Mediterranean (Mariño 1992).

Accelerated coastal erosion

While coastal erosion is inherent in the nature of the operation of most coastal systems, there is increasing evidence that a number of coastal areas are at risk from accelerated coastal erosion within the Mediterranean Basin. Accelerated erosion can occur because of a number of factors. Among these, two are most significant:

1. relative sea-level change which can promote major environmental changes, such as changes in coastline configuration, wave energy, wave patterns, and the nature of coastline materials;
2. loss of sediment supply to the coastline – affecting water and sediment discharge. This often occurs because of structural changes within the river systems feeding the deltas.

These effects are well seen on the coastlines of the Ebro and Nile deltas. A number of studies indicate that the shoreline of the Nile delta began to experience erosion in several areas at the beginning of this century, and parts of the delta (e.g. the Burullus area) have experienced significant accelerated erosion since the tenth century (Broadus 1993; Fanos *et al.* 1995; Frihy 1988, 1992; Stanley 1996; Stanley and Warne 1993a). Dramatic erosion has occurred on some beaches of the Nile delta. This is greatest at the tips of the Rosetta and Damietta promontories, with shoreline retreat up to 58 m/yr. Accelerated erosion commenced primarily because of the construction of nine barrages along the main course of the river at the beginning of the twentieth century, but increased greatly after the construction of the High Aswan Dam in 1964. This resulted in much of the river sediment which formerly reached the delta being entrapped in Lake Nasser. In addition, river sediment which reached the Nile plain was primarily retained in an extremely dense network of irrigation and drainage channels. The drastically reduced amount of sediment which now reaches the sea is discharged primarily from lagoon outlets and several canal mouths and is removed by strong, easterly-directed coastal and innermost shelf currents. This erosion is greatest adjacent to distributary mouths, with shoreline retreat rates of up to 60 m per year (Frihy 1988, 1992; Smith and Abdel-Kader 1988). Erosion, coupled with salinisation and pollution problems, is inducing a marked decline in agricultural productivity and loss of land and coastal lagoons (Stanley and Warne 1993a).

Although Stanley (1996) cites the Nile delta as an extreme example of a depocentre which has been completely altered by human activity (i.e. from an active prograding delta to a locally eroding coastal plain), this is increasingly becoming the 'norm' for Mediterranean deltas. The Ebro delta, for example, also shows signs of extensive accelerated erosion, resulting from sea-level rise aggravated by human-induced erosion and reshaping of the delta coastline due to dam construction in the upper and middle reaches of the river (Mariño 1992). Erosion effects are currently seen mainly on the front and southern lobes of the delta, affecting the wetlands and the natural areas which may eventually disappear. A study by Palanques *et al.* (1990) showed that the amount of sediment supplied by the Ebro River is presently less than 5 per cent of that supplied 50–60 years ago. At present, only the finer particles escape the dams, leaving the remainder trapped behind them. The present supplies and fluxes of suspended sediment are significantly different from those which occurred before management of the river and coast began.

Collapsing ecosystems

Marine and on-shore pollution has taken a heavy toll in terms of damage to fauna and flora in the Mediterranean Sea and coastal zone. Collapsing ecosystems are manifest in a variety of ways – plagues of jellyfish, 'red tides' of algal bloom, and high morbidity levels in marine life-forms. Pearce (1995) indicates that pollution has weakened the coastal ecosystem to the point that alien species are able to take over from native ones. For example, *Caulerpa toxifolia*, a bright green weed commonly found in the tropics and previously unknown in the Mediterranean, is spreading rapidly, after what is thought to have been accidental introduction in 1984. Altogether it is estimated that there are about 300 alien species in the Mediterranean, two-thirds of them discovered since 1970. Pearce emphasises that this form of biological pollution is irreversible and represents a major environmental hazard, which 'may prove to be one of the major ecological problems of the coming century' (Pearce 1995, p. 27).

This type of biological pollution is aggravated by sewage discharge, which has the effect of increased eutrophication, which encourages algal growth. In the Mediterranean, this is manifested in the build-up of dinoflagellate 'red tides' and the formation of mucus-like foam secreted by diatoms. Dinoflagellates release chemicals that are toxic to higher marine life forms and dinoflagellate blooms have proliferated so much over the past 15 years that shellfish have accumulated enough toxin to render them unfit for consumption. Large-scale formation of foam secreted from diatoms has even more devasting effects. The foam can foul beaches for several weeks and, by removing oxygen from sea water, tends to impart considerable stress on surrounding environments.

Eutrophication is now commonplace in the Mediterranean, particularly in the northern Adriatic Sea. It tends to be particularly bad near sewage outlets (Newberry and Siva Subrmaniam 1978). For example, in the Lac de Tunis lagoon, which receives municipal wastes from Tunis, up to one-third of the lagoon is covered in algae during the summer, effectively killing off marine life. Similarly, at Izmir Bay, on the Aegean coast of Turkey, an annual discharge of 500 000 tonnes of raw sewage has led to severe periodic eutrophication and increased pollution-related illness among the resident population.

Over-fishing also has a significant effect on the marine ecosystem. For example, the Mediterranean currently yields about 2 million tonnes of fish per year, which is 50–75 per cent above the sustainable catch (Pearce 1995). In addition to these catches, larger marine life is at risk. Several thousand dolphins and whales die each year after being accidentally caught in drift nets. This selective accidental or deliberate harvesting of marine life has left several species at risk, including Atlantic bluefin tuna, hake, red mullet and sole, and may indirectly encourage proliferation of species further down the food chain (e.g. jellyfish).

People

The presence of people, in large numbers and in high concentrations, ultimately raises the possibility for environmental stress on coastal systems. The potential for enhanced stress is most marked in the Mediterranean, where people have utilised the sea and its coastal margin for several thousands of years. In fact it is fair to say that the coastal margin has been central to the development of civilisation in the area, with many of the great cities of the Mediterranean situated on or near the coast, often on the coastal lowland deltas where water and fertile agricultural land were readily available. This indigenous population continues to grow, at alarming rates. The combined population of the countries of the Mediterranean Basin, at 352 million in 1985, is predicted to rise to around 550 million by the year 2025 (Viles and Spencer 1995). With concomitant increase in urbanisation, more and more of these people will live along the coastal fringe.

The pressure on resources is exacerbated by the annual influx of tourists into the region. Over the past few decades the number of tourists visiting the Mediterranean has been increasing rapidly, as we saw in Chapter 14. These visitors are predominantly 'coastal tourists', drawn mainly from Northern Europe by the desire for sun, sand and sea (Tangi 1977). Local – domestic and intra-Mediterranean – tourism is also increasing fast. Blue Plan scenarios project between 379 and a staggering 758 million tourist arrivals by 2025 (Grenon and Batisse 1989).

These 'sunseekers' tend to occupy a relatively small proportion of the shoreline and as a whole do not venture very far inland. This, coupled with the near tideless conditions of the sea, has promoted the very rapid and concentrated development of hotels, entertainment centres and other forms of accommodation, mainly on the sandier, more unstable, parts of the Mediterranean shoreline. The influx effectively doubles the resident coastal population of around 133 million (Milliman *et al.* 1992). But because most tourists visit the Mediterranean between May and September the resident population in certain coastal zones may expand ten- or twenty-fold.

Pollution

Although the Mediterranean loses deep water relatively enriched with mineralised or recycled nutrients, and is reasonably well adapted to avoid excessive eutrophication and pollution, the high and increasing concentrations of people at the coastal margin in the Mediterranean, coupled with long mean residence times of 80–90 years for sea water before renewal, pose a major problem in terms of pollution from industrial and domestic effluents.

Most of these pollutants emanate from the northern coastal margin (Fig. 17.8) of the Mediterranean (mainly Spain, France and Italy) where people and industrial activity are concentrated. Contaminants enter the Mediterranean from rivers (principally via the Nile, Rhône, Ebro and Po), direct discharges (both land-based and off-

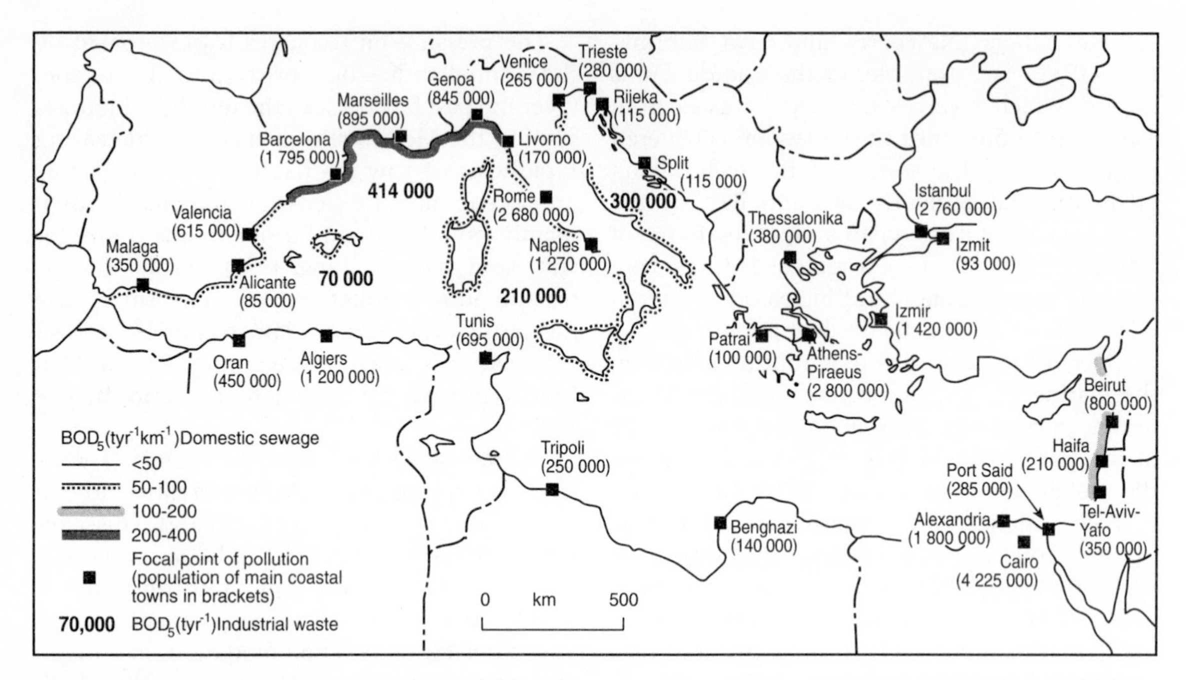

FIGURE 17.8 Domestic sewage and industrial waste discharge into the Mediterranean. No detailed information is provided for the southern coastal margin (domestic waste) or the Aegean Sea, Levantine basin and African waters (industrial waste).

Source: After Clark (1989)

shore), atmospheric deposition and through water exchange, primarily with the Atlantic Ocean and Black Sea (Stanners and Bourdeau 1995). More than 500 million tonnes of raw (i.e. untreated) sewage – representing about 80 per cent of all sewage generated by human activity – are poured into the Mediterranean each year (Pearce 1995). This is in addition to 120 000 tonnes of mineral oils, 60 000 tonnes of detergents, 3800–4500 tonnes of lead, 5000 tonnes of copper, 3600 tonnes of phosphates, and 100–120 tonnes of mercury (UNEP 1989).

Pollution inputs are in some instances balanced by the exchange of dissolved trace metals between the Mediterranen and the Atlantic. Recent calculations suggest a net export of cadmium, lead and copper to the Atlantic. Although some previous estimates for riverine inputs are now thought to be rather high, these are compensated for by the increasing potential for pollution by atmospheric deposition. For example, in the northwestern basin, atmospheric inputs of cadmium, lead and copper (in the dissolved phase) are higher than riverine inputs (CEC 1990). Finally, numerous other types of non-biodegradable pollutants (e.g. plastics) are either accumulating in the sea or being redistributed along the shoreline (Golik and Gertner 1992).

Health problems

Pollution within the confines of the Mediterranean is a major threat to the environmental health of the permanent inhabitants, visitors and marine and coastal ecosystems. However the health consequences of pollutants are not always readily apparent, because of long incubation periods for some diseases, and a confusion between cause and effect, which produces a richly varied morbidity pattern (Briscou 1977).

Problems include bacterial, viral and parasitic diseases which are common to all climates and populations: candidiases, straphylococcal skin diseases, conjunctival, nasopharyngeal and ear infections, streptococcicoses, gastrointestinal infections, intestinal virus diseases and acute pulmonary disorders. Alongside these infections, there are some which are more specific to, or more firmly implanted on the Mediterranean coast, including the typhoid and paratyphoid infections caused by *Salmonella*, bacillary and amoebic dysentery, virus diseases (e.g. poliomyelitis, type A epidemic hepatitis), and various helminthic diseases transmitted directly by ingestion of parasitic eggs resistant to external conditions (Briscou 1977).

Risks to health are enhanced on the coastal margin because these diseases are generally either water- or food-borne, transmitted by polluted water, fruit, vegetables or shellfish – often consumed raw. They can also be transmitted by insects (mainly flies) and birds. Severe health risks also arise from direct organic and chemical pollution. Human beings are exposed to pollutants and disease through several pathways of infection. For example, in addition to ingestion, infection can occur due to inhalation, skin contact and insect bites.

Because of these infection pathways, the main risk groups of populations tend to be those who make heavy use of the sea for occupational or recreational purposes – seafarers, tourists and coastal populations. The risk of infection increases dramatically where hygiene is unsatisfactory due to overcrowding or poor sanitary conditions. For example, salmonellae are carried in abundance in sewage, and where raw sewage is pumped into the sea, coastal waters and any shellfish raised in the area are contaminated. Consequently, the most frequent cases of infection due to salmonellae is through consumption of shellfish (Briscou 1977).

Some of these agents can have remarkably long residence times in water before destruction. Salmonellae can survive long enough in the marine environment to infect unimmunised and unvaccinated bathers, especially the young and old. Other pathogenic agents, such as *vibrio cholerae* micro-organisms, can survive for at least 80 days in seawater. They commonly survive and even multiply in shellfish over periods of many weeks.

The risk to coastal populations is further enhanced because of the large seasonal increases in population, which dramatically increase the amount of wastewater produced at the coast, and because of overcrowding, which creates ideal conditions for person-to-person contagion. Kocasoy (1995) highlighted this problem in a study of two coastal resorts in Turkey, where the resident population increases about tenfold during the main tourist season. Kocasoy established that even small increases in the degree of microbial pollution in seawater resulted in considerable increases in the number of water-borne disease incidences. He also established that foreign tourists tend to have a much higher susceptibility to infection relative to local populations.

In addition to health risks from diseases, there are risks from uptake of pollutants, particularly metals, through consumption of seafood and contact with polluted water. For example, seafood consumption tends to be the main source for uptake of mercury, which, if ingested in high dosages, can lead to severe poisoning. Mercury is only one constituent of a complex cocktail of pollutants present within Mediterranean seawater which can have harmful effects. Recent studies on occupationally-exposed groups (i.e. fishermen) on the island of Favignana, off the coast of Sicily, indicate that although mercury concentrations never reach risk levels, a significant proportion of the sample population show mercury concentration values above the security level, even though there are no large industrial complexes in this area (Valentino *et al.* 1995). Other studies demonstrate that environmental risk increases with proximity to industrial and residential centres (Nicolaou and Andreadakis 1995) and river outflow points (Herut and Hornung 1993). These pollutants tend on the whole to be dispersed alongshore, reflecting and depending on local coastline configuration and wave conditions.

Coastal litter

Litter is now a significant problem along the Mediterranean coastline. Measurements of persistent litter on 13 beaches in Spain, Sicily, Turkey, Cyprus and Israel between 1988 and

1989 by Gabrielides *et al.* (1990) showed that plastic items are the most abundant, followed by wood, metal and glass items. Spoiling by litter derives from a number of sources:

- indiscriminate, occasional spillage by coastal inhabitants, particularly tourists;
- dumping at sea, which is subsequently directed on-shore by wave processes;
- point-source spoiling due to dumping of domestic waste.

Gabrielides *et al.* (1990) found that the quantity of litter present on a beach is inversely related to its geographical distance to a population centre and directly related to the number of visitors frequenting it. Seasonal fluctuations in coastal litter are caused by storm waves (which wash the litter landward, leaving the actual beach clean during winter) and by bathers who pollute it during summer. Their study suggested, based on the nature of the garbage, that most Mediterranean coastal litter is land-based.

Ironically, although coastal litter is aesthetically displeasing, it generally does not present a direct health threat to coastal inhabitants because most of it is relatively inert (Briscou 1977). In addition, most people tend, where possible, to avoid areas which are heavily littered. Avoidance of coastal litter by tourists presents a major incentive to dispose of this type of waste and many coastal authorities have programmes to remove litter on a daily basis, at least during the tourist season.

MANAGING MEDITERRANEAN COASTAL PROBLEMS

Management of coastal zone environments in the Mediterranean Basin, as with many of the other global 'commons', is a multi-disciplinary, cross-sectional, and cross-national problem (Kutting 1994). This, by its very nature, makes it a complex task, and one which must be executed at a range of scales – local, regional, national and international. How do you manage a vast area susceptible not only to the effects of its inhabitants, but also to passing international trade? How do you manage an area like the Mediterranean, which is politically, socially, culturally and economically incoherent, when to manage effectively requires strategic coherence at all levels within and between collaborating groups? Also, to manage often assumes, somewhat wrongly, that we are omniscient – that the problems are readily identifiable and solvable.

Problem differentiation also poses a number of difficulties. How do you establish, or decide, which particular coastal areas or environments are most at risk, in terms of the potential for environmental degradation? Is it sufficient to assess the threat potential to human occupants alone? Attempts to manage also presuppose that we have full knowledge of the factors that contribute most to environmental degradation or change, which is not always the case.

Notwithstanding these imponderables, we can identify some areas that are sensitive to intrinsic and extrinsic change: in Mediterranean coastal settings these include most of the low-lying deltas and associated wetlands. A number of coastal plains are also susceptible to increasing risk, largely by intensification of agricultural or economic activity. In these areas, the foreseeable risks stem primarily from erosion, sea-level change and pollution, which may be manifested through habitat loss or change, soil loss, or chemical degradation of soil-water.

In order to illustrate the difficulties and potential conflicts in attempting to manage environmental problems in the Mediterranean Basin, we now look at the problem at three levels: inter-regional, regional and local. First we examine the Mediterranean Action Plan and then we present two case studies, a regional study of the Adriatic Sea and a detailed local study of the Venice lagoon.

Managing inter-regional coastal problems: the Mediterranean Action Plan

While defining and understanding the environmental issues facing coastlines is crucial, the need to provide an effective executive structure for what is essentially a global commons (by this we mean a physical entity which is not owned, either singularly or collectively; cf. Hardin 1968) is equally important. To this end, Mediterranean countries have examined ways to manage their inter-regional environmental problems, primarily to save coastal and marine environments from pollution. In 1975, under the auspices of the Regional Seas Programme (RSP) of the United Nations Environment Program (UNEP), 16 Mediterranean countries ratified the Action Plan for the Protection of the Mediterranean (Mediterranean Co-ordinating Unit 1985). This called for:

- a number of legally binding treaties to be drawn up and signed by Mediterranean governments (Measure 1);
- the creation of a pollution monitoring and research network (Measure 2); and
- a socio-economic programme that would reconcile environment and development (Measure 3).

These measures have been achieved to varying degrees and illustrate many of the difficulties of achieving coordinated inter-regional coastal zone environmental management.

Measure 1: Conventions and protocols

In Barcelona in 1976, 16 countries signed a convention to 'take all appropriate measures ... to prevent, abate and combat pollution ... and to protect the marine environment'. Between 1976 and 1983, a number of Conventions and protocols were adopted:

- to prevent dumping of pollutants at sea from ships and aircraft;
- to compel governments to cooperate in cases of emergency (e.g. oil spillage at sea);
- to control pollution from land-based sources.

The first two protocols were adopted immediately. However, the third protocol was only adopted in 1980 and not implemented until 1983, following lengthy negotiations on the basis of Phase I MED POL data. Furthermore, although by the end of 1984 there were 102 projects carried out by 62 research groups in 16 countries, only 7 countries had, by mid-1985, signed the agreement with UNEP and implemented their National Monitoring Programme (Mediterranean Co-ordinating Unit 1985, p. 11).

Measure 2: Mediterranean pollution monitoring

Measure 2 has been achieved with a great measure of success. Immediate priority was given to a sound assessment of the state of the Mediterranean, through establishment of MED POL – the Mediterranean Pollution Monitoring and Research Programme (Keckes 1977). During Phase I of MED POL (1976–80), 83 laboratories in 16 countries were equipped to monitor the quality of water, sediments and marine fauna and flora, using standardised analytical methods. The projects adopted during Phases I (Table 17.1) and 2 (Table 17.2) were critical in establishing common principles and guidelines setting out how, where and when to measure pollution levels in order to understand and assess their effects (Mediterranean Co-ordinating Unit 1985).

On the whole, these activities represent a detailed and in-depth attempt to address severe coastal and marine environmental problems within the Mediterranean Basin. The results of MED POL Phases I and II are found in many excellent publications (Gabrielides *et al.* 1990; Haas and Zuchman 1990; Jeftic 1992; Jeftic *et al.* 1992; Nicolaou and Andreadakis 1995; Saliba 1990a, 1990b; UNEP 1989).

Measure 3: The Blue Plan

Measure 3 was implemented in 1979, through the launch of the Blue Plan (Grenon and Batisse 1989). The main aim of the Blue Plan was to facilitate evaluation of the relationship between environment and development in the Mediterranean, and assist Mediterranean countries to make appropriate practical decisions for the protection of their marine and coastal environments (Mediterranean Co-ordinating Unit 1985, p. 12). By 1985, the Blue Plan was moving towards its basic objective to produce realistic models for sustainable, integrated social and economic development of the Mediterranean Basin, with the establishment of a series of Priority Action Programmes (Table 17.3).

TABLE 17.1 Projects implemented under Phase 1 of MED POL, 1976–80

Project	Description
1	Baseline studies and monitoring of oil and petroleum hydrocarbons in marine waters.
2	Baseline studies and monitoring of metals in marine organisms.
3	Baseline studies and monitoring of DDT, PCBs and other chlorinated hydrocarbons in marine organisms.
4	Research on the effects of pollutants on marine organisms.
5	Research on the effects of pollutants on marine communities/ecosystems.
6	Problems of coastal transport of pollutants.
7	Coastal water quality control.
8	Biogeochemical studies of selected pollutants in open waters.
9	Role of sedimentation in marine pollution.
10	Pollutants from land-based sources.
11	Intercalibration of analytical techniques and common maintenance services.

Source: Mediterranean Co-ordinating Unit (1985)

Despite a well-founded programme, Pearce (1995) argues that the actions of MAP have not matched objectives. In 1985, the Mediterranean countries set a series of objectives for 1995 (Table 17.4) to clean up the sea and protect its environment. Many of these have not been realised as yet. In particular, it has proved difficult to assess the level of progress in the implementation of MAP objectives, largely because contributing countries will not or cannot provide the necessary information. Perhaps this is not surprising as it may reflect the considerable difficulties faced in developing a cohesive management framework within an area with a complex administrative framework (see Chapter 1, Fig. 1.3). The geopolitical

TABLE 17.2 Projects implemented under Phase 2 of MED POL, 1981–90

Project	Description
1	Monitoring sources of pollution, providing information on the type and amount of pollutants released directly into the environment.
2	Monitoring of nearshore areas (including estuaries) under the direct influence of pollutants from identifiable primary (outfalls, discharges, dumping) or secondary (rivers) sources.
3	Monitoring of offshore areas.
4	Monitoring of the transport of pollutants into the Mediterranean through the atmosphere.
5	Developing and testing sampling and analytical techniques for monitoring marine pollutants.
6	Development of reporting formats for the *Measure I* protocols.
7	Formulation of the scientific rationale for Mediterranean environmental quality criteria.
8	Epidemiological studies related to environmental quality criteria.
9	Guidelines and criteria for the application of the land-based sources protocol.
10	Research on oceanographic processes.
11	Research on toxicity, persistence, bioaccumulation, carcinogenicity and mutagenicity.
12	Eutrophication and associated plankton blooms.
13	Pollution-induced ecosystem modifications.
14	Effects of thermal discharges on coastal organisms and ecosystems.
15	Biogeochemical cycles of specific pollutants.
16	Pollution-transfer processes.

Source: Mediterranean Co-ordinating Unit (1985)

TABLE 17.3 Priority Action Programmes of the Blue Plan

Programme	Description
1	Directories of Mediterranean institutions and experts in the fields of aquaculture, water resources management and renewable energy.
2	Water resources development of islands and coastal areas.
3	Integrated planning and management of coastal areas.
4	Rehabilitation and reconstruction of historic settlements.
5	Land-use planning in earthquake zones.
6	Solid and liquid waste management, collection and disposal.
7	Promotion of soil protection as an essential component of environmental protection in the Mediterranean coastal zones.
8	Development of Mediterranean tourism harmonised with the environment.
9	Environmental aspects of Mediterranean aquaculture.
10	Renewable energy resources.
11	Environmental impact assessment in the development of the coastal zone.
12	Balance between the hinterland and gravity coastal areas – problems and experiences in planning and management policies.

Source: Grenon and Batisse (1989)

constraints (and achievements) surrounding environmental cooperation were outlined in Chapter 8.

Managing a regional coastal problem: the Adriatic Sea

Much of the northern Adriatic Sea region is considered to have the most serious water quality problem in the whole Mediterranean (CEC 1990). The main environmental problems in this area stem from three processes.

1. The formation of phytoplankton blooms appearing particularly in the northern and central Adriatic. These blooms can cover an area of 1000 km^2 and, because of the resulting anoxic conditions, they are harmful to much of the marine ecosystem.
2. The formation of gelatinous clusters, particu-

TABLE 17.4 Mediterranean Action Plan objectives, 1985–95

Objective	Description
1	Measures to reduce the amount of contaminated water discharged by ships into the sea, including facilities at ports to offload dirty ballast and oily residues.
2	Sewage treatment works for all cities with populations exceeding 100 000 people, plus appropriate sewage outfall pipes into the sea or treatment plants for towns with more than 10 000 people.
3	Environmental impact assessments for development projects (e.g. new ports, holiday resorts, etc.).
4	Improved marine navigation safety to minimise the risk of collisions at sea.
5	Protection of endangered marine species.
6	Identification and protection of 100 historic coastal sites.
7	Identification and protection of 50 marine and coastal conservation sites.
8	More protection against forest fires and soil loss to minimise runoff from rivers.
9	Reduction of acid rain.

Source: Pearce (1995)

larly in the northern Adriatic. At times these can cover 10 000 km^2 (as in 1989) and can suffocate benthic organisms due to sedimentation of mucilaginous material.
3. Build-up of toxic algal blooms, particularly in the northern and central Adriatic.

These problems are associated primarily with increased inputs of anthropogenic nutrients from land into the coastal waters of the northern Adriatic Sea.

In response to this problem, two main approaches were adopted by the Italian authorities:

1. definition of a research programme;
2. development of a master plan for the restoration of the marine ecosystem.

Under Law 57/90 (Article 1) a research programme was approved in June 1991 which was designed to eliminate unnecessary programmes, maximise effectiveness of the research efforts aiming to be consistent with real needs, ensure optimal use of existing infrastructures, allow the development of a sound master plan, and promote cooperation with national and international programmes. This programme set goals and priorities for action for seven topics:

1. to understand how hydrographic factors are changing spatially and temporally and understand the hydrodynamic and biogeochemical processes and fate of contaminants entering the marine environment;
2. to understand the sources, fates and effects of nutrients entering the marine environment as a result of human activities;
3. to understand the sources, fates and effects of toxic substances entering the marine environment as a result of human activities;
4. to understand the sources, fates and effects of biological substances entering the marine environment as a result of human activities;
5. to understand the effects of modifying marine habitats as a result of human activities;
6. to document trends in the status of the marine ecosystem;
7. to understand the implications of marine pollution to human health.

By mid-1994, three main research activities had been scheduled:

1. the development of knowledge-based systems for the scientific management of the Adriatic ecosystem;
2. establishment of standard methods to determine which marine pollution-based pathogens are causing human ill-health effects and to identify indicators of microbial contamination of seafood and marine waters;
3. a support framework to the research.

The estimated costs to achieve some of these initiatives for the management of the Adriatic are enormous, whilst the budget for maintenance of the programme is modest. For example, the financial support needed to take 'down process' measures to control point sources of phosphorous in the Po valley alone is estimated at 2500 MECU. The allocated funds for the master plan to provide resource managers and decision-makers with scientific information about pollution phenomena is running on allocated funds of 5 MECU per annum in 1992 (Chiaudini and Premazzi 1995).

Management of a local coastal problem: the Po delta and Venice lagoon

The Po delta and Venice lagoon epitomise the sensitivity, local environmental conflicts and challenges that face much of the littoral zone, and in particular the coastal deltas of the Mediterranean. It is a deltaic and lagoonal setting, characterised by active geological subsidence, frequent river and tidal flooding, storm surges and coastal shifting (Sestini 1992, p. 429). It is composed of complexes of beach-dune barriers, lagoons, saltwater and freshwater marshes and reclaimed lands separated by numerous elevated channels which together form the interlocking deltas of the Isonzo, Tagliamento and the Po. Of the lagoons, that which encloses the city of Venice covers some 546 km^2 and is divided into three basins separated by former alluvial ridges. The inland margin comprises salt marshes and freshwater lake basins which vary in depth from 1–3 m in the open lagoon to 15–20 m at the three outlets at Lido, Chiogga and Malamocco (Fig. 17.9). Much of the lagoon has been altered by human activities and little of the original tidal marshes and mud flats remain. The inland shore is bounded by the levées of the River Sile, by road embankments and discontinuous sand dunes. On the seaward side, the 1–3-m-high barrier islands have been much modified by the construction of 1.5–4.0-km-long jetties and sea walls that were begun as early as the eighteenth century. Groynes built along the barriers have produced low-angled, protected beaches in places, but elsewhere the beach face is steep and indicative of active erosion in areas starved of sediment.

The area has a long history of human occupation and environmental modification, both on the coast and in the hinterland. Since prehistory, the Po delta has been the main access route from the Adriatic to the northern plains and Alpine zone of Italy. The importance of this route continued through Roman times with the development of major mainland settlements at sites such as Ravenna. However, by the post-Byzantine period, social and political uncertainty led to the development of settlements on the more defensible islands of the lagoon, including Torcello, Murano, Burano and Venice itself. From these beginnings in approximately the sixth century, Venice rapidly developed into a major maritime and trading power. But as the city grew, so did the need to manage the lagoonal environment, to stabilise the islands, to improve mainland agriculture and to maintain access to the sea via passages between the barrier islands. The scale of these modifications was, however, relatively minor and it was not until the early decades of the twentieth century that rapid and far-reaching changes were initiated. These included the progressive reclamation of large areas of swamps and marshes, the recreational use of the coastline and, most importantly, the development of the coastal plain as a major industrial centre. As early as 1920, the decision was taken to create a new port and industrial centre near Mestre. How-

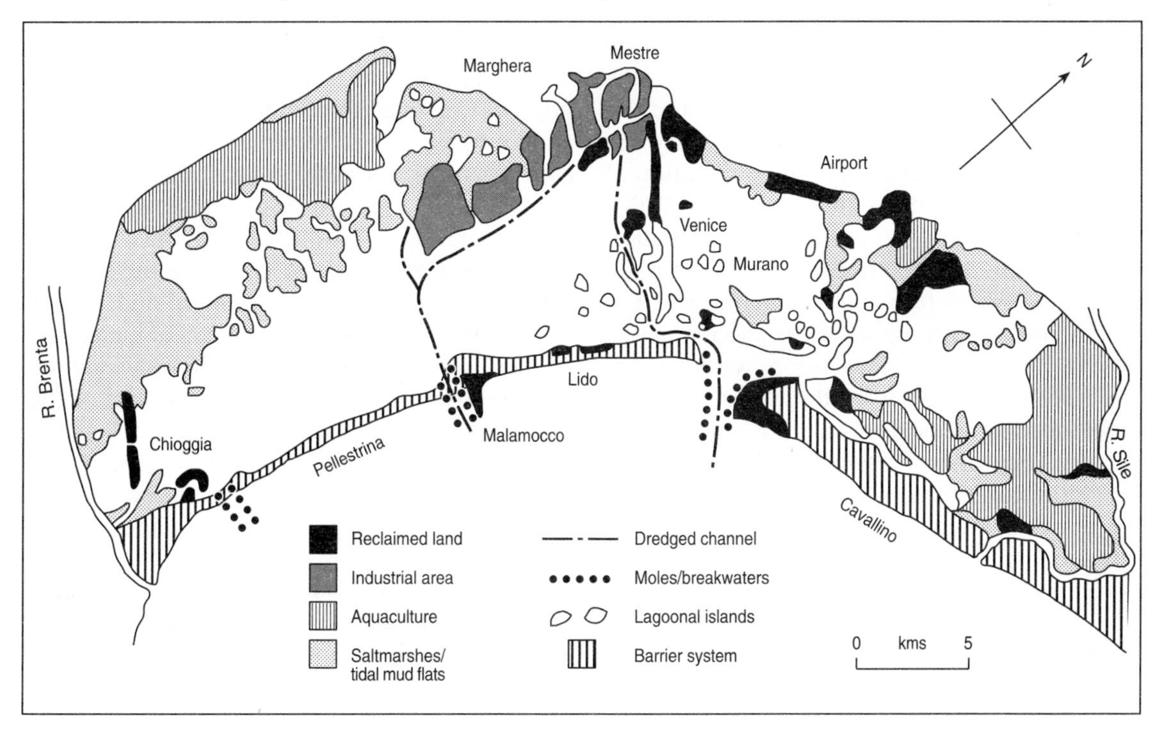

FIGURE 17.9 Location map of Venice lagoon.

ever, it was after the Second World War that economic growth really took off, with the nearby discovery of natural gas and the expansion of chemical, metallurgical and consumer goods industries. This includes the large petrochemical facility comprising 200 factories employing 40 000–50 000 workers and three major industrial zones at Marghera. In support of this development, there has been the reclamation of some 800 ha for agriculture (Ghetti and Batisse 1983).

Extensive urbanisation has also taken place around the fringe of the lagoon and there has been the progressive dredging of two major navigational channels, one of which runs directly through Venice. It can be considered, therefore, that 'the north Adriatic coastal plains have reached a stage of advanced landuse, with complexly inter-related agricultural, industrial and tertiary activities that are tied to national and supranational markets, and are supported by a well developed communications network' (Sestini 1992, pp. 430–1).

Despite these developments, the resident population of the coastal plain remains at less than 2 million, due mainly to a low birth rate and continued out-migration. Indeed, the lack of employment opportunities in Venice and the poor infrastructural provision, particularly for the young, resulted in a fall in the population of Venice itself from 165 000 in 1951 to only 79 000 in 1989. This fall in the permanent population has been more than compensated by the daily influx of workers from the mainland and the annual invasion of up to 12 million tourists. Most of the latter are day visitors whose numbers have been swollen by the liberalisation of travel opportunities from central and eastern Europe. They often therefore make a minimal individual contribution to the economy of the city, but place a major strain upon facilities such as waste disposal, transport and, through these, on factors such as air and water pollution.

Degradation of the Venetian environment

In response to the long-term degradation of the environment and rapid industrialisation, the Venice authorities have pioneered awareness of the environmental damage that rapid economic development can inflict on the coastal margin. Because of the high international profile and historic significance of Venice, the city has also seen one of the most concerted efforts of any area in the Mediterranean to protect and rehabilitate both the natural environment and the cultural heritage contained within it. Most damage can be linked primarily to industrial development this century and can be divided into a number of interrelated categories.

First, there is the *physical structure of the lagoon*. Sestini (1992) has suggested that most of the shoreline is in a state of retreat due to reduced sediment inputs resulting from upstream river management and sand mining from river beds. Over the lagoon as a whole, natural subsidence has been exacerbated by over-extraction of groundwater in the middle part of this century and locally, subsidence has been further accentuated by the dredging of the navigation channels. However, the lack of gradient on these artificial channels has subsequently encouraged siltation as well as altering the pattern of tidal flow in the lagoon (Ghetti and Batisse 1983). Moreover, the widening of the entrances to the lagoon (that at the Lido is now 900 m) has greatly facilitated the inflow and outflow of sea water and together these changes have 'radically altered the configuration of the lagoon causing a large number of natural channels to silt up, destroying a number of low mud banks ... and changing the spreading pattern of tidal waters' (Ghetti and Batisse 1983, p. 14). This has had the effect of increasing the sea water component within the lagoon and, to an extent, turning it into an arm of the sea.

The scope for accommodating and counteracting these changes has been reduced because the natural flexibility of the coastline is now greatly constrained by fixed structures designed to protect agricultural and tourist developments and access to the ports. There are also conflicts between the different economic and political interests involved and administrative wrangles between local authorities, central government and various ministries as to who is responsible for the management of the environment (Sestini 1992).

The second dimension is that of *water quality*. There is no doubt that the Venice lagoon experiences massive inputs of agricultural, industrial and urban pollution which clearly exceed its 'self-cleaning' capacity through tidal flushing. Lee *et al.* (1990) identified a large range of pollution sources which include:

- Treated industrial effluent from the Marghera complex.
- Cooling water from the industrial zone.
- Treated domestic sewage and industrial waste.
- Untreated domestic sewage from the historic city of Venice.
- Runoff containing industrial, urban and agricultural pollutants coming down waterways feeding into the lagoon.
- Influxes of pollution from the Adriatic on high tides.
- Accidental spillages into the lagoon.
- Deposition of atmospheric pollutants.
- Release of pollutants from sediments already in the lagoon, for example, lead.

The unacceptable conditions that had developed by the mid-1970s were highlighted by Sestini (1992) who pointed out that urban pollution principally comprised domestic sewage. He estimated that annually some 3970 t of nitrogen and 1096 t of phosphorus were being discharged into the coastal lagoons, although 50 per cent of the latter may have derived from non-sewage pollutants. The large quantities of organic matter discharged into the lagoons had caused oxygen depletion in many areas, and under anaerobic conditions ammonia and sulphides were producing environments amenable to pathogenic bacteria that made fish farming impossible. Industrial pollution came mainly from Porto Marghera and included high concentrations of aliphatic hydrocarbons, aromatic polynucleids, cyanides and metals (Cu, Zn, Ni, Cr, Fe). The metals, in particular, accumulate in sediments, organic matter and living tissue and in this way can concentrate upwards through the food chain. Industry was also responsible for thermal pollution, especially in summer, which can lead to thermal stratification and anoxic conditions. Finally, agricultural pollution was contributing large quantities of nitrates, phosphates, herbicides and pesticides into the lagoons.

Since the 1970s there have been some improvements. In a survey of the Venice lagoon carried out by Lee *et al.* (1990), zinc levels were seen to have improved between 1976 and 1985, anionic detergent levels for 1986/7 were back to those for 1976/7 and faecal coliform levels had declined steadily over the previous 20 years. Over the same period there had been reductions in certain forms of nitrogen and phosphorus and overall levels of nitrate and nitrite had not increased. Despite these improvements, however, they acknowledged that organic, bacterial and chemical pollution had remained high in the canals of Venice since cleaning stopped in the 1950s and that some water quality parameters had actually deteriorated since the 1970s.

Third, there is the 'external' influence of *marine pollution*. Outside the lagoon, water quality has also deteriorated, due in part to pollutants washed out of the lagoons. Previously, extensive mixing in the autumn and winter produced water quality comparable to the rest of the Mediterranean. Now, eutrophic conditions are common near the shoreline as nutrient-rich runoff encourages algal blooms of diatoms (mainly at the end of winter) and dinoflagellates (August–October). Thus, by the end of summer so-called 'red tides' of algal blooms can extend up to 60 km from the coast producing malodorous, anoxic conditions and mass fish kills which are a serious threat to the region's tourist industry.

Subsidence and flooding comprise the fourth element in the sad story of Venice. Subsidence is common on deltaic coastal plains due to tectonic sinking of depositional basins and compaction of clays and peats. Until the 1950s natural subsidence of approximately 2.7–5.0 cm per annum in the region of Venice was compensated for by active sedimentation (Sestini 1992). However, since the 1950s a combination of reduced sedimentation, draining of marshes and de-watering of aquifers by over-abstraction of groundwater produced significant additional subsidence. Locally within the deltaic region subsidence reached 30 cm per annum and beneath the Po delta the piezometric level fell by 40 m. Around the Venice lagoon, additional subsidence between 1952 and 1969 was 14 cm at Porto Marghera and 10 cm at Venice. This may seem insignificant until one realises that most of the city is only some 80 cm above mean sea level and that a tide of +1.3 m will flood 80 per cent of the city.

The combined effects of natural and accelerated subsidence are seen in the reversal of hydraulic gradients in canals and rivers and increased coastal erosion. However, when subsidence is compounded by a eustatic rise in sea level, the most obvious effect has been the increased threat of flooding. This was drawn to international attention by the events of 4 November 1966, when Venice was inundated by a 1.94 m flood. Regular invasions of the *acqua alta* have become a feature of Venetian life. Individual floods are triggered by a number of regional factors that include 'atmospheric tides', in which onshore winds such as the Sirocco and low atmospheric pressure can add up to 90 cm and 20 cm respectively to astronomic tides. There are also 'hydrodynamic tides', related to oscillatory movements of water within the elongated Adriatic possibly linked to the Sirocco, and marine currents that can raise sea level under the impact of sudden variations in atmospheric pressure. Theoretically these influences could raise sea level by up to 2.5 m and it is fortunate that the flood of 1966 did not additionally coincide with a spring tide (Ghetti and Batisse 1983).

Although the damage to buildings and artefacts was immense, the floods of the 1960s did at least raise awareness of the problems facing the region and stimulate national and international remedial efforts. Venice represents one of, if not the, greatest concentration of historic buildings anywhere around the Mediterranean. It was the threat to these structures by flooding that was the principal stimulus to international concern for Venice. Many of the buildings are of brick and stucco and are protected from rising damp by foundations of dense, white Istrian limestone that acts as an effective damp course. Unfortunately, the Istrian stone rarely rises more than a few centimetres above water level and these courses are overtopped during floods and many lower floors of houses are now uninhabitable (Fig. 17.10).

At the national level, the Italian government finally, in 1973, passed the 'Law of Venice', which released the equivalent of £200 million to be used to build an aqueduct to Marghera, new sewage schemes and flood prevention works, to fund research into the meteorology of tidal surges and to restore and stabilise the historic centre of Venice (O'Riordan 1975). These measures have succeeded in halting artificial subsidence, but as Sestini (1992) has noted, the effects of previous mismanagement still remain. This includes not only a lingering susceptibility to flooding but also a legacy of unusable groundwater beneath the city contaminated with saltwater and industrial pollutants.

Po delta and Venice lagoon: future prospects

Efforts to stabilise the Venetian environment, and in some cases to reverse damage caused by previous economic developments, continue. Tidal barriers have now been erected across the three entrances to the lagoon. In addition, as well as sea walls along the barrier beaches, plans were drawn up in 1991 to redress the effects of

FIGURE 17.10 Abandonment of 'ground'-floor buildings, Grand Canal, Venice

past erosion by constructing a rock barrier some 300 m offshore from the most vulnerable section of coast between the Malamocco and Chioggia inlets. Local legislation has been enacted to reduce atmospheric pollution within Venice, complemented by compliance with European-wide limits on emissions of sulphur and nitrogen. Problems of groundwater over-abstraction and subsequent subsidence have been successfully addressed. There remains, however, the potentially catastrophic threat of environmental damage linked to possible climatic changes associated with global warming. Sestini (1992) identified the northern Adriatic coastal environment as a particularly sensitive area and envisaged that a regional rise in temperature of 0.5–1.5°C by 2025 could significantly increase the frequency of abnormal events. These could include: hot dry summers, unusually dry and mild winters, erratic precipitation (including storms and floods), marine storms, tidal surges and eutrophication of coastal and lagoon waters. If sea level were to rise between 12–20 cm by 2025 it is unlikely to produce extensive flooding, but beach erosion would increase and tidal flats and reed beds would degrade under stronger tidal currents and more estuarine conditions. Finally, if sea level were to continue to rise between 2025 and 2050 by a further 40–50 cm, Sestini (1992) forcasts serious degradation of Venice and other lagoon towns, a decline of beach resorts and restrictions on harbours. A milder winter climate would increase runoff during autumn and winter from the Alps, with resultant risks of erosion and sedimentation, but lower river flows in longer, hotter summers would threaten irrigation and increase risks of water pollution.

THE 'COMMONS' PROBLEM IN THE MEDITERRANEAN

For the past two decades, the countries of the Mediterranean have, in response to severe pollution problems, cooperated to manage the Mediterranean Sea. However, it is difficult to see what lies ahead. The management approaches adopted raise issues about co-governance of transboundary environmental problems and the problem-solving capacity of the countries of the Mediterranean, acting either individually or in concert (Skjaerseth 1993). The example of managing a regional area (i.e. the Adriatic Sea) illustrates the limited ability of national governments to attain their quality objectives unilaterally. One might argue that the action plan adopted by the Italian authorities is itself an attempt to maintain 'operational sovereignty' over the management of the Adriatic, which is perhaps unrealistic. On the other hand, the action plan could be considered effective if it forms part of 'negotiated orders' for joint action, which are repeated at each of the different levels of collective decision-making inherent in MAP. Clearly, solutions within the area of the Adriatic require this in the form of cooperative regional management approaches. However, their action plan could also be seen to be an 'optimum response' by a nation–state during a time when many of the potential cooperating partners are affected by civil conflict and fragmentation of their 'nation' states. Certainly, the degree of unrest within the area of the eastern Adriatic raises serious doubts about the potential for long-term effective cooperation to manage regional problems as envisaged in the Adriatic Initiative of the four Coastal States (Italy, the former Yugoslavia, Albania, Greece). For example, political fragmentation of nation–states may inhibit future cooperation as in the case of the former Yugoslavia, while Albania is a good example of a nation–state which has had only limited participation in MAP activities over the past two decades.

The 'global commons' nature of the Mediterranean Sea raises questions about the unevenness of environmental impact. By and large most of the pollution in the Mediterranean is generated along its northern industrialised coastal margin, and transmitted throughout the sea area. Thus the costs and impacts of a limited number of countries are transferred to less industrialised and generally less wealthy nation–states along the southern coastal margin. This raises a number of searching questions:

- Who is or should be responsible for these problems?
- Who are the victims?
- Who pays (polluter or recipient)?

This unevenness of environmental impact ultimately raises the possibility for increasing conflict rather than cooperation between the countries who share the Mediterranean 'commons'. The threat of conflict may be enhanced by the uncertainty and elusiveness of some of the environmental risks which we have outlined in this chapter. Our analysis indicates that some of the causative factors (e.g. people) that contribute to the environmental problems in the Mediterranean (e.g. water pollution) are undoubtedly set to increase in magnitude, thus increasing the threat potential for the future.

Blowers and Glasbergen (1996, p. 265) ask 'what political circumstances do we need for environmental management and planning to become central activities in contemporary society?' Here we would ask:

- Do these propitious political circumstances exist today in the Mediterranean?
- Are they likely to exist in the near future?

Certainly, over the past two decades it seems that the countries of the Mediterranean have met the basic political conditions – a high level of collective, coordinated action. There are, however, signs that this cohesiveness is deteriorating. By the early 1990s the MAP was close to collapse, with the major European contributors failing to pay their dues to the secretariat and many research activities cancelled. Pearce (1995) indicates that, by April 1994, unpaid contributions amounted to $3.7 million, equivalent to 7 months of the budget of the secretariat. Other difficulties have been identified. Cohesiveness has also been affected by instances of war and civil conflict within the Mediterranean, some of which are still continuing.

Part of the 'success' of MAP has been the importance of expert credibility in the definition of environmental problems within the Mediterranean (Liberatore 1996). However, even this is being undermined at an intellectual as well as a practical level. This credibility is threatened if the work of cooperating scientific groups cannot be adequately supported and sustained, both in terms of programme continuance and relevance. Questions have arisen about why it is necessary to evaluate the 'objective nature of "episteme knowledge" and the legitimate role of expert communities in proposing camps of action' (Liberatore 1996, p. 63). It could be argued that it is more important to try to evaluate the social and political processes which affect the efficacy of cooperating and coordination of environmental problems. However, Liberatore (1996) clearly highlights one of the problems of dealing with environmental problems in the long term, that of changes in fundamental approaches to problem-solving due to growth of our knowledge base or changes in the intellectual approach to the problems.

What is clear is that, in attempting to manage the Mediterranean, we have to define the nature of the environmental problems and formulate certain solutions for which there are many scientific uncertainties. Liberatore (1996, p. 65) suggests that natural science findings need to be 'translated' in political and economic terms to become policy issues. However, in translation, we should not lose sight of these uncertainties, which require on-going resolution. Dr Mostafa Tolba, Executive Director of UNEP, stated in his opening address to the Seventh Ordinary Meeting of the Contracting Parties (to MAP) in Cairo in October 1991 that

> Our Common goal is to protect and rehabilitate the Mediterranean as a basis for the sustainable development of all countries around it. Treaties and plans alone will not do that. Actions will. A very basic question we have to answer is how far has the Mediterranean benefited from the (1976) Barcelona Convention, its various protocols and its Action Plan.

This statement is epitomised to some extent in the intentions of the Nicosia Charter, adopted in June 1990 by the Commission of the European Communities and 12 Coastal States, which set priorities for the period to 2025. Paramount among these was the mobilisation and support of the public of the Mediterranean for environmental protection. After all, it is people who create the conditions for the abolition of economic and political obstacles and the taking of effective measures.

REFERENCES

ALBAIGES, J., AUBERT, M. and AUBERT, J. 1985: The footprints of life and of man. In Margalef, R. (ed.), *Western Mediterranean*. Oxford: Pergamon Press, 317–51.

BLOWERS, A. and GLASBERGEN, P. 1996: *Environmental Policy in an International Context: 3. Prospects*. London: Arnold.

BOUSQUET, J.C. 1977: Quaternary strike-slip faults in southeastern Spain. *Tectonophysics* **52**, 277–86.

BRISCOU, J. 1977: The health situation around the Mediterranean. *Ambio* **6**, 342–5.

BROADUS, J.M. 1993. Possible impacts of, and adjustments to, sea level rise: the cases of Bangladesh and Egypt. In Warrick, R.A., Barrow, E.M. and Wigley, T.M.L. (eds), *Climate and Sea Level Change*. Cambridge: Cambridge University Press, 263–75.

CAPUTO, C., ALESSANDRO, L., LA MONICA, G.B., LANDINI, B. and LUPIA PALIERI, E. 1991: Past erosion and dynamics of Italian beaches. *Zeitschrift für Geomorphologie*, Supplementband **81**, 31–9.

CEC (Commission of the European Communities) 1990: *Eutrophication-Related Phenomena in the Adriatic Sea and in Other Mediterranean Coastal Zones*. Brussels: Water Pollution Research Report 16.

CHIAUDANI, G. and PREMAZZI, G. 1995: Recent technical and scientific initiatives for the management of the Adriatic Sea problems. In Nicolaou, M.L. and Andreadakis, A.D. (eds), Pollution of the Mediterranean Sea. *Water Science Technology* **32**, 357–63.

CLARK, R.B. 1989: *Marine Pollution*. Oxford: Clarendon Press.

DEFANT, A. 1961: *Physical Oceanography*. Oxford: Pergamon Press.

EMERY, K.O. and KUHN, G.G. 1982: Sea cliffs: their processes, profiles and classification. *Geological Society of America, Bulletin* **93**, 644–54.

FANOS, A.M., KHAFAGY, A.A. and DEAN, R.G. 1995: Protective works on the Nile Delta coast. *Journal of Coastal Research* **11**, 516–28.

FRIHY, O.E. 1988: Nile delta shoreline changes: aerial photographic study of a 28-year period. *Journal of Coastal Research* **4**, 597–606.

FRIHY, O.E. 1992: Sea-level rise and shoreline retreat of the Nile Delta promontories, Egypt. *Natural Hazards* **5**, 65–81.

GABRIELIDES, G.P., ALZIEU, C., READMAN, J.W., BACLI, E., ABOUL DAHAB, U. and SALIHOGLU, I. 1990: MED POL survey of organotins in the Mediterranean. *Marine Pollution Bulletin* **21**, 233–7.

GHETTI, A. and BATISSE, M. 1983: The overall protection of Venice and its lagoon. *Nature and Resources* **19**, 7–19.

GOLIK, A. and GERTNER, Y. 1992: Litter on the Israeli shoreline. *Marine Environmental Research* **33**, 1–15.

GRENON, M. and BATISSE, M. 1989: *Futures for the Mediterranean Basin: The Blue Plan*. Oxford: Oxford University Press.

GUILLÉN, J. and PALANQUES, A. 1993: Longshore bar and trough systems in a micro-tidal, storm-wave dominated coast: the Ebro delta (Northwestern Mediterranean). *Marine Geology* **115**, 239–52.

HAAS, P.M. and ZUCHMAN, J. 1990: The Med is cleaner. *Oceanus* **33**, 38–42.

HARDIN, G. 1968: The tragedy of the commons. *Science* **162**, 1234–48.

HARVEY, A.M. 1990. Factors influencing Quaternary alluvial fan development in southeast Spain. In Rachocki, A.H. and Church, M. (eds), *Alluvial Fans: A Field Approach*. London: Wiley, 247–69.

HERUT, B. and HORNUNG, H. 1993: Trace metals in shallow sediments from the Mediterranean coastal region of Israel. *Marine Pollution Bulletin* **26**, 675–82.

HINDRICHSEN, D. 1990: *Our Common Seas: Coasts in Crisis*. London: Earthscan.

HOPKINS, T.S. 1985. Physics of the sea. In Margalef, R. (ed.), *Western Mediterranean*. Oxford: Pergamon Press, 100–25.

JEFTIC, L. 1992: The role of science in marine environmental protection of Regional Seas and their coastal areas: the experience of the Mediterranean Action Plan. *Marine Pollution Bulletin* **25**, 66–9.

JEFTIC, L., MILLIMAN, J.D. and SESTINI, G. (eds), 1992: *Climatic Change and the Mediterranean, Vol. 1*. London: Edward Arnold.

JELGERSMA, S. and SESTINI, G. 1992: Implication of a future rise in sea level on the coastal lowlands of the Mediterranean. In Jeftic, L., Milliman, J.D. and Sestini, G. (eds), *Climatic Change and the Mediterranean. Vol. 1*. London: Edward Arnold, 282–303.

KECKES, S. 1977: The Coordinated Mediterranean Pollution Monitoring and Research Program. *Ambio* **6**, 327–8.

KELLETAT, D. 1985: Bio-destructive und bio-konstructive formelemente an den Spanischen Mittelmeerkusten. *Geoökodynamik* **6**, 1–20.

KOCASOY, G. 1995: Waterborne disease incidences in the Mediterranean region as a function of microbial pollution and T_{90}. In Nicolaou, M.L. and Andreadakis, A.D. (eds), Pollution of the Mediterranean Sea. *Water Science Technology* **32**, 257–66.

KOMAR, P. 1973. *Beach Processes and Sedimentation.* Englewood Cliffs, NJ: Prentice Hall.

KUTTING, G. 1994: Mediterranean pollution: international cooperation and the control of pollution from land-based sources. *Marine Policy* **18**, 233–47.

LE CAMPION-ALSUMARD, T. 1979. Le biokarst marin: rôle des organismes perforants. In *Actes du Symposium International sur l'érosion karstique.* Aix en Provence: UIS, 133–40.

LEE, N., MATARRESE, G., MIANI, P., FOSSATO, V.U., DEL TURCO, A., MORETTO, L. and CARCASSONI, B. 1990: *Porto Marghera, Venice and its Environment.* Marghera: Ente Zona Industriale di Porto Marghera.

LIBERATORE, A. 1996. The social construction of environmental problems. In Glasbergen, P. and Blowers, A. (eds), *Environmental Policy in an International Context: 1. Perspectives.* London: Arnold, 59–83.

MARGALEF, R. 1985. Introduction to the Mediterranean. In Margalef, R. (ed.), *Western Mediterranean*, Oxford: Pergamon Press, 1–16.

MARIÑO, M.J. 1992. Implications of climatic change on the Ebro delta. In Jeftic, L., Milliman, J.D. and Sestini, G. (eds), *Climatic Change and the Mediterranean. Vol. 1.* London: Edward Arnold, 304–27.

MEDITERRANEAN CO-ORDINATING UNIT 1985. *Mediterranean Action Plan.* Athens: UNEP.

MILLIMAN, J.D. 1992: Sea-level response to climate change and tectonics in the Mediterranean Sea. In Jeftic, L., Milliman, J.D. and Sestini, G. (eds), *Climatic Change and the Mediterranean. Vol. 1.* London: Edward Arnold, 45–57.

MILLIMAN, J.D., JEFTIC, L. and SESTINI, G. 1992: The Mediterranean Sea and climatic change: an overview. In Jeftic, L., Milliman, J.D. and Sestini, G. (eds), *Climatic Change and the Mediterranean. Vol. 1.* London: Edward Arnold, 1–14.

MOSES, C.A. and SMITH, B.J. 1994: Limestone weathering in the supra-tidal zone: an example from Mallorca. In Robinson, D.A. and Williams, R.B.G. (eds), *Rock Weathering and Landform Evolution.* London: Wiley, 433–51.

NEWBERRY, J. and SIVA SUBRMANIAM, A. 1978. Middle East: sewerage projects for coastal towns of the Libyan Arab Republic. *Quarterly Journal of Engineering Geology* **11**, 101–12.

NICOLAOU, M.L. and ANDREADAKIS, A.D. (eds), 1995: Pollution of the Mediterranean Sea. *Water Science Technology* **32**, 1–385.

O'RIORDAN, N.J. 1975: The Venetian ideal. *Geographical Magazine* **47**, 416–26.

PALANQUES, A., PLANA, F. and MALDONADO, A. 1990: Recent influence of man on the Ebro margin sedimentation system, northwestern Mediterranean Sea. *Marine Geology* **95**, 247–63.

PANZAC, D. 1983: Espace et population en Egypte. *Méditerranée* **50**, 71–80.

PEARCE, F. 1995: Dead in the water. *New Scientist* **145**, 26–31.

RIOS, J.M. 1978: The Mediterranean coast of Spain and the Alboran Sea. In Nairn, A.E.M., Kanes, W.H. and Stehli, F.G. (eds), *The Ocean Basins and their Margins, Volume 4B, The Western Mediterranean.* New York: Plenum Press, 1–16.

SALIBA, L.J. 1990a: Coastal land use and environmental problems in the Mediterranean. *Land Use Policy* **7**, 217–30.

SALIBA, L.J. 1990b: Making the Mediterranean safer. *World Health Forum* **11**, 274–81.

SESTINI, G. 1992: Implications of climatic change for the Po Delta and Venice Lagoon. In Jeftic, L., Milliman, J.D. and Sestini, G. (eds), *Climatic Change and the Mediterranean. Vol. 1.* London: Edward Arnold, 428–94.

SKJAERSETH, J.B. 1993: The 'effectiveness' of the Mediterranean Action Plan. *International Environmental Affairs* **5**, 313–34.

SMITH, M. 1991: Dams that cut off the lifeblood of the wetlands. *The Independent*, 25 November, 15.

SMITH, S.E. and ABDEL-KADER, A. 1988: Coastal erosion along the Nile delta. *Journal of Coastal Research* **4**, 245–55.

STANLEY, D.J. 1988: Subsidence in the northeastern Nile delta: rapid rates, possible causes, and consequences. *Science* **240**, 497–500.

STANLEY, D.J. 1996: Nile Delta: extreme case of sediment entrapment on a delta plain and consequent coastal land loss. *Marine Geology* **129**, 189–95.

STANLEY, D.J. and WARNE, A.G. 1993a. Nile delta: recent geological evolution and human impact. *Science* **260**, 628–34.

STANLEY, D.J. and WARNE, A.G. 1993b: Sea level and initiation of Predynastic culture in the Nile delta. *Nature* **363**, 435–8.

STANNERS, D. and BOURDEAU, P. (eds), 1995: *Europe's Environment: The Dobrís Assessment*. Copenhagen: European Environment Agency.

TANGI, M. 1977: Tourism and the environment. *Ambio* **6**, 336–41.

UNEP 1989: *State of the Mediterranean Marine Environment*. Athens: MAP Technical Report Series No. 28.

VALENTINO, L., TORREGROSSA, M.V. and SALIBA, L.J. 1995: Health effects of mercury ingested through consumption of seafood. In Nicolaou, M.L. and Andreadakis, A.D. (eds), Pollution of the Mediterranean Sea. *Water Science Technology* **32**, 41–7.

VILES, H. and SPENCER, T. 1995: *Coastal Problems: Geomorphology, Ecology and Society at the Coast*. London: Arnold.

18

CONCLUSION: FROM THE PAST TO THE FUTURE OF THE MEDITERRANEAN

LINDSAY PROUDFOOT AND BERNARD SMITH

MEDITERRANEAN IDENTITY REVISITED

Geographical identities are notoriously elusive: difficult to delimit and even more difficult to portray and to understand. The Mediterranean is no exception. The very term itself contains an underlying ambiguity which at once invites and denies straightforward analysis. The physical definition of the Mediterranean Basin perhaps poses the fewest problems, although it is still far from easy to achieve. The littoral regions bordering the Mediterranean Sea itself; the catchments draining into it; or even the regions characterised by the 'Mediterranean Climate' beloved of past school textbooks: all conjure up a widely recognisable – and to a great extent mutually reinforcing – picture of what 'the Mediterranean' as a physical entity might be.

At various points in the book other semantic and conceptual elaborations of the term 'Mediterranean' were introduced. In Chapter 1 the essence of the region was portrayed as the integrating notion of 'Mediterraneanism': the close blend of the physical, the cultural and the visual – the Mediterranean as a sensual experience (cf. Houston 1964, p. 707) to be seen, felt, described, painted, lived. For Perry, in Chapter 3, the concept of Mediterraneanism is partly climatic, above all the predictable biseasonality of the regime, and partly linked to proximity to the sea. In Chapter 16 Faulkner and Hill referred to 'Mediterraneity' as the seasonal climatic variability which presents plants with a winter precipitation maximum (but also cold stress) combined with a summer drought. Whilst these climatic parameters are important, it is also vital to appreciate how they 'produce' Mediterranean landscapes and ways of life. The danger of physical determinism is present here, but few can ignore the environmental influences (amongst other influences, of course) on crop types, rural dwellings, the forms of some settlements, or a human habit such as the siesta.

But closer inspection reveals these seemingly reassuring congruities to be less certain than they at first appear. Thus the simple climatic stereotype of 'hot dry summers and warm wet winters' learnt by rote by generations of schoolchildren proves to be altogether more variable and capable of greater extremes than the definition itself suggests – as Braudel (1972) and St Paul both came to realise! The same is also true of the geology, landforms and soils which occur within the basin (Chapters 2 and 4). Whether defined in terms of its catchment area or more narrowly by its coastal littoral, the 'physical' Mediterranean proves to be so diverse as to prompt the remark that its only unifying physical feature is the Mediterranean Sea itself.

A MARGINAL PLACE?

Within the environmental diversity decribed above, one unifying feature of the Mediterranean Basin may, however, be its marginal suitability for human occupation. It is ironic that an area that has been settled for so long should be in many respects so hostile. A defining characteristic, as noted, is the occurrence of seasonal drought, but a shortage of fresh water is only one aspect of the resource limitations that produce the marginality of the Mediterranean. To hydrological constraints must be added shortages of fertile, usable land, limited soil resources, a fragile vegetation cover, erosion risk and frequent seismic and volcanic activity. To these terrestrial problems must be added the sensitivity of the Mediterranean Sea itself to human impact through pollution and resource depletion ranging from over-fishing to the introduction of foreign species inimical to indigenous marine life.

As a primarily 'youthful' geological region (Chapter 2) much of the basin is mountainous and supports only shallow residual soils. Productive land is invariably restricted to narrow coastal plains and compact intermontane basins, and even in these areas the accumulation of fertile soils has often been at the expense of deforestation and erosion of adjoining uplands. Typically the soils of the hills and mountains are on limestones, uplifted during the Cainozoic from the floor of the Tethys Ocean that was the precursor to the present-day Mediterranean. The combination of relatively pure limestones and a Mediterranean climate is not conducive to rapid or extensive soil formation. Indeed, classic soils of the Mediterranean such as the *terra rossas* are considered by many to be relicts from previous humid subtropical climates during the Quaternary and/or to comprise a significant component of dust deposited from winds blowing out of the Sahara (Chapter 16). Thus, once soils are removed there is little possibility of regeneration or a return to the mature forest cover that characterised most hillslopes prior to human intervention. The key to soil erosion has been the destruction of this forest cover in conjunction with the natural erosivity of Mediterranean environments produced by steep slopes and intense seasonal rainfall. Destruction of the Mediterranean forest has its origins in prehistoric and classical periods (Chapter 16), and it is no accident that, for example, the sequence of sclerophylous forest–maquis–garrigue–steppe is firmly entrenched in the geographer's lexicon as the archetypal example of vegetation degradation. This destruction has now progressed to the point where, for most people, severe degradation is viewed as the climatic norm. Our perception of the Mediterranean is therefore littered with images of isolated pine trees clinging to pockets of soil amongst jagged outcrops of white limestone. Within this framework, the efforts of farmers to terrace impossibly steep slopes, and the well-ordered farms of the plains are seen as gallant attempts to master a harsh environment, rather than as the final acts in the degradation of a once stable landscape.

It is upon this foundation of a resource base severely denuded by centuries of over-exploitation, that the dramatic cultural, economic and political changes of the second half of the twentieth century have been played out. Environmental stress has risen exponentially in response to: the rapid growth in indigenous populations, particularly in the poorest countries, and their rising economic expectations; massive influxes of temporary visitors; widespread urbanisation and industrialisation; and unprecedented intensification of agriculture on land that has not been built upon. To these increased human pressures must now be added the certainty that the very climate of the region is changing in response to global warming. The seriousness with which this prospect is viewed is indicated by the research effort already expended (e.g. Jeftic *et al.* 1992, 1996). Most researchers agree that the likelihood of a warmer Mediterranean can only add an extra dimension of pressure to already overstretched environments and economies.

RESOURCE SHORTAGE AND ENVIRONMENTAL STRESS

Nowhere will this pressure be more critical than in the provision of fresh water and the safe disposal of domestic, industrial and agricultural wastes. Any regional warming will not only increase evaporation, but is also likely to increase the length of summer drought periods while reducing the quantity of snowmelt that is so important for the recharge of lowland reservoirs. In the light of these problems, and the fact that many Mediterranean countries already consume more than their annual rainfall, it is significant that water supply is the one issue which brought both Arabs and Israelis to the negotiating table long before any political *rapprochement* took place. Already water is a limiting factor on economic development in many Mediterranean countries (Chapter 15), and it has dictated political relationships between countries to the point where threats of war have been made if access to supplies is prejudiced. Conflicts are not restricted to nation–states, and an interesting by-product of increased regional autonomy in Spain is the reluctance with which scarce supplies are now shared between the regions.

Disputes within regions and states over scarce water are paralleled and reinforced by competition for other resources, most particularly land. Typically, competition is for traditionally cultivated land or one of the few areas of remaining forest and is between intensive agriculture (especially horticulture), residential use, commercial development (especially tourism) and industrialisation. To these must be added a growing, though still rarely successful, international and grassroots movement for the conservation of key environments. Nowhere is the conservation issue more urgent than in the threats posed by economic development to the coastal wetlands of the Mediterranean. These are crucial not only to the ecological well-being of their immediate surroundings, but because of their role as stopover points on global bird migration routes. They also serve to illustrate how competition for land is concentrated within the narrow coastal zone (Chapter 17) and the degree to which successful competition is not simply a question of replacing one land-use with another. Land-use change almost always accelerates resource depletion with ramifications at and beyond the specific site of the change. Thus, regional water resources are contaminated by sewage and industrial effluents, saline incursions occur where coastal aquifers are lowered and pesticides and fertilisers increasingly find their way into drinking water supplies. On the land, over-irrigation can lead to salination of soils and in extreme cases once-fertile land is turned to stone by the formation of calcretes. Finally, atmospheric pollution from industry and, increasingly, vehicles has made Mediterranean cities such as Athens and Cairo barely habitable during the hot summer months. Atmospheric pollution and indiscriminate construction are also threatening another great resource base of the Mediterranean in the shape of its cultural heritage. Not only is this destruction a crime against society, but it is also economically counterproductive. At a time when the Mediterranean must broaden its tourist base in response to the challenge of 'long-haul tourism', it would be a folly to destroy one of the main reasons why many visitors come to the region.

Fortunately for the Mediterranean Basin, the fragility of its environment and the pressures upon it have been recognised for some time (King 1990). As early as 1975 the issues at stake were formulated in 'The Blue Plan', initiated under the auspices of the United Nations in Barcelona. This sought to:

> make available to the authorities and planners of the various countries in the Mediterranean information which will enable them to formulate their own plans to ensure optimal socio-economic development without causing environmental degradation ... and to help the governments of the states bordering the Mediterranean region to deepen their knowledge of the common problems facing them, both in the Mediterranean Sea and its coastal regions.
>
> (Grenon and Batisse 1989, p. vii)

The background and aims of the plan are comprehensively laid out in Grenon and Batisse (1989) and anyone attempting a review of environmental issues in the Mediterranean has to be indebted to their pioneering study. What remains to be seen, however, is the extent to which the awareness of the issues and problems that have been created are translated into actions by governments and responses by developers. To date, the omens do not appear auspicious. Despite some international co-operation on problems such as controlling oil spillages, there is little sign of concerted international economic action designed to manage the environment in a sustainable fashion. It is true that there have been many local initiatives but such undertakings remain 'a drop in the ocean'.

From cultural diversity to contemporary geopolitics

The human identities, cultures and state ideologies which border upon the Mediterranean are equally diverse, both in their current configuration and historical provenance. They, too, defy easy delineation. Historically, the Mediterranean Sea acted as both routeway and barrier, at times dividing the cultures and peoples around its shores, at others providing the means whereby they might influence – or conquer – each other. Invariably, however, these confrontations and collaborations formed part of broader processes of human interaction which, though they may have focused upon the physical Mediterranean or, indeed, originated within it, also extended far beyond its confines. Consequently, whether in the fragmented, cellular world of Hellenic Antiquity, or under the over-arching imperial structures of the Roman Empire (Chapter 5), or during the uneasy confrontation between medieval Christianity and Islam (Chapter 6), or faced with the impact of European modernity and capitalism (Chapter 7), the human landscapes of the Mediterranean proved to be susceptible to a series of external cultural influences, each of which left its trace on these varied 'Rhodian shores' (cf. Glacken 1967). The Parthenon, the Dome of the Rock, the minarets of Hagia Sofia or Nicosia, the walls of Rethymnon or Valetta, the regular fieldscapes of the Tunisian *tell*: all bear witness to the multi-layered cultural identities which have taken root in the Mediterranean soil, as successive generations have created *place* out of *space*, formed and reformed in their own image.

These legacies continue to resonate in the present day, and together serve to ensure that the Mediteranean continues in its historic role as a cultural 'shatter-belt', where fundamentally opposed political and religious ideologies collide in a world where sensitivities are already sharpened by ethnic differences and social and regional inequalities (Chapters 8 and 9). Inevitably, our portrayal of these differences has been an exercise in hermeneutics, reflecting our own cultural conditioning and standpoint. Thus Westernising, Eurocentric perspectives tend to stress those tensions and problems which impact most strongly on European identities, economic interests and political security (Chapter 10); hence the European Union's growing concern with illegal South–North immigration from the Maghreb and Mashreq to Spain, France and Italy, and its attempted involvement in the long-running political instability in former Yugoslavia.

Such concerns are understandable. Arguably, the demographic pressures in Algeria, Morocco and Tunisia which have given rise to illegal immigration constitute one of the most potent problems currently facing Mediterranean states, particularly since these pressures are unlikely to diminish in the near future (Chapter 11). Similarly, the speed of Yugoslavia's collapse and the savagery of the subsequent internecine warfare in Bosnia and Croatia provide an eloquent reminder both of the relative recency of 'modern' geopolitics in the eastern Mediterranean and the intensity of the ancient ethnic and religious rivalries which sometimes underpin these.

Yet other cleavages also exist which expose different lines of fracture within the Mediterranean world. Among the region's Islamic states, for example, the earlier distinction between pro-Western and Arabist regimes has been increasingly blurred both by the collapse of the former USSR as the latters' patron, and the rise of Islamic fundamentalism. This perhaps over-convenient 'catch-all' phrase masks the considerable diversity displayed by these movements, which vary from the militant radicals of the Algerian *Front Islamique du Salut* to the members of the pro-Islamic *Welfare Party*, now leading the current constitutionally elected coalition government in Turkey. Nevertheless, the fact remains that virtually none of the 'moderate' (i.e. Westernising and more or less secular) Islamic states have remained immune to the effects of fundamentalist movements seeking in some way to reconstitute government and society according to Koranic precepts. Their effect has been profound, most noticeably in Algeria, where to date almost 50 000 people have died in the civil war which followed the military take-over in the wake of the government's annulment of the 1991–92 election, which the *Front Islamique* was almost certain to win. In Tunisia, the fundamentalist parties provide the only coherent opposition to President Ben Ali's regime and therefore have been banned. In Egypt, since 1992 Islamic fundamentalists opposed to President Mubarak have carried out a series of attacks in which over 400 people, including 26 tourists, have been killed. In Turkey, possibly the most secular of all Mediterranean Islamic states, the long sequence of pro-Western governments finally came to an end in June 1996, with the election of the first Islamist-led government since the foundation of the secular state in 1923. While signalling its intention to continue to govern Turkey as a democratic, secular and social state based on the 'principles of Ataturk', the new government has also made it clear that Turkey's traditional pro-Western stance will be tempered by increasing co-operation with Islamic central Asian and Balkan countries with which it has 'historic and spiritual links'. Subsequently, Prime Minister Necmettin Erbakan has strengthened Turkey's economic ties with Iraq, and engaged in sabre-rattling with its more immediate neighbour, Syria, over the disputed province of Antakya.

Elsewhere in the Mediterranean, other long-standing 'zones of political cleavage' are also showing signs of movement, but not always in ways which portend future stability. Thus to the south, while the relationship between Israel and its Arab neighbours continues to slowly evolve towards (but currently retreat from) the possibility of a permanent peace, with the 'cold peace' between Egypt and Israel now supplemented by the latter's more thorough co-operation with Jordan in a 'warm peace', the newly established Palestinian State, created as an essential part of this process, seems inherently unstable and riven by internal tensions from birth. To the west, the traditional emnity between Greece and Turkey continues to prove itself capable of potentially violent expression, whether along the Attila Line which has divided Cyprus since the Turkish invasion of 1974, or in the disputed waters off Turkey's western Ionian coast.

Neither are the 'developed' states of the north-western Mediterranean entirely immune from these sorts of problems: both Spain and France continue to face violent demands for regional autonomy from Basque and Corsican separatist groups. France, too, faces the growing problem of Islamic fundamentalist violence among its immigrant population. But in every case, whether in Corsica, Algeria, Gaza, Cyprus or elsewhere, these disputes presuppose a fundamental mismatch between the idea of the state – or its construction – and the aspirations and identities of a significant proportion of its people. Almost invariably, as argued here, the causes are in part at least historic, whether they are to be traced back to the imperial geographies constructed by British and French colonialists in the nineteenth century, or further to the much earlier misalignments between political authority, religion and ethnicity which characterised the Ottoman Empire and its predecessors.

THE CONTRADICTIONS OF DEVELOPMENT

On to this fractured and precarious political template a whole raft of developmental processes has been fastened, bringing with it the problems and opportunities which have formed the subject matter of a large part of this book. The growth of package tourism has clearly been among the most far-reaching of these changes (Chapter 14), but there have been others, including EU membership or association (Chapter 10), the exploitation of fossil fuels (Chapter 9) and mass urbanisation (Chapter 12). The effects of each have varied regionally and socially. In the case of tourism, the consequences have included the creation of post-modernist urban geographies along significant stretches of the Spanish, French, Italian, Greek and Turkish coasts, the destruction of traditional value systems, ecologies and environments, and the exacerbation of existing regional imbalances or the creation of new ones. EU membership or prospective membership has led to significant and sometimes socially traumatic realignments within various regional agrarian economies, and has failed to deter existing state rivalries, notably between Greece and Turkey. The continuing exploitation of oil reserves in Libya and Algeria continues to add to the growing problem of environmental pollution within the Mediterranean, a trend exacerbated particularly on the south shore by the failure of infrastructural modernisation to keep pace with urbanisation.

Underlying all these developmental processes lie relationships of dependency between the 'developed' and 'less developed' parts of the Mediterranean world which have changed remarkably little since the growth of the world capitalist system in the early nineteenth century. Just as then, when the industrialising economies of Northern Europe exerted their economic power to establish colonial relationships of exploitation with various North African and Levantine provinces of the decaying Ottoman Empire, so today in the post-colonial period, many 'less developed' states along the southern and eastern shores of the Mediterranean continue to stand uneasily dependent on international financial aid mediated through institutions like the World Bank. The difference, however, is that whereas two hundred years ago, European economic power legitimised the assertion of European cultural supremacy, today recipient states like Morocco, Tunisia or Algeria confidently assert their own cultural and political identities, even though these may be challenged from within. It is in this conundrum – the dissonance between cultural identity, economic modernisation, political motivation and social aspiration – that the future of the Mediterranean lies.

References

BRAUDEL, F. 1972: *The Mediterranean and the Mediterranean World in the Age of Philip II.* 2 vols. London: Collins.

GLACKEN, C. 1967: *Traces on the Rhodian Shore.* Berkeley and Los Angeles: University of California Press.

GRENON, M. and BATISSE, M. 1989: *Futures for the Mediterranean Basin: The Blue Plan.* Oxford: Oxford University Press.

HOUSTON, J.M. 1964: *The Western Mediterranean World.* London: Longmans.

JEFTIC, L., KECKES, S. and PERNETTA, J.C. (eds), 1996: *Climatic Change and the Mediterranean. Vol. 2.* London: Arnold.

JEFTIC, L., MILLIMAN, J.D. and SESTINI, G. (eds), 1992: *Climatic Change and the Mediterranean. Vol. 1.* London: Edward Arnold.

KING, R. 1990: The Mediterranean – an environment at risk. *Geographical Viewpoint* **18**, 5–31.

INDEX

Page references in **bold** refer to illustrations and tables.